DEEP WATER

James Bradley

DEEP WATER
THE WORLD IN THE OCEAN

HARPERONE

An Imprint of HarperCollins*Publishers*

HarperCollins books may be purchased for educational, business, or sales promotional use. For information, please email the Special Markets Department at SPsales@harpercollins.com.

First published in the UK by Scribe Publications, 2024.

Published by arrangement with Scribe Publications.

Typeset in Adobe Caslon Pro by Midland Typesetters, Australia

Library of Congress Cataloging-in-Publication Data has been applied for.

ISBN 978-0-06-339017-1

24 25 26 27 28 LBC 5 4 3 2 1

This book was written on the unceded lands of the Gadigal People of the Eora Nation. Gadigal have cared for the beaches, waterways and woodlands of the part of what is now known as Sydney that I call my home for tens of thousands of years. They are its original storytellers, and their songs and ceremonies are woven through its fabric. I acknowledge their deep and continuing connection to Country and pay my respects to Elders past and present.

For Annabelle and Theo

'How inappropriate to call this planet "Earth", when clearly it is "Ocean".'

Arthur C. Clarke

CONTENTS

1

BEGINNINGS

'How does it start the sea has endless beginnings.'
Alice Oswald, *Nobody*

ON CHRISTMAS EVE 1968, as *Apollo 8* completed its fourth orbit of the Moon, its crew looked up to see the Earth rising above the lunar surface. Over sounds of laughter and delight, astronaut William Anders asked for a roll of film, and, pressing his Hasselblad camera to the capsule's hatch window, snapped a photo of something no human had ever seen before.

The photo Anders took – dubbed 'Earthrise' – captured our planet suspended above the lifeless surface of the Moon, its blue and white and dusky green and brown vivid and alive against the inky void of space. Together with the 'Blue Marble' photo taken a few years later by the astronauts on *Apollo 17*, it changed the way we saw our planet, revealing not just its wonder and fragility, but also that seen from without it is not a collection of continents and countries, but a living, interconnected whole. As Anders famously remarked, 'We set out to explore the Moon, and instead discovered the Earth.'

More than fifty years later, 'Earthrise' has not lost its extraordinary power. Indeed, in many ways its impact today is even greater than when it was first taken. Seen in the context of the catalogue of destruction humans have inflicted on the Earth over the past half century, the image of our planet suspended like a solitary oasis in the blackness of space takes on an almost elegiac charge.

This recognition of the fragility of the Earth's systems is complemented by another, which is that seen from space, our world is a blue planet. Perhaps this should not be a surprise; after all, the oceans cover more than 70 per cent of the planet's surface, and the largest of them – the Pacific – is bigger than all the continents combined. But 'Earthrise' makes these sorts of statistics graspable, the vast blue of the South Atlantic and the swirling bands of cloud above it dwarfing the faded browns and greens of Africa's western edge.

Psychologists have dubbed the transformative feelings of awe and wonder many of the Apollo astronauts experienced at the sight of the Earth from space 'the overview effect'. This same feeling of connectedness is likely to be familiar to anybody who spends time in the water. In my twenties and thirties I spent as much time as I could in the ocean, mostly catching waves off the beaches that divide Sydney's eastern suburbs from the sea beyond. Having come to surfing late I was next to useless on a board, and while fun, boogie boarding always seemed a little undignified. And so, instead, my brother, a few friends and I took to using hand-skis to catch the swells that roll in from the Pacific.

Hand-skis are precisely what they sound like – curved pieces of wood or plastic about the size of a dinner plate that are strapped to your hand. Using them is intensely exciting: the hand-ski acts as a hydrofoil, lifting your head and shoulders out of the water and sending you shooting across the face of the wave, your body suddenly as sleek and manoeuvrable as that of a dolphin or a seal.

The catch is that the best rides aren't possible in the mess of the shore break. And so we took to swimming out into the deeper water. Out there, we would take on waves almost twice our height – great green and blue slabs of curving water that, if caught right, would send us hurtling beachward until we slipped out the side or were swallowed by the spume. It was exhilarating because it offered a reminder of the sheer power of the ocean: cut it too close and the weight of the wave's collapse would detonate behind you like something seismic – *an atom bomb*, we used to call it, only half-jokingly – but if you mistimed an exit or got caught in the dump zone, that excitement could quickly give way to fear, and sometimes, real danger: I still have scars on my back from being thrown onto the rocks by a particularly massive midwinter swell.

Yet what really interested me about those sessions in the surf wasn't the adrenaline rush of riding the waves, but the quieter moments. Out beyond the break, as the swell moved beneath me, slow and steady as breath, it was sometimes possible to feel the intimation of something larger, a sense of time's depth, of the great pulse of the world's cycles. And although it was never easy to articulate, it often seemed that something transformative inhered in those moments – as if to give oneself over to the ocean's largeness and the movement of the wind and the waves might offer a glimpse of another way of being.

This sense of boundlessness is a common theme among those who spend time in the water. Australian author and surfer Fiona Capp writes of the 'inkling of infinity' experienced in the face of the ocean's immensity and power, observing that 'few images capture this primal "at oneness" better than that of the surfer crouched inside the crystal, womb-like tube of a breaking wave; an image made all the more exquisite by our knowledge of the wave's imminent destruction'. Capp notes this moment of suspension and inevitable ejection from the 'all-embracing amniotic realm' of the wave's oneness is also embedded in the Hawaiian word for surfing, *he'e nalu*, or 'wave-sliding'. '*He'e* means "to run as a liquid" or "flee through fear", while *nalu* refers to the surging motion of a wave or the slimy liquid on a newborn child.' Free diver James Nestor says something similar, suggesting that 'letting the ocean envelop you' can be 'a spiritual practice, a way of using the human body as a vessel to explore the wonders in the Earth's inner space'.

In correspondence with his friend, Sigmund Freud, the writer Romain Rolland dubbed this apprehension of the infinite and sense of oneness with the universe 'oceanic feeling', and argued it was the source of religious awe. Freud was more circumspect, contending (in a suggestive echo of the Hawaiian association with gestation and birth) that if it did exist, it was a vestigial remnant of early infancy, and our still egoless and undifferentiated self's inability to distinguish between the world and ourselves.

More recently scientists have found evidence that such feelings of expansive awareness have a neurological origin, associated with the interplay between the neural network our brain uses when focusing on the external world and the network that governs internal processes relating to self-reflection, self-awareness and emotion. Usually these two networks operate independently, but studies of the brains of Buddhist monks suggest that the feelings of expansive awareness and connection that arise during meditation are the result of the two operating simultaneously, effectively lowering the barrier between the self and the environment. Something similar happens in activities such as surfing and skydiving, where the intense focus on the immediate collapses the boundary between the self and the world, and with psychedelic drugs, such as LSD and psilocybin, once again resulting in feelings of connection and oneness with the universe.

Birth, dissolution, oneness: perhaps it is little wonder that we so often invoke the water or the ocean when we speak of origins, or time. Surely it is not a coincidence that when the American writer John McPhee sought a metaphor to encompass the vastness of geological time, a time frame in which the permanence of the geological comes unfixed, and flows like water, so mountain ranges rise and fall and continents drift across the face of the planet, he reached for the marine metaphor of 'deep time'. Or that as Austrian philosopher and writer Ivan Illich reminds us, so many human Creation stories begin with an act of division, with the conjuring of form from the primal unity of water. Genesis speaks of a world without light or form, yet before the world there is already the deep; similarly there are early Islamic texts that describe the seven earths as balanced on the back of a great whale, which swims in an endless ocean. Time, water, the ocean: the three are inextricably connected.

*

WHILE LIFE ON EARTH would be impossible without the ocean, its origin is surprisingly mysterious. To appreciate why, it is necessary to first understand how the water that fills it came to be, and to do that we must go back almost 14 billion years, to the immediate aftermath of the Big Bang. In the inconceivable heat and violence of those first infinitesimal fractions of a second, the fundamental forces that shape our Universe unfolded, followed by elementary particles such as quarks and leptons, which in turn began to form protons and neutrons.

At first these particles remained separate – at many thousands of millions of degrees the early Universe was so hot it was impossible for atoms to form, so they instead existed in a plasma-like fog, or soup. As the Universe continued to expand it also continued to cool, until finally, around 380,000 years after the Big Bang, temperatures fell to only a few thousand degrees Celsius, making it possible for protons and neutrons to begin to combine with electrons to form stable atoms. Approximately three quarters of these early atoms were hydrogen – with its single proton and electron the simplest element, and then, as now, the most abundant element in the Universe – the rest were mostly helium, along with trace numbers of lithium and deuterium atoms.

For tens of millions of years these atoms were spread through the Universe in a thin cloud. But over time they began to condense into clumps, and these clumps grew larger and denser, to collapse in on themselves, until finally the pressure at their centres increased so much that they ignited. In the nuclear furnace at the hearts of these first stars, atoms began to fuse together.

Initially most of what these early stars created was helium. But as the stars aged, more complex elements began to be produced. Sometimes these new elements were as heavy as iron, but most of what was produced was oxygen, which gradually became the third most abundant element in the universe. As these early stars died, and these elements were released, this oxygen reacted with the even more abundant hydrogen to form water molecules. This began relatively soon after the Universe's creation: in 2016 scientists at the Paranal Observatory in Atacama, Chile, observed the chemical signature of water in a galaxy 12.88 billion light years from Earth, the light from which began travelling towards us less than a billion years after the Big Bang. And water is still being made today: in the immense stellar nursery of the Orion Nebula, enough is created every day to fill Earth's oceans sixty times over.

Some of these water molecules were present in the cloud of gas and dust from which our Solar System formed. This process started a bit over 4.5 billion years ago, when some disturbance caused part of the cloud to begin to collapse inwards. As the gas and dust at the centre grew denser the cloud began to rotate, drawing more and more matter into itself. Most of this matter clumped into the centre to form the Sun; the rest spread out into a thin protoplanetary disc: as this disc swung around the newborn Sun the matter comprising it also began to cohere, clumping together into larger and larger agglomerations that gradually became planetesimals, and eventually planets.

The composition of these planets was determined by their distance from the Sun. In the inner Solar System, where temperatures were too high for water and volatiles such as methane, carbon dioxide and ammonia to condense into solid form, they formed out of metals and silicates, creating small, rocky worlds such as Mercury, Venus, Earth and Mars. Further out, beyond what astronomers call the frost line, where water was able to freeze, they formed out of gases borne outwards on the solar wind, creating the brooding ice and gas giants of the outer Solar System with their icy rings and confusions of frozen moons.

In the immediate aftermath of its formation the Earth was an unimaginably violent place. Its surface was a highly unstable sea of magma, temperatures exceeded 1000 degrees Celsius, radioactivity was annihilatingly high, and the atmosphere was a toxic miasma of ammonia and methane. Over time a crust began to form, but even once it did it was ruptured by volcanic eruptions and frequent impacts by meteors and other debris left over from the formation of the planets. The most catastrophic of these collisions occurred around 4.44 billion years ago, when a Mars-sized object known as Theia slammed into the Earth, ejecting the material that became the Moon.

For many years it was assumed that during this phase of its existence, known as the Hadean, the Earth was so hot that any water that might have been captured during its formation would simply have boiled away. Instead scientists argued the bulk of our planet's water must have arrived after the surface had cooled, delivered to Earth by a barrage of water-bearing asteroids and comets between 3.8 and 4 billion years ago. In recent years, however, a more complex story has begun to emerge. The first part of that story stems from the analysis of the water contained in comets. All naturally occurring water contains trace amounts of deuterium, an isotope of hydrogen that contains a single neutron as well as a proton. Because the precise amount of deuterium varies depending upon where the water originated, it can be used as an atomic fingerprint or isotopic signature.

Earth's water has one deuterium atom for every 6400 atoms of the far more common protium isotope of hydrogen. But data from probes that have intercepted comets in space suggests that the water contained in many of them has much higher levels of deuterium, making it unlikely the bulk of our planet's water arrived on comets.

Another part of this story emerges out of geological samples found in the Jack Hills in Western Australia. Located 300 kilometres inland from Shark Bay, the rocks that comprise the Jack Hills are ancient, having formed some 3.6 billion years ago. Yet within them are minute zircon crystals that are even older, dating back an astonishing 4.4 billion years.

These crystals are the oldest known remnants of our planet's primordial crust – so old, in fact, that they formed a mere 160 million years or so after the Earth itself. Analysis of their chemical composition shows they must have grown somewhere wet, suggesting not just that there must have been water present on Earth not long after its creation, but that the Earth's crust must have cooled sufficiently for this water to condense much earlier than previously thought.

The presence of liquid water doesn't imply this early Earth was particularly hospitable: surface temperatures would have been close to 200 degrees, and the water was only prevented from boiling away by the extremely high atmospheric pressure. But what it does suggest is that at least some of Earth's water was present in the rocks and dust from which the planet formed and, instead of boiling away, somehow survived not just the heat of the Hadean, but the collision that formed the Moon.

Further support for the idea that some of Earth's water was here from the very beginning has emerged from analysis of meteorites known as enstatite chondrites. A class of very ancient meteorites left over from the formation of the Solar System, and therefore similar to the material from which the Earth formed, enstatite chondrites turn out to contain significant amounts of hydrogen of a similar isotopic signature to that in Earth's mantle. The recent discovery of huge amounts of water distributed through the Earth's mantle more than 400 kilometres below the surface also offers evidence our planet's water arrived early: locked up in a bluish rock called ringwoodite, this reservoir contains enough water to fill the oceans several times over, and may indicate the existence of a planetary water cycle, in which water deep within it very slowly permeates upwards to the surface in a process known as mantle rain. Even more intriguing is the discovery of what appear to be fossils of microbes dating back

4.3 billion years in ancient rocks in Canada. Closely resembling organisms that live on hydrothermal vents in the deep ocean today, these fossils not only offer evidence that water was present a mere 200 million years after the Earth's formation, but raise the tantalising possibility that life on our planet began much earlier than was previously believed possible.

Although these findings suggest that much of Earth's water may have been present in the rocks from which the planet formed, they cannot account for all of it. The incidence of deuterium in enstatite chondrites is slightly lower than that of Earth's water, meaning at least some of Earth's water must have arrived later in comets or meteorites. This process continues today – as comets swing in towards the Sun they leave a fine sprinkling of ice and dust in their wake. As our planet sweeps its way around the Sun it sweeps up this dust and snow; each year many tonnes of it fall to Earth, a gentle rain of cometary material that is absorbed into our atmosphere, our rivers and oceans. Our bodies.

EARTH IS NOT THE ONLY WORLD with oceans. On Jupiter's moon, Europa, an ocean perhaps 100 kilometres deep and containing more than twice as much water as there is on Earth is believed to lie between the moon's rocky centre and its frozen surface. Kept liquid by the heat created by Jupiter's colossal tidal force, the waters of this subterranean sea are believed to be warm and salty, like Earth's oceans, and our blood. On Saturn's moon, Enceladus, a similar ocean exists beneath an ice sheet some 30 or 40 kilometres thick; its waters escape through ice volcanoes at the moon's south pole in huge geysers that rise hundreds of kilometres into space. Most of the water released in these plumes drifts back to the surface of Enceladus and freezes again. That which does not is lost to space, where much of it ends up in Saturn's rings.

Although the oceans on Europa and Enceladus are the most famous, they are not the only instances of such worlds among the many moons and dwarf planets scattered through the outer reaches of the Solar System. Jupiter's largest moon, Ganymede, may hide an ocean even larger than Europa's bound up in concentric layers of ice and liquid water, and it is possible liquid water also lurks under the cratered surface of Callisto, the fourth of the Galilean moons. Similarly, around Saturn, Titan, the surface of which is so cold it has lakes of liquid methane and sandhills of frozen hydrocarbons, likely conceals an ocean as salty as the Dead Sea on Earth, as do Dione and tiny Mimas. Further out Uranus's Ariel and Neptune's Triton may also harbour sub-surface seas. Even Pluto – so far from the lifegiving heat of the Sun that its light takes five-and-a-half hours to reach it – may possess an ocean beneath its icy surface.

These frozen worlds out beyond the frost line bear little resemblance to Earth and our neighbours in the inner Solar System. Yet even in here, close to the Sun, there is evidence of ancient oceans. In the last years of the nineteenth century, the astronomer Percival Lowell believed he had glimpsed a network of lines on the surface of Mars that faded and reappeared in time with the Martian seasons. Convinced these lines were artificial structures, he theorised they were the work of a dying civilisation, part of a system of canals and oases designed to channel the last water on a drying planet from the poles to the lands in the temperate regions.

Although we now know Lowell was wrong and that Mars is barren and implacably hostile, four billion years ago it possessed abundant surface water, probably concentrated in two major bodies: a vast, shallow ocean that covered the northern third of the planet, and another, smaller sea in the south. What happened to these Martian seas is not fully understood. What we do know is that Mars lacks the magnetic fields that protect Earth, meaning its atmosphere must have been gradually stripped away by the solar wind. As atmospheric pressure fell, Mars' surface water was sublimed away; that which was not lost to space sank into the surface, combining with chlorate and perchlorate salts to form the brines that even today seem to flow occasionally on the Martian surface, or collect in icy reservoirs far underground.

Venus was also once very different. Although closer to the Sun than Earth, Venus is so similar to our planet in size and composition that the two might almost be sisters. The clouds that envelop its surface, however, meant it remained tantalisingly unknowable until the first spacecraft visited it in the 1960s.

Faced with this mystery, astronomers were free to speculate. In the early twentieth century the Swedish scientist Svante Arrhenius (who was also one of the first to recognise that changes in the Earth's climate are connected to variations in the concentration of carbon dioxide in the atmosphere) suggested Venus's clouds were water vapour, and that beneath them the planet was both extremely warm and extremely humid. Predicting its surface would be covered by swamps and oceans, he imagined conditions rather like those on Earth during the Carboniferous era (when much of the deposits of coal and oil that have fuelled the transformation of our own planet were laid down), and speculated its seas and jungles might be home to abundant but simple plant life, fast-growing and quickly decaying.

By the 1930s this idea of a humid, swampy Venus had become a commonplace in popular culture. Edgar Rice Burroughs, creator of John Carter of Mars and the racist fever dream of *Tarzan*, set a series of novels there, imagining a cloud-covered world of warm oceans and dripping forests, populated by humans and degenerate ape-men and pygmies, while Burroughs' competitor Otis Adelbert Kline pictured a world of teeming oceans and jungles 'grotesque to Earthly eyes' and crowded with immense tree ferns 'more than seventy feet high'. And as late as the 1950s writers like Robert Heinlein and

Ray Bradbury were writing stories and novels that imagined a wet, swampy Venus.

It is telling that Arrhenius compared the conditions on his imaginary Venus to those in the Congo; certainly the fantasies of dripping jungles, strange dinosaur-like creatures and alien princesses his vision engendered in the work of Burroughs and others are deeply coloured by the fantasies of colonialism and its racist underpinnings (Arrhenius himself was a committed eugenicist). Irrespective, this image of a steamy, tropical planet persisted in the popular imagination until the 1960s, when probes finally visited Venus, and found not lush rainforests and warm seas, but a horrifying hellscape where surface temperatures average more than 450 degrees Celsius and an incredibly dense atmosphere composed almost entirely of carbon trapped beneath heavy clouds of sulphuric acid. Atmospheric pressure is ninety-two times that of Earth's at sea level, or equivalent to almost a kilometre beneath the ocean, so dense in fact that early probes collapsed soon after they entered the atmosphere.

Venus's unimaginable hostility was a shock. But there were more surprises to come. For as scientists learned more about our nearest neighbour it became clear Venus was not always a hellscape. Instead, early in its existence, Venus boasted conditions similar to those on Earth, up to and including the presence of liquid water in the form of oceans. At some point however it grew warmer, and the water began to evaporate, combining with the carbon dioxide and methane to cause a runaway greenhouse effect that boiled away 99.95 per cent of Venus' water and created the conditions we see today.

Exactly why this happened is not clear. Until recently it was assumed it was connected with a gradual increase in the Sun's intensity. But some now speculate Venus may have continued in a relatively Earthlike state until about 750 million years ago, at which point a sudden spike in carbon dioxide – perhaps as the result of a massive volcanic eruption – triggered the runaway greenhouse effect. If this is correct, that means Venus may have had oceans until less than a billion years ago. By that time Earth's oceans were home to multi-cellular organisms: is it possible those on Venus were as well? Or that life swims in the depths of the oceans beneath the ice of Enceladus or Europa?

If it's possible to see in the stories of Mars and Venus different versions of the Earth's history, alternative scenarios in planetary evolution that offer a disquieting vision of our own future, perhaps it is also possible to discern something more. For much of the twentieth century both planets provided a theatre for the colonial imagination, a stage upon which its lurid racial fantasies could be enacted in stories about bestial aliens and doomed civilisations. And while scientific reality has largely banished the pulpier elements of these imaginings, they still echo through the vision of Martian colonisation promoted by Elon Musk and others: for what else are their fantasies of new frontiers but the colonial vision on a planetary scale?

NO MATTER WHERE IT CAME FROM, there is no question that water has shaped our planet since its molten surface first began to cool, 4.5 billion years ago. Initially that water would have issued from cracks in the planet's crust as steam, condensing in the choking atmosphere before falling to the surface as rain and boiling away again, only to gather and fall once more, over and over again. But as the surface slowly cooled over the next 100 million years, this water began to collect in basins and declivities.

As these bodies of water grew they formed lakes and, finally, shallow seas. In this early stage of Earth's evolution the continents had yet to emerge; as a result water is likely to have covered most, if not all, of the planet's surface. Unconstrained by large landmasses and other barriers the tides would have washed around the world in a single bulge, pulled upwards by the gravity of the Moon, which was not yet the pale, silvery disc we are familiar with, but instead, in a reminder of the violence of its creation, a looming, volcano-scarred ball that swung around the planet at less than a quarter of the distance it does today.

The salt that fills the oceans also began forming during this period, as rain saturated with carbonic acid dissolved the newly formed rocks of the surface, leaching away minerals and washing them out to sea. In time this process slowed and stabilised at a saltiness of about thirty-five parts per million – about the same as today – but not before it had transported so much salt into the oceans that if it was dried out it would cover the entire Earth in a layer 50 metres thick.

Exactly when life emerged, and whether that happened in a shallow pool, at a hydrothermal vent on the ocean floor or somewhere else, is not clear. Yet by 3.7 billion years ago the simple, single-celled organisms were established enough to have begun to agglomerate into matlike structures that lay spread across the seabed or floated on the ocean's surface. Primitive photosynthesis followed 3.4 or 3.5 billion years ago, and by 2.4 billion years ago cyanobacteria were producing oxygen, provoking a radical shift in the history of life that led to the development of multicellular organisms, plants and, eventually, perhaps three quarters of a billion years ago, the first animals.

It is unlikely we will ever discover the identity of the first of these primitive animals, but molecular evidence suggests that sponges and comb jellies were the first lineages to cleave away, followed by the cnidarians, ancestors of today's jellyfish, corals and anemones. What we do know is that by 575 million years ago, during the Ediacaran Period, the Earth's oceans were thick with strange, soft-bodied life, ranging from the gardens of frondlike *Charnia* and sea pens to the elegantly ribbed ovals of the *Dickinsonia*, which looked a little like an inflatable leaf, or a cross between a trilobite and a surfmat, and the weird discs and triskelions of the trilaterally symmetrical Trilobozoa. Many of these organisms were so strange that scientists are not always sure how to classify them – despite their plantlike appearance the *Charnia* seem to have been animals, as were the sea pens – many of which seemed to have lived rooted in place, and to have drawn their sustenance from the water or the thick microbial mats that covered much of the ocean floor.

About 539 million years ago most of these Ediacaran lineages abruptly disappeared, seemingly in one of the first of the mass extinction events that have punctuated our planet's history. But as life rebounded in the warm, shallow seas of the Cambrian it took on an astonishing diversity of new forms. As new, burrowing animals spread, the microbial mats that had covered the ocean floor for billions of years largely disappeared; in their place the spongelike Archaeocyatha formed huge, reeflike structures, and hard-shelled arthropods such as the vaguely shrimplike Radiodonta and the trilobites exploded in number. Before long in geological terms, some of these began to emerge from the ocean, at least for short periods. Fossilised footprints show centipede-like creatures the size of lobsters known as euthycarcinoids were skittering around on shore about 530 million years ago. The appearance of plants on land came later – the oldest known fossils of land plants date back around 470 million years, but while their arrival paved the way for the development of centipedes and other primitive arthropods, it would not be until about 400 million years ago that the first fish crawled onto the beaches of the ancient continent of Gondwana.

These first fish resembled the lungfish that even today can be found in rivers in northern Australia, Africa and South America, possessing lungs as well as gills, and fleshy, lobed fins instead of the delicate fanlike fins we are more familiar with today. From them would spring not just the amphibians, reptiles, dinosaurs and birds, but also the ancient synapsids that would give rise to the mammals and, eventually, ourselves.

*

ALTHOUGH WE DO NOT KNOW when early humans first encountered the ocean, there is no question the experience altered them in profound ways. As these ancient species spread outwards across Earth's waters and along its shores their journeys opened new horizons and demanded new technologies and new forms of social organisation.

Much of our knowledge of these early movements must be inferred, yet the scattered traces we have map out a series of waves of expansion that begin with the migration of *Homo erectus* out of Africa a little over 2 million years ago, and which saw them reach central China around 2.1 million years ago, and Java by 1.3 million years ago. Most of this journey could have been made on land, including to Java: the cooler conditions that prevailed in this period meant sea levels were lower, and large sections of the Indonesian archipelago were part of a single landmass known as Sundaland. But while Sundaland connected Java, Sumatra and Borneo to the Asian mainland, even when sea levels were at their lowest a deep channel separated it from Flores and the other Wallacean islands to the east.

Despite this, ancient humans were present on Flores by a million years ago. To get there these humans – the ancestors of the diminutive *Homo floresiensis*, or 'hobbits', and almost certainly *Homo erectus* – would have had to cross a wide stretch of treacherous water. It's possible this was the result of some kind of accident – perhaps individuals were swept across the channels during a storm or some other natural disaster. But the presence of other descendants of *Homo erectus* on other islands in Indonesia and the Philippines suggest it is more likely to have been an intentional process, involving not just the construction of watercraft, but also a capacity for planning and communication not previously associated with *Homo erectus*.

Homo erectus were not the only ancient hominids who seem to have been able to undertake such journeys. There is evidence Neanderthals may have been moving between the islands of the eastern

Mediterranean, and that their cousins, the mysterious Denisovans, whose genes are found in Indigenous Australians, Papuans and many of the peoples of the Pacific, also crossed the deep channels between Sundaland and what are now the islands of Wallacea several hundred thousand years ago.

Modern humans undertook even more ambitious voyages along the seaways of the Stone Age world. Although we do not know for sure what route the ancestors of Australia's Indigenous peoples took through Sundaland and Wallacea to reach what is now Australia and New Guinea more than 60,000 years ago, the journey would have involved multiple hops of tens of kilometres, and at least one long crossing of 100 kilometres or so. Likewise there is evidence humans had colonised islands in the Bismarck Islands north of New Guinea almost 40,000 years ago, and the Solomon Islands 28,000 years ago, a process that would have required crossing 180 kilometres of open ocean. Further north, in the waters separating Japan and Taiwan, people were moving between the Ryukyu Islands 35,000 years ago, and between the Talaud Islands, between the Philippines and Indonesia, some 20,000 years ago. And while it involved fewer crossings of open sea, the movement of humans into the Americas also took place on water rather than through the ice-locked interior of the continent, a long migration that followed the ribbon of biologically bountiful seaweed forests now dubbed the kelp highway down the western coast of North America somewhere between 15,000 and 20,000 years ago.

These journeys did not just disperse modern humans to almost every corner of the planet, they irreversibly altered us. As the archaeologist Jon Erlandson has observed, 'the human species, from the beginning, was an animal of the edge' that made its home along rivers and other places where ecosystems meet. But on the boundary between land and ocean, we found an unparalleled richness of resources in the form of fish and shellfish that allowed us to develop in new and transformative ways. As the historian John R. Gillis puts it, on the seashore 'the human learning curve accelerated, making possible generational continuity and the foundations of a complex, ongoing social and cultural order'. It was on the ocean's edge that the first sedentary communities took shape, and where many technologies crucial to our species' development were first invented. And it was through the act of journeying that we learned to plan and organise.

The ocean and its denizens also helped ignite the secret fire of our imagination, paving the way for the creation of art and symbolic thinking. Half a million years ago a *Homo erectus* seated by the Solo River in what is now Java took a shark tooth and carved a series of zigzag patterns into its surface with a shell. These markings were not random: they are the earliest known example of the intuitive, symbolically charged pattern-making we know as art.

Although human beings have used shells to create tools for hundreds of thousands of years, shaping them into fishhooks and employing them as blades to butcher meat and cut skins and cord, their curving surfaces and complex interplay of pattern and colour also appeal to us in other, less easily quantifiable ways. Shells were being collected and carried long before humans made paintings, or etched images in stone, their *koan*-like forms offering a medium onto which we could project meaning and value, accelerating the development of our minds. And while it is likely we were painting our bodies and faces long before we began to create ritual objects or jewellery, the oldest known example of ornamentation – thirty-three beads created by boring holes through shells dating back 142,000 years that were found in a cave on the Atlantic coast of Morocco – underline the importance shells held for our ancestors.

The presence of the shell beads in Morocco (and indeed similar evidence that Neanderthals were painting shells and making clam necklaces on the shores of southern Spain more than 100,000 years ago) reveal that these ancient humans were capable of complex symbolic thought, and that they inhabited societies that employed markers of identity and status. But it seems likely the significance of shells did more than signal power. In many cultures, shells – and cowries in particular – are associated with fertility and birth, a connection rooted in the way their rounded forms echo the fullness of the pregnant belly, and the similarity between their openings and the folds of female genitalia. And, just as our ancestors placed shells in graves more than 70,000 years ago, cultures around the world have long treated shells as markers of the boundary between life and death, using them to decorate the bodies of the dead, or to create burial mounds. This widespread usage is a reminder of the strange liminality of shells, the way they exist midway between sea and land, living and non-living, animal and object, and of their mysterious inwardness, the infinite nature of their spiral form.

AS HUMAN SOCIETY CONTINUED to expand across the globe the ocean helped create increasingly complex networks of trade and cultural interchange, enabling the development of new forms of social organisation. By 20,000 years ago people in the Bismarck Archipelago were trading obsidian with each other, and had begun moving wild animals such as the cuscus – a marsupial that is still hunted for food in New Guinea – between islands. This indicates not just systems of exchange, but also the ability to deliberately reshape their environment by introducing new sources of food and tools. These signs of early cultivation predate similar processes in the Fertile Crescent and elsewhere by many thousands of years.

Trade has also thrived in the Indian Ocean for thousands of years. At first this was mostly coastal, as the cultures that grew up on the coasts of Africa, the Arabian Peninsula and the Indian subcontinent, and along the shores of the Red Sea and the Persian Gulf, traded with their neighbours further afield. But as merchants learned to harness the annual cycle of the trade and monsoon winds they established a web of commerce that connected the east coast of Africa and Egypt to India and Sri Lanka, the Bay of Bengal and the cultures of Southeast Asia, which transported spices and other goods from India, Pakistan and the Arabian Gulf to Mogadishu and Dar es Salaam. By 4000 years ago crops such as sorghum and millet had been transplanted from Africa to Gujarat, on the west coast of India, and sesame and pepper had made their way from India to Egypt (along with the black rat). Meanwhile references in the *Rig Veda* suggest Indian sailors were experienced at deep-sea navigation well before 3000 years ago. Not long before the beginning of the first millennium, the Greek navigator and merchant known as Hippalos described the sea road made by the monsoon winds from the Red Sea to India. A few decades later the unknown Egyptian Greek merchant who wrote the work we know as the *Periplus of the Erythraean Sea* offered a detailed description of the trade routes that extended from the Red Sea along the Horn of Africa and south along the African coast, as well as eastwards as far as the Ganges.

In the enclosed waters of the Mediterranean the movement of people and goods along and between the coasts helped spur the development of increasingly powerful cultures and societies. By 4800 years ago the Minoans had founded a maritime trading empire centred on Crete; only a few hundred years later the Phoenicians and Greeks rose as well. And further west, the coastal peoples of Atlantic Europe were connected by a complex network of 'seaways'. Known as the *aster mara* in Gaelic, the *veger* in Norse, and the *hwael-weg*, or 'whale's way', in Old English, these routes, shaped by wind and current and steered by reading the waves and the stars and the movement of fish and birds, formed a cat's cradle connecting northern Spain with the coasts of England, France, Ireland and settlements on the Orkneys, Shetlands and elsewhere, as well as connecting the Atlantic to the Baltic and beyond. The threads of these seaways disseminated languages and beliefs, songs and stories, new technologies for making tools and shaping stone and metal, and precious items ranging from jewellery to carved bone and amber.

Yet the most remarkable story concerns the settlement of the Pacific. While many Polynesian cultures describe a migration outwards from an ancestral island known as Hawaiki, archaeological evidence suggests their ancestors moved west through the islands of Micronesia before striking east, out through the Pacific. These journeys were unparalleled in human history, expeditions across many hundreds of kilometres of open water that involved weeks at sea and required those making them to find islands in the middle of an ocean that covers a third of the planet's surface. Along the way Polynesians and other Pacific Islanders developed an extraordinary seafaring culture that allowed them to read the stars, sky, waves and currents, as well as understanding the movements of birds, whales and fish with such sophistication they could follow the road of the winds as far as Rapa Nui, or Easter Island, more than 2000 kilometres from the nearest island.

For these Pacific cultures their islands were not isolated outcrops in a vastness of ocean; instead they were what the historian Thomas Gladwin describes as a 'constellation linked together by pathways on the ocean', or to use Tongan and Fijian anthropoligist and writer Epeli Hau'ofa's eloquent formulation, 'a sea of islands', in which 'peoples and cultures moved and mingled'. Encoded in these systems of navigation and exchange was a way of being in the world that was not grounded in the top-down abstraction of charts and maps but conceived of as a system of relationships in which embodied knowledge and experience take precedence. Something similar is true of the shift in understanding involved in seeing the world through the lens of the ocean: no longer is it possible to understand the development of societies in isolation. Instead the map dissolves and reforms to reveal the fluid patterns of migration and exchange that connect human societies across time and space. It also demands a recognition of the complex interplay between human activity and the environment – the way we are shaped by the world around us as much as it is shaped by us.

THERE IS ANOTHER BEGINNING here as well. In 1347 galleys bearing Genoese traders arrived in Constantinople from the port of Kaffa in the Crimea. With them they brought a disease that had been slowly percolating its way across Eurasia: the bubonic plague. As the sickness took hold in Constantinople more Genoese ships arrived in Sicily, their crews already dead; after the ships were looted the disease spread into the local population, and from there to Venice and Genoa. By the end of January 1348 the plague had taken hold in Pisa, and entered France through the port of Marseille. By the beginning of summer it was burning its way through Portugal and Spain and England, as well as northwards and eastwards through Germany and Scandinavia.

The devastation the plague left behind it was almost unimaginable. Between a third and a half of the European population died, leaving entire villages empty and cities and towns half-deserted. The dead were buried in pits, or left to rot where they fell. Agricultural production collapsed as the cycles of the farming year were disrupted by the lack of workers capable of sowing the fields or harvesting crops. Trade largely ceased, and social and intellectual life was deeply affected. Many of those who survived turned to religion for succour, often giving way to displays of extreme piety such as flagellation in an attempt to curb God's wrath. Social unrest also increased: in the absence of an explanation for the catastrophe, Jews, lepers and gypsies were accused of spreading the disease, leading to mass expulsions and pogroms that saw thousands of Jews burned alive in Basel, Strasbourg and elsewhere.

But the implications of what was often called the 'Great Pestilence' extended beyond the immediate. The feudal systems that had underpinned European society for half a millennium relied upon the existence of a large peasantry to keep labour costs low. In the wake of the plague, labour was scarce, and as the peasantry grew aware of their new power they began to demand greater control over their lives and better conditions, leading to growing discontent and violent uprisings.

At first the old order sought to keep a lid on these tensions by passing repressive laws: in England wages were frozen and peasants were banned from leaving the land, while in France agitators were savagely suppressed. But it was not enough. No longer able to rely on profits at home, Europe's rulers began to look outwards, across the ocean, for new sources of wealth – chief among them African gold and enslaved Africans. And as they did, they laid the groundwork for an entirely new kind of economy.

The template for this new economy first came into view in the early fifteenth century, when Portuguese sailors landed on Madeira. Located 500 kilometres to the west of Africa and just under 400 kilometres north of the Canary Islands, Madeira and its neighbouring islands were not just uninhabited but untouched by humans, and so densely covered with forest that, in the words of Portuguese chronicler Azurara, there was not 'a foot of ground that was not entirely covered with great trees'. Within a few decades Portuguese settlers had begun to transform the islands, clearing the dense woodlands and carving watercourses in order to exploit the rich, volcanic soil. Almost immediately their carelessness had unintended and disastrous consequences: fires lit to clear trees quickly burned out of control, placing settlements and lives at risk, while in a rehearsal for the devastation Europeans would visit upon Australia four centuries later, the island of Porto Santo had to be abandoned after rabbits (introduced by the man who would become Christopher Columbus's father-in-law) exploded in numbers, devouring crops and leaving the settlers in despair. Over time the colony stabilised, growing rich enough to export small amounts of wheat and other goods back to Portugal. But that changed in 1452, when Prince Henry granted permission to establish a sugar mill on the island.

Sugar is one of the hidden engines of world history. Originally cultivated in New Guinea and the surrounding islands more than 6000 years ago, sugar cane was carried north by Austronesian traders through the islands of Indonesia and the Philippines, and through the Andaman Sea and the Bay of Bengal to China and India, as well as eastwards into the Pacific by the ancestors of the Polynesians. The process for refining the juice of the cane was developed in India around 2500 years ago (the word 'sugar' seems to be drawn from the Sanskrit word for granulated sugar, *śarkarā*, which originally meant grit, or gravel); from India it spread first to China, Persia and East Africa before arriving in the Mediterranean sometime in the first millennium. By the fourteenth century sugar cane was being grown and processed in Cyprus, Sicily and Southern Spain.

Although sugar was much sought after in Europe, limited supply meant it remained a delicacy, often costing as much by the pound as spices such as nutmeg, cinnamon and cloves. Henry's decision to establish a sugar industry was taken with one eye to this market, and it was a gamble that quickly paid off. Underwritten by the backing of Genoese merchants and bankers, the industry expanded at a dizzying rate. Within five years the island was producing more than 70,000 kilograms of sugar a year, which was borne outwards across the ocean to Portugal, Genoa and Venice in the east, and England and Flanders in the north. A decade and a half later it had risen to around 175,000 kilograms a year, and by the early years of the sixteenth century production had increased almost tenfold. As more plantations and mills opened to meet Europe's insatiable appetite for sugar the population on the islands soared as well, increasing from about 800 in 1455 to as many as 20,000 at the end of the century.

This also gave bloody birth to a new economic model, driven by the unique nature of sugar production. Sugar cane is an incredibly destructive crop that guzzles water and strips nutrients from the soil. The window for harvest is also narrow, usually only a matter of days, meaning the cane must be cut fast, and processed quickly before it spoils.

Meeting these demands required enormous amounts of labour. And so the operators of Madeira's sugar plantation began to import enslaved people. Slavery was not new in Southern Europe – by 1450 the Portuguese slave-trading station at Arguin on the Mauritanian coast was already supplying as many as 1000 enslaved people a year, some of them to sugar plantations in Valencia and elsewhere – yet the boom on Madeira caused the trade to expand, people flowing in almost as fast as sugar flowed out. Some of these enslaved people came from North Africa, but many more were Guanches, the Indigenous inhabitants of the Canary Islands, which had been invaded by Europeans half a century earlier. The growing number of enslaved labourers created a workforce of interchangeable bodies that was deployed to meet the demands of mass production by digging channels to funnel water, felling trees to fuel the sugar vats, and cutting, processing and transporting cane and molasses.

Nor was it just human lives that were fed into the infernal engines of Madeira's sugar mills. The transformation of cane into sugar consumers huge amounts of energy – boiling and distilling a single kilogram of sugar on Madeira required at least 50 kilograms of timber. At the height of the frenzy 500 hectares of forest were being felled to fuel sugar production every year.

Unsurprisingly this assault on Madeira's ecology was not sustainable. By the 1520s the trees that had once covered the islands were largely gone, and the industry collapsed as quickly as it had coalesced. In its wake it left a devastated ecology of razed forests, unchecked erosion and ruined soil. To borrow economic historians

Raj Patel and Jason W. Moore's formulation, 'Europe's wealthy ate the sugar, and the sugar ate the island.' Similarly the Guanches and their culture were almost almost entirely wiped out, victims of what the historian Mohamed Adhikari describes as 'unrestrained mass violence . . . near-total confiscation of land and near-total enslavement and deportation'. It was the first instance of a pattern of colonial-settler genocide that would be repeated many times over the next half a millennium and which continues even today in many parts of the world.

From Madeira, the sugar frontier metastasised outwards. First was the island of São Tomé, off the coast of Africa, where the model was refined, and the ad hoc mixture of indentured labour and Indigenous slavery developed on Madeira was replaced with fully racialised slavery of Black Africans, followed by Brazil and the Caribbean. As the demand for slave labour grew, its nexus began to move southwards, into West Africa, spurring the development of the Atlantic slave trade. This trade was made possible by the two systems of wind and ocean currents that dominate the North and South Atlantic, turning in opposite directions on each side of the equator like two great wheels. Using the clockwise motion of the system in the North Atlantic, goods flowed from England and Northern Europe to Africa, where they were exchanged for human beings, who were then carried west, across the Atlantic to the Caribbean and the Americas, before the ships returned to England and other Northern European ports laden with the sugar, tobacco and cotton produced on the slave plantations. Meanwhile, in the Southern Hemisphere, anti-clockwise currents carried Portuguese slavers back and forth between West Africa and Brazil.

The human cost of this extended atrocity is difficult to comprehend. Between 1501 and 1867 more than 12.5 million Africans were transported to the New World, and millions more were killed in raids and skirmishes with slave-traders. Almost two million of those who survived capture died at sea, their bodies tossed overboard. As many again died in 'seasoning' camps in the Caribbean, where the enslaved people were subjected to a regime of violence and deprivation designed to strip them of their identity and enforce their subordination to their enslavers. Those who survived were trapped in lives of such unending brutality that life expectancies were as low as seven years. This abduction of millions of people, and the extended violence of their capture and transportation to the coast, also ravaged African societies, leaving them weakened and retarding their economic development by holding back population growth at precisely the moment population was soaring in Europe.

Africans were not the only people who suffered as a result of European expansion into the New World. Although estimates vary, it seems likely there were approximately 60 million people living in the Americas prior to the arrival of Europeans in 1492. Only a century later, this number had fallen to a mere six million. Many died of diseases such as smallpox and measles, but many also died at the hands of the invaders, or in mines and on plantations. This genocidal violence, and the collapse in human population it triggered, was so catastrophic it altered the planet's climate. Ice cores show that as carbon-hungry forests reclaimed farmland, concentrations of atmospheric carbon dioxide fell by seven to ten parts per million, a decline that seems to have played a part in intensifying the global dip in temperatures known as the Little Ice Age.

The human costs of the introduction of European diseases were shadowed by a now largely invisible ecological catastrophe. As ships bore European crops and animals westwards to the Americas, and American crops such as tomatoes, potatoes and tobacco back to Europe, they brought into contact ecosystems that had been separate for almost 200 million years. Dubbed the Columbian exchange, this transfer of plant and animal species fundamentally

altered ecosystems in the Americas, simplifying and homogenis-
ing them. New plants took root, edging out indigenous species,
while introduced insects devastated crops, rats competed with native
rodents, cats preyed on birds and other small animals, and live-
stock such as cattle and horses reshaped grasslands. Even the soil
was transformed, as European earthworms, aided by their ability
to drag leaf litter down into the soil, outcompeted and largely
replaced their American cousins. In the forests of New England and
Midwest North America, where earthworms did not exist before
the arrival of Europeans, this altered forest ecosystems by removing
the leaf litter upon which many species depended, transforming
forests from places with rich layers of leaves into drier, more open
environments.

WHILE THE TRANSFORMATION of the human and ecological landscapes of Africa and the Americas by maritime trade was
perhaps the most visible instance of the impact of European expansion, it was only one part of a far larger – and longer – process. For
alongside the movement west across the Atlantic, Spanish and
Portuguese ships were also heading east, across the Indian Ocean to
India, the Indies and even China in search of spices and other forms of
wealth to exploit. By the late fifteenth century these networks of trade
finally encircled the globe, as Spanish galleons began to ship silver
extracted from the mines of Potosí, high in the Bolivian Andes, across
the Pacific to the Philippines and China to exchange for luxury goods
such as silk and porcelain. As European empires jostled for control
of these new sources of revenue the companies that controlled them
grew ever more powerful and wealthy: by the second half of the seventeenth century the Dutch East India Company employed 50,000
people and controlled a fleet of 150 merchant ships and forty warships
as well as ports and cities across Sri Lanka, Indonesia and Southeast

Asia; a century later the British East India Company controlled half the world's trade.

The profits of this process were staggering. It is impossible to separate the differing trajectories of Europe, Africa and the Americas from the fifteenth century on from the expansion of colonial empires and slavery: the increasing wealth in Western Europe from 1500 onwards is almost entirely accounted for by the rise of nations with access to the Atlantic Ocean, such as the Portuguese, Spanish, English and Dutch. This in turn is inseparable from the role of slavery, which not only underpinned the growth of economies in the Americas – through the seventeenth century close to 70 per cent of export commodities produced in the Americas were the product of slave labour, a figure that rose to 80 per cent by the eighteenth century – but thereby fuelled the growth of European nations. And this dependence did not end with the industrial revolution: the French economist Thomas Piketty calculates that between 1770 and 1810, the value of enslaved people in the United States exceeded that of all domestic capital combined; similarly by the 1850s as much as 75 per cent of the cotton used in European textile mills came from the American South, a reminder that slavery wasn't confined to agricultural production, but was a vital cog in the machine of industrialisation that drove the expansion of American and European economies. Meanwhile European nations, and Britain in particular, used their power to actively suppress industrialisation in Asia and elsewhere to secure the development of their own manufacturing. India, in particular, suffered as a result, as textile production and other sectors were held back, preventing it from developing its own industries, and helping lay the foundations for the unequal trajectories of development that shape the world today.

These systems of oceanic trade and exploitation also transformed European societies. The wealth generated in the plantations of the Americas and Asia created a powerful new merchant class and helped spur the development of new systems of property rights and constraints on state power. Sugar and other new sources of nutrition transported back across the Atlantic allowed workers to be fed cheaply, hastening the growth of urban centres and fuelling the long workdays of those who laboured in the mills and factories of Europe's rapidly industrialising cities. The journalist and historian Howard W. French has even argued that the coffee houses of the eighteenth century that gave birth to the public sphere that disseminated the values of the Enlightenment owe their existence to the availability of sugar and coffee brought back to Europe from its colonies.

European modernity, in other words, is built on the bodies of enslaved Africans and the others whose lives were used to fuel its rise, and by the systems of oceanic trade that enabled their capture and exploitation. Yet its intellectual underpinnings – and, in particular, the values of individual liberty and equality that lay at the centre of Enlightenment thought – were also shaped by the encounter between Europe and the Americas, and more particularly the criticisms of the cruelty, destructiveness and manifest inequalities of European society made by Indigenous Americans such as the Wendat states-man Kondiaronk. This indigenous critique played a vital role in shaping the arguments of philosophers such as Voltaire, Diderot and Rousseau, many of whom wrote works in which they used invented Indigenous characters – in Voltaire's case one explicitly modelled on Kondiaronk – to observe their own societies. Indeed the archae-ologist David Wengrow and the late anthropologist David Graeber go so far as to speculate that the origins of the Western gaze – the supposedly objective, dispassionate and scientific frame of reference that precludes other ways of seeing – does not arise from Europeans looking out, but 'rather in European accounts of precisely these imag-inary, sceptical natives: gazing inwards, brow furrowed, at the exotic curiosities of Europe itself'.

*

ALTHOUGH MORE THAN 500 YEARS separate us from the sugar boom on Madeira, the template created there looks startlingly familiar. Agile, inventive, fantastically destructive, the velocity and violence of its expansion and subsequent collapse, and its human and ecological cost, have been repeated over and over again. No less significantly, the experience on Madeira also makes it clear that capitalism was not born in the mills and factories of eighteenth-century Britain. Its antecedents lie further back and are bound up in the process of transformation that takes place when capital encounters new resources. Just as importantly, it reveals that slavery and the genocide of Indigenous cultures such as the Guanche were not incidental to the creation of the modern world, or to capitalism; instead they are written into their very foundations.

The birth of capitalism did not just transform the global economy. It also altered the way those who inhabit its structures see the world. For most of human history cultures around the world have understood human beings as part of a larger web of being in which all things – from biological entities such as animals and plants to the stones of the Earth and the waters of the rivers and the sea – are seen as alive and imbued with spiritual presence. To live within such a world is to recognise the essential interrelatedness of things, and our kinship with everything around us. In such cultures people take what is needed; to take more is usually seen as an act of violence that does not just harm the land but shames the perpetrator. As historian Carolyn Merchant puts it, 'as long as the Earth was considered to be alive and sensitive, it could be considered a breach of human ethical behaviour to carry out destructive acts against it'. Or in the words of the Potawatomi biologist Robin Wall Kimmerer, 'Never take the first plant you find, as it might be the last – and you want that first one to speak well of you to the others of her kind'.

In stark contrast, capitalism privileges materiality: extracting value to create goods that can be traded and sold. In so doing it ignores and renders valueless the complex webs of relationship between human and non-human beings, and the meanings inherent in the landscape. And, more deeply, it demands that which is to be exploited be transformed into things. As part of the same sleight of hand that grants corporations personhood, not only are animals and plants, rivers and forests all robbed of agency and meaning, even many human beings are reduced and made less than human.

Nowhere is this more visible than in the treatment of the enslaved, who were stripped of their identity and dignity and used as a resource. But the same process lies behind the treatment of the Indigenous peoples of the Americas, Australia and elsewhere, who were dispossessed and murdered, their systems of knowledge disregarded and suppressed. This same rendering as something less is fundamental to the cycles of exploitation and violence that have underpinned the transformation of the planet – the felling of forests for fuel and to build ships to move the enslaved, the slaughter of whales to make oil for lamps, lubricant for the machines in the factories and nitroglycerine for explosives. Or to put it another way, dispossession, dehumanisation and disregard for other living beings are as much a part of the forces that have brought an end to the relatively benign climactic conditions that have prevailed since the end of the last Glacial, some 11,700 years ago, and ushered in what many regard as a new geological epoch known as the Anthropocene, as the reckless burning of fossil fuels.

None of this would have been possible without the oceans. Their sea routes connected continents and cultures, allowing Europeans to keep pushing outwards into new frontiers, allowing goods to flow back and forth, and enabling the systems of exploitation and inequality that shape the modern world.

'VIEWED FROM THE DISTANCE of the Moon, the astonishing thing about the Earth . . . is that it is alive,' wrote the physician and poet Lewis Thomas in 1974. 'Aloft, floating free beneath the moist, gleaming membrane of bright blue sky . . . the only exuberant thing in this part of the cosmos. It has the organized, self-contained look of a live creature, full of information, marvellously skilled in handling the Sun.'

If Thomas's words were inspired by 'Earthrise', they were not directed at it. Instead they were prompted by the work of English scientist James Lovelock and his Gaia hypothesis. First formulated by Lovelock in the late 1960s, and later developed into a complete theory by American biologist Lynn Margulis, the Gaia hypothesis proposes not just that life has shaped the Earth since it first appeared, but that the Earth itself is a synergistic, self-regulating system, in which biological and non-biological systems interact to produce and sustain the conditions necessary for life.

Lovelock's idea was not new. Two thousand years ago some Stoics believed the Earth was a living being. The first-century Roman geographer Pomponius Mela wondered whether the tides might be the water being drawn back and forth by the oscillation of its breath. Instead Lovelock's thinking recognised the same awareness of interconnection and value that lies at the heart of many Indigenous cultures, by treating the world as a single, interconnected system in which all things play a part. To borrow the terminology of the writer Amitav Ghosh, the systems we inhabit are not inert: they have their own meaning and qualities that exceed and transcend the sum of their parts. A reef is not simply a collection of corals and fish, any more than a forest is simply a collection of trees: it is an extraordinarily complex system made up of many billions of organisms, and imbued with meanings both human and non-human. Energy and information ripple through a reef, memories of past shapes and existences inheres in its physical structure, and the lives of the creatures that inhabit it are shaped and made meaningful by their part in it. Nor is a reef bounded or unitary; instead it is part of a larger oceanic system, connected to the waters around it by the movement of animals and nutrients, and to the planet as a whole by the systems that drive those processes.

Gaia embodies a return to these older ways of understanding the world. Just as the ocean's immensity allows us to see past ourselves and intuit a larger frame of reference, seeing the planet as an interconnected whole alters our understanding of our relationship to the world around us. No longer is it possible to imagine ourselves as outside, or separate from the world. For we are part of it, one tiny part of an immense living system.

IN FIVE BILLION YEARS' TIME, the Sun will burn through the hydrogen at its core and begin to expand, swelling outwards and transforming into a red giant. As it grows, it will swallow Mercury and Venus, finally reaching beyond the current orbit of the Earth. Our planet may escape being consumed, released into a larger orbit by the weakening of the Sun's gravity – but even if it does, life will already be gone, perhaps as soon as a billion years from now, when the Sun's heat will have increased enough to trigger a runaway greenhouse effect similar to that which transformed Venus 700 million years ago. Subjected to the Sun's heat and radiation, Earth's oceans will be stripped away, broken down into their constituent atoms and borne away on the solar wind.

We are connected to that deep future, just as we are connected to the deep past. The water that will be boiled away billions of years from now is the same water that once filled the shallow seas in which life began and evolved, the same water that flows through the oceans and the atmosphere and the bodies of every living thing on Earth today. Its shared liquidity is a reminder of time's impossible depth and of the extraordinary, wondrous complexity of the web of processes that sustain life on our planet. But it also connects us to the events that have shaped the world we now share, its manifest inequalities and many histories of violence and grief. The scientist and author Rachel Carson wrote of the ocean's mysteries, of the way its meanings haunt and elude us, instead offering 'an uneasy sense of the communication of some universal truth that lies just beyond our grasp'. But perhaps this truth is not so elusive, perhaps this truth lies in a recognition of the forces that have brought us to this moment, and of the impossibility of moving forwards without a reckoning with them.

2

BODIES

*'I am fluent in water. Water is fluent in my body –
it spoke my body into existence.'*
Natalie Diaz, 'Exhibits from the American Water Museum'

AS OFTEN AS I CAN, I drive the twenty minutes from my home to the beach. When I was younger I tended to go in the late afternoon, often picking up a friend or two along the way and heading for Bondi or Tamarama; these days it tends to be in the early morning, a brief window before the day begins, and usually Coogee, which isn't my favourite beach but has the advantage of being closest to my house.

In the summer months it is necessary to fight for a parking spot; as a result I often prefer sunny days in the winter, when the beach is quieter. There is something magical about the beach on those mornings. On the days when the winter swells thunder in, you can hear them rumbling from streets away. But more often the water gleams in the low sun, clear and clean, the waves slowly washing onto the chilly sand.

On those days the water is often so cold that when I dive in my teeth ache. If there are waves I race to catch them; if it's calmer I strike out, swimming fast in order to get my blood moving. Either way it always takes a minute or two until my body adjusts and I feel my breath coming easily again, but it never matters. The Australian Olympic swimmer Murray Rose once said a swimmer needs to have 'a feeling for the water', and while I would never compare my prowess to his, I understand what he means. Entering the water never feels like passing into an alien medium; instead, surrendering to its weightlessness is like returning to something intuitive, a way of being remembered below the level of language or conscious thought. While one of my earliest memories is a fragmentary recollection of almost drowning in a pool under a jetty, swimming seems so natural to me that I cannot remember a time when I could not.

This affinity between humans and the ocean is deep, and ancient. We spend the first nine months of our lives suspended in the liquid of the womb; once we emerge our tissues are made of water, our blood salt as the sea. Studies consistently demonstrate proximity to water, or even just the colour blue, reduces anxiety and improves psychological wellbeing. And while at some point in our evolutionary history humans and most primates lost the innate ability to swim that most other mammals possess, our bodies remember the water in other ways: simply submerging our faces immediately triggers a series of hard-wired responses, our heart slowing as it shunts blood away from the extremities to our core to preserve oxygen and protect crucial organs. Sometimes known as the mammalian diving reflex or the master switch of life, this physiological mechanism is common to all mammals, and is actively employed by free divers, who use their mastery of it to control their bodies and dive to great depths without suffering the ill effects of hypoxia and nitrogen narcosis.

Some human communities have even evolved adaptations that improve their abilities in the water. The Bajau people of Southeast Asia make their home on the waters of the Philippines, northeast Indonesia and Malaysia. Although many now dwell on land, the Bajau traditionally lived on houseboats, sustaining themselves by fishing and diving, and only coming ashore to trade. Because of their maritime existence, they are as at home in the water as out, spending many hours every day swimming and diving, sometimes descending to depths of 70 metres and remaining underwater for minutes at a time. These abilities are the result of years of training and expertise, but a 2018 study revealed the Bajau also possess a series of genetic adaptations that seem to equip them to spend time underwater. The most significant of these involves alterations to a gene associated with higher levels of thyroid hormones, and may be the reason the Bajau's spleens tend to be larger than those of other human populations, enabling an increased supply of oxygenated blood cells. But there is also evidence of adaptation in genes that control the dive reflex, and another that affects the levels of carbon dioxide in the blood. Fascinatingly this gene appears to have come from the Denisovans, whose genetic legacy is also responsible for the ability of Tibetans to thrive in the attenuated atmosphere of the Himalayas.

There is also evidence to suggest some of the Moken people, who live among the islands of the Andaman Sea, have developed the ability to change the shape of the lens in their eyes and reduce the size of their pupils when they dive, allowing them to see as clearly underwater as above. Unlike the abilities of the Bajau this capacity is not innate: instead it seems to be developed in childhood, perhaps as a result of long periods swimming and diving. Nonetheless it testifies to the ways in which close contact with the water can shape human beings in profound ways.

Some scientists have even gone so far as to suggest swimming *made* us human. In 1960 the marine biologist Alister Hardy published an article arguing our ancestors may have returned to the water at some point in our evolution, pointing to our relatively streamlined bodies, hairlessness, dive reflex and layer of subcutaneous fat as evidence for such a phase. Hardy proposed this phase might account for our bipedalism, noting that it is much easier to stand upright when supported by water. His ideas were later developed and refined by the Welsh writer Elaine Morgan, who argued this aquatic ape theory offers an alternative to the patriarchal fantasy of human evolution that focuses on 'the Tarzanlike figure of the prehominid who came down from the trees, saw a grassland teeming with game, picked up a weapon and became a Mighty Hunter'.

Although Morgan's ideas continue to haunt the pseudoscientific fringes of evolutionary theory, the evidence for an aquatic phase in our development is tenuous at best. Yet while there are alternative explanations for almost all the characteristics the aquatic ape theory purports to explain, there is little question many ancient hominids were at home in the water. The various feats of seafaring undertaken by *Homo erectus* suggest skill and experience on the water, something that is unlikely to have been possible if they were not able to swim.

Archaeological remains on Crete and other islands in Aegean and Ionian seas suggest our Neanderthal cousins were seafarers as well. But whether they were sailors or not, the presence of the shells of a species of clam only found 3 to 4 metres below the surface in Neanderthal encampments on the Italian coast make it clear Neanderthals were swimming and diving in the waters of the Mediterranean 90,000 years ago. This finding is backed up by the presence of bony growths in the ear canals of many Neanderthal skulls; in modern humans these same excrescences are most often seen in surfers and swimmers who spend extended periods of time submerged in cold water. Likewise remains found on Gibraltar, where the last Neanderthals survived until a mere 28,000 years ago, show they not only ate fish and shellfish but caught and butchered seals and dolphins – all practices that would be extremely difficult were they not comfortable in the sea.

Modern humans have also been at home in the water for tens of thousands of years or even longer. The acceleration of human social evolution triggered by our encounter with the ocean depended upon us being able to take advantage of the abundant supply of fish, shellfish and other resources: it seems implausible that took place without us being able to swim. Nor is it easy to imagine a scenario in which our ancestors spread outwards through the islands of Indonesia, the Philippines and the Japanese archipelago, as well as into Australia and along the Kelp Highway into the Americas, without them being proficient in the water.

Despite this, the first direct evidence of modern humans swimming comes not from the coast, but somewhere else altogether. In 1933, in the region of the Gilf Kebir plateau in southwestern Egypt known as the Wadi Sura, the Hungarian explorer and archaeologist (and inspiration for Michael Ondaatje's novel *The English Patient*) László Almásy stumbled upon a cave in which images of human figures mingle with paintings of animals such as hippopotamuses and giraffes. The cave is not the only one with rock art – indeed the name 'Wadi Sura' is sometimes loosely translated as 'Valley of Pictures', and other caves nearby feature images of other animals, as well as archers and other human figures dating back 6000 to 9000 years – but the paintings Almásy found were different. For there, etched in red across the pale stone, the human figures glide and undulate, their bodies horizontal and their arms extended out in front of them, in what is – unmistakably – the act of swimming.

Almásy's discovery was made even more remarkable by the fact the Gilf Kebir lies in the Libyan Desert region of the Sahara, one of the driest places on Earth. Faced with this conundrum Almásy made a remarkable intellectual leap, asking whether it was possible this dry region had once been rich with water.

This idea was treated as fanciful at the time yet, as our understanding of our planet's climatic history has grown more sophisticated, Almásy's intuition has been borne out. The Earth is tilted slightly on its axis, meaning that at any given time one hemisphere is receiving more solar radiation than the other. But this tilt is not constant; the planet's inclination oscillates slightly, moving from a fraction more than 22 degrees to just over 24.5 degrees and back every 40,000 years or so. Some 14,500 years ago, this process caused the African Monsoon to shift northwards slightly, bringing rain to the Sahara. These monsoonal rains filled the valleys and lowlands of the Sahara, transforming the desert into a lush landscape of interconnected lakes and wetlands, teeming with fish and shellfish and animals such as hippopotamuses, elephants, crocodiles and giant tortoises. By 10,000 years ago, there were humans here as well. While little is known of the people who lived in what is now the Gilf Kebir, at the same time in the Gobero, just under 2000 kilometres to the southwest and now part of the Ténéré region of the Sahara in Niger, the people known as the Kiffian were catching fish – including huge Nile perch up to 2 metres in length – using slim spears with beautifully carved heads and harpoons shaped from bone, as well as producing elegant pottery and jewellery shaped from hippo tusks.

Yet the figures that stroke their way across the walls and ceilings of the Cave of Swimmers speak to the planet's future as well as its past. For 6000 years ago, when the Earth's rotation shifted slightly once more, the Sahara dried out rapidly, transforming the lush wetlands that had sustained the culture that produced the underground galleries of the Wadi Sura into the barren landscape we know today in the space of only two centuries, a humbling reminder of how abruptly and how dramatically the climate can change, and of the consequences when it does.

IRRESPECTIVE OF WHERE it happened first, there seems little doubt that swimming has been a common practice in cultures around the world for many thousands of years. Along the eastern and western coasts of Africa, people were skilled and enthusiastic swimmers; similarly in Australia, the Americas, Southeast Asia and much of India, people seem to have been at home in the water. The same was true in ancient Egypt, where hieroglyphs from almost 5000 years ago show people swimming, and the gods themselves were believed to stage competitions of strength and prowess in the water. In the Persian Gulf, divers have been harvesting pearls from the sea floor for at least 7000 years; some even seem to have pioneered the first diving gear in the form of goggles made of leather and polished turtle shell. Meanwhile the ruins of the palace at Mohenjo-daro, built around 4500 years ago in modern-day Pakistan, contain a pool that appears to have been used for swimming or bathing. The Chinese *Shijing*, or *Book of Odes*, contains a reference to swimming that dates back almost 3000 years, while in the Americas Mayan reliefs show swimmers performing something that looks remarkably like breaststroke more than 2000 years ago.

While many, if not most, cultures seem to have been at home in the water several thousand years ago, one group was not. There is evidence that, despite the icy conditions, humans in western Eurasia spent time in the water during the last Glacial, and continued to do so after the ice sheets retreated. The same sorts of bony growths that are found in the ears of Neanderthals have been found in skeletons of modern humans dating back 11,500 years or more from locations in France and Germany. Further north people also relied on aquatic resources, as demonstrated by the discovery of fishhooks dating back 10,000 years on the shores of Norway and Sweden, and a more than 10,500-year-old willow-bark net recovered from a peat bog in Finland. Yet from about 8000 years ago people in northern Eurasia seem to have stopped spending time in the water.

Exactly why this shift occurred is not certain, but it seems to have coincided with the arrival of agriculture, which reduced the reliance of these ancient peoples upon aquatic resources. Over time this cultural aversion to swimming spread through Europe and the Middle East, and later eastwards, into China, so that by 4000 years ago, cultures from West Asia to Europe had largely abandoned the water. From the *Epic of Gilgamesh* to the Torah the water features as a force of destruction, and immersion leads to death by drowning. And, no less importantly, the ability to swim became a signifier of difference, a way of distinguishing between us and them.

It would take centuries for attitudes to begin to change once more. The shift this time came from the south. Unlike their neighbours to the north, Africans and Egyptians had never stopped swimming. As the power and influence of these cultures grew, their enlightened attitude to the water began to spread north, altering attitudes in Central Asia and Southern Europe. But while swimming gradually became more common in these regions, the practice was not universal. Among both the Greeks and the Romans the ability to swim was largely confined to the **wealthy and** privileged. Nor was it regarded as something natural; it was seen as a skill that had to be acquired through training, and therefore as a marker of education and sophistication: Plato speaks of those who are 'unlettered and unable to swim', and similar prejudices echo through the works of Herodotus and others.

As the Roman Empire's influence waned in the West, that began to change. Although swimming continued to be a mark of distinction, a skill possessed by kings and queens, most Europeans abandoned the water, often coming to regard the ocean with fear and horror. The reasons for this were complex but seem to have had more than a little to do with the rise of the Church and its discomfort with the body and female sexuality. Especially in Britain, swimming came to be regarded as a fool's activity, and water was transformed from a symbol of purity into a source of suspicion and fear, becoming, as the writer Charles Sprawson put it, a substance 'devilish rather than divine . . . a breeding ground for rats, a source of plague and disease'. The legacy of this rejection of the water persists today, visible in the tendency of Western culture to treat the ocean and other waterways as blank spaces, cultureless voids whose main purpose is to delineate terrestrial spaces, rather than recognising the complex interrelationship of land and water that underpins many cultures elsewhere.

Despite the European retreat from the water, other cultures retained a deep affinity with it. This was especially true on the coast of West Africa, where many of the rich and complex societies that thrived along the Atlantic coast in the Middle Ages were oriented outwards, towards the ocean. In many of these cultures, land and water were not separate: instead the ocean shores and the spreading deltas and rivers that fed into them formed what the historian Kevin Dawson describes as 'waterscapes', in which land and water were seamlessly bound together. These waterscapes were an integral part of daily life. The abundant fish and shellfish provided sustenance and goods for trade, just as the annual movements of life through coastal waters helped structure the rhythms of social life. The ocean and inland waterways also connected communities and cultures as canoes bore people and goods back and forth along the coast and into the interior.

This aqueousness also took on other, deeper forms. In many West African cultures water and the ocean had multiple meanings, existing as both the medium of everyday life and as a sacred, supernatural space, inhabited by spirits and deities whose voices were audible in the sound of the breaking waves and running water. The surface of the ocean was not just the boundary between the air and the water, but the *kalunga*, the shoreline between the world of the living and the world of the ancestors. Thus to die was to pass downwards, through this boundary, and into the world of the ancestors; in some places it was customary to bury the dead with a small model of a canoe that symbolised this passage, or with cowries, which served not just as money, but as powerful charms and tools of divination with which to buy passage across the *kalunga*. As more and more people were abducted by slavers, the Atlantic Ocean itself came to form this boundary, so those enslaved and taken westwards to the New World were sometimes believed to have been taken to the land of the dead.

In such a world swimming was an integral part of life. Both men and women learned to swim at an early age: 'Once the children learn to walk by themselves, they soon go to the water in order to learn how to swim', observed Dutch merchant Pieter de Marees in the early seventeenth century. Skilled swimmers gathered oysters and cowries from the sea floor; these served as sources of food and lime for building, as well as repositories of spiritual and material value. Still others guided canoes and other boats along rivers and through the breaking waves along the coast. The skills involved in the latter were of particular importance: the scarcity of natural harbours on the Atlantic coast of Africa meant the ability to navigate the surf zone played a vital role in connecting coastal communities to fisheries and coastal trade routes. Many West Africans also surfed for pleasure, either bodysurfing or employing small wooden boards, a variety of longboards and single-person canoes to ride the waves, as well as guiding larger fishing canoes and other craft through the surf.

These abilities astounded European travellers from the outset. During one of the first recorded encounters between Europeans and sub-Saharan Africans, off Senegal in 1445, Portuguese sailors captured several local fishers; when the men escaped, their captors marvelled at their skill in the water, declaring they 'dived like cormorants'. Similarly, one fifteenth-century Venetian slave-trader related the story of two Senegalese men who carried a message 5 kilometres through violent seas and then returned with a reply before concluding 'that the Blacks of that coast are the greatest swimmers which there are in the world'. Later writers would declare Africans were 'amphibious', and able to 'swim beneath the water like a fish', while slave owners in the New World compared the abilities of enslaved Africans to otters, beavers, dolphins and more.

While these descriptions speak to the astonishment felt by Europeans, they are also part of a darker process of exoticisation and dehumanisation. European travellers in Africa constantly emphasised the differences between Africans and Europeans, contrasting their own refinement to the libidinousness and savagery of Africans in the most sickening terms. Some asserted that African women did not experience pain during labour, others likened Africans to apes. Comparing the aquatic abilities of Africans to those of animals, or claiming that, like animals, African children were born knowing how to swim, thus became one more way of othering Black Africans and treating them as less than fully human.

This was not the first time swimming had been racialised: a less rigid but similar distinction seems to have existed in the Muslim world, where the aquatic skills of enslaved East Africans stood in opposition to those of their masters. Nor was it the last: in the Americas the aquatic aptitude of the Aztecs and others was also seen as a sign of their animal nature, as Cortés's chronicler Bernal Díaz's observation that 'like crocodiles or seals', Aztec warriors 'swim as easily as they walk on land' or sixteenth-century descriptions of Brazilians swimming like ducks and porpoises attest. Similar observations were made about Indigenous Australians and Pacific Islanders, among others. In other words, Europeans subsumed the ability to swim into the poisonous edifice of racism that under-pinned European colonialism – another way of distinguishing what they regarded as their natural superiority from what they saw as the backwardness and bestiality of the people upon whose bodies European wealth was built.

BACK IN EUROPE, attitudes to swimming were in flux once again. On the one hand, concern about the moral and physical risks of swimming – or, given how few people could actually swim, the very idea of swimming – was tangible, and often backed up by official edict. As late as 1571 the vice-chancellor of the University of Cambridge banned swimming altogether; scholars caught breaching the rule faced flogging, time in the stocks, or even expulsion. Yet simultaneously the Classical notion of swimming as a mark of refinement had begun to resurface. In 1531 English diplomat Thomas Elyot published *The Boke Named the Governor*, which advocated that swimming should form part of the education of young noblemen. This was followed in 1538 by German scholar and linguist Nikolaus Wynmann's *Colymbetes, sive de arte natandi dialogus et festivus et iucundus lectu* (*The Swimmer, or a Merry Dialogue on the Art of Swimming*) and, more importantly again, Cambridge theologian Everard Digby's 1587 treatise, *De Arte Natandi*.

Digby's book, which was translated into English by Christopher Middleton as *A Short Introduction for to Learne to Swimme*, makes some surprising claims, including that humans excel at swimming 'above all fowls of the air, fishes of the sea, beasts of the earth, or other creatures whatsoever' (this unlikely proposition is justified by the fact that swimming is the only gift God bestowed upon fish, whereas men can dive to the bottom of the sea, carry things, swim on their backs and just generally cavort in the water in a way animals and fish cannot). But Digby also argues eloquently for the importance of learning to swim, not just as a way of guarding against the danger of drowning, but as a way of cleansing the body, cooling off in summer and promoting longevity.

Digby advocates four strokes – breaststroke, backstroke, a form of sidestroke and an exhausting-looking doggy paddle – as well as various 'decorative feats', which include using the hands to hold two birds out of the water while swimming a form of backstroke. Possibly more importantly, Digby included a series of woodcuts, that offered a visual guide to show how one might swim. And while these images might seem curious to a contemporary eye and their practical application is limited at best, their energy and playfulness evoke a delight that is immediately familiar and suggest a way of imagining the body in water that presages a new way of inhabiting its fluidity.

Over the next century or two multiple treatises expounding the virtues of swimming were published in England and Europe, and schools such as Eton and Harrow began to train young men to swim as a safeguard against drowning. But it was not until the mid-eighteenth century that the activity really came back into vogue in Europe and America.

One of the most significant figures in this rehabilitation of swimming's reputation in Europe and North America was Benjamin

Franklin. Franklin learned to swim as a boy in Boston, where he took pleasure in diving into the water as well as rowing boats and canoes. He continued to exercise this proclivity after arriving in England. In London the eighteen-year-old Franklin taught several of his friends to swim. Then in 1726 he demonstrated his prowess in the water to a boatload of acquaintances by diving into the Thames and swimming several kilometres from Chelsea to Blackfriars, 'performing on the Way many Feats and Activity, both upon and under the Water, that supriz'd and pleas'd those to whom they were Novelties'.

Franklin's passion for swimming was part of a wider revival of interest in the benefits of physical exercise during the eighteenth century, a process that saw the educated and wealthy take up a range of activities. In the case of swimming this was hastened by medical theories that extolled the virtues of immersion, especially in seawater. But as its popularity grew, swimming also began to be associated with ideals of masculine endeavour and martial virtue. In 1786 Barthélemy Tarquin opened a swimming pool in the Seine with the aim of teaching naval and marine cadets to swim, an innovation that spread to other European nations in the aftermath of Napoleon's military successes.

This conception of swimming as a symbol of male courage was given even greater currency in the early nineteenth century by Lord Byron. In 1810, the 22-year-old Byron and a young marine, Lieutenant William Ekenhead, attempted to replicate the mythical Leander's fateful swim across the Hellespont to the tower of his lover, the priestess Hero. Setting off from Sestos, Byron and Ekenhead struck out towards Abydos, but abandoned the attempt after the current carried them so far seaward they were afraid they would be swept out into the open sea. A week later they tried again, and succeeded, reaching the southern shore in just over an hour.

Byron was soon boasting of his achievement in correspondence, verse and conversation: he was notorious for pretending to be vague about exactly how far he had swum and would often pause in the middle of an account of his feat to ask his valet, William Fletcher, to remind him. 'Three miles and a half,' the long-suffering Fletcher would reply, presumably to the amazement (or amusement) of those present, before Byron dove back into his tale.

Over the next few decades the popularity of swimming surged in Europe and America. In London and other metropolitan centres pools began to be constructed, partly to promote hygiene and cleanliness, partly as places of recreation. People began racing as well, and champions emerged, as did an entire industry of swimming instructors and coaches to assist those who wished to improve their skills.

As Europeans celebrated swimming as a civilised and enlarging pursuit, its racialised dimensions persisted and multiplied, perhaps most visibly in attitudes to particular strokes. Although there were a number of odd variations, many of the strokes used in these races and carnivals would have been familiar to a modern observer. Nonetheless there was one very notable absence: nobody swam the overarm or front-crawl stroke modern swimmers know as freestyle. Instead, in Britain, Europe and America, breaststroke was king. This may come as a surprise to some: while breaststroke remains a favoured stroke in parts of Europe and Asia, in Australia it is often regarded as a slower and weaker (and therefore less manly) stroke than freestyle. Despite this it would be a mistake to underestimate its power. Byron swam breaststroke across the Hellespont, and Captain Matthew Webb, who in 1875 became the first man to swim across the English Channel, also swam breaststroke, battling jellyfish stings and difficult currents to complete his 66-kilometre route in just under twenty-two hours.

In contrast, overarm or crawl strokes have been in use in other parts of the world since ancient times. Images in the memorial temple of Rameses II show ancient Egyptians using a form of crawl more than 3000 years ago, and similar images dating from around 2900 years ago show Assyrians swimming overarm across a river (assisted in one case by an inflated goat bladder). Crawl strokes were common in the Pacific as well: in the sixteenth century Magellan observed Micronesians swimming overarm, while one of the last records left by the ill-fated French explorer Lapérouse contains an account of Easter Islanders doing the same, together with a comment about the speed and power of their swimming. Historical accounts make it clear that Africans, Indigenous Australians and Native Americans also routinely swam versions of the crawl, and that it had begun to be adopted by at least some American colonists by the first half of the nineteenth century.

The popularity of breaststroke in Europe may have been because Europeans tended to swim lakes and rivers, rather then the turbulent waters of the ocean, where crawl strokes come into their own. Or it may have reflected a cultural suspicion of immersing one's face in the water. But it was also intimately bound up in the European need to categorise the swimming abilities of Africans and Indigenous peoples as primitive and subhuman. Whereas European swimming and the strokes it favoured were seen as scientific and civilised, the swimming styles of other cultures – and, in particular, overarm strokes – were vulgar and animalistic. It is not accidental that nineteenth-century authors such as Theodorus Mason insisted that swimming should be 'smooth and graceful', or that breaststroke was sometimes referred to as 'White' swimming.

This was vividly illustrated in 1844, when two Ojibwe men, who had been brought to London by the American adventurer, artist and promoter George Catlin, provided the first known demonstration of overarm swimming in England. The pair, known by their stage names Flying Gull (Wenishkaweabee) and Tobacco (Sahma), swam two lengths of London's Holborn Baths before a crowd of several hundred. In their first race the two men competed against each other, swimming 130 feet (just under 40 metres) in less than thirty seconds before facing off against the noted English swimmer Harold Kenworthy.

Despite the two men's speed, the *London Times* described their style of swimming as 'totally un-European. They lash the water violently with their arms, like the sails of a windmill, and beat downwards with their feet, blowing with force and performing grotesque antics.' It also – predictably – celebrated the reassertion of English superiority in the match-up with Kenworthy, who beat 'the Red Indians . . . with the greatest of ease'.

This notion of breaststroke as emblematic of European civilisation persisted well into the second half of the nineteenth century. Adventurous swimmers in England and elsewhere were experimenting with forms of crawl (or 'the Indian stroke', as it was often known) as early as the 1850s, and by the 1870s many people had begun to adopt the one-armed sidestroke known as the Trudgen stroke. But crawl strokes did not really begin to supplant breaststroke among Europeans until the beginning of the twentieth century, when Sydney swimming coach Dick Cavill introduced the stroke that would become modern freestyle: the Australian crawl.

While the Australian crawl took the world by storm and made Cavill a star, its origins suggest a different kind of history. Legend has it that the originator of the stroke was not Cavill but a young Solomon Islander, Alick Wickham. Wickham, who was employed as a houseboy, often swam in weekend races at Sydney's Bronte Baths, usually employing the stroke he had learned as a child in the Roviana Lagoon in the Solomons, the *tapa-tapala*. According to this version of the story, swimming coach George Farmer spotted Wickham swimming *tapa-tapala* during one such session and exclaimed 'Look at that kid crawl!' Intrigued, Cavill and his family adopted and refined Wickham's style, developing the basic form of the stroke – and its name – we know today.

Australian historians Gary Osmond and Murray Phillips argue that this version of the stroke's origin does not accord with the historical record, and that Cavill may have picked up the technique earlier, while on a trip through the Pacific. Either way, this overarm stroke is not and never was the 'Australian crawl'; if it was anything it was the Solomon Islands, or possibly the Roviana Lagoon, crawl. The story of the crawl's invention and appropriation is better understood as another example of the British disdain for 'native swimming' and the erasure of the influence of Indigenous swimming techniques more generally. It also serves as a reminder that swimming was always enmeshed in the racial politics of its time and in the systems of colonialism and nationalism that gave shape to them. Channel swimmer Matthew Webb encapsulated these assumptions with brutal succinctness when he pronounced in 1875, 'None of the black people I have ever known approach a first-class English swimmer.'

The use of swimming as part of a racialised national mythology was not confined to Europe. In Japan, a place where the water and swimming have long been deeply enmeshed in social and cultural life, swimming was exalted as one of the military arts. Those proficient were expected to be able to swim *hira-oyogi*, or classical Japanese breaststroke, and to master feats: *tachi-oyogi-shageki*, a special eggbeater kick that allowed those who had mastered it to aim and shoot a gun or a bow while treading water, and, *inatobi*, a kick powerful enough to launch those who had mastered it out of the water and into a boat like a dolphin.

The incorporation of swimming into the idea of the nation and its symbolic power reached its apotheosis in Nazi Germany in the 1930s. The Germans had long regarded swimming as a central part of their military tradition, tracing its origins as a symbol of German vitality and heroism back to Germanic tribes such as the Suebi, whose prowess in the water was noted by Julius Caesar and others. The Nazis married these fantasies of Aryan vigour to Classical ideals of bodily perfection, and imbued them with mythic power, transforming the act of swimming – and more particularly the swimmer's body – into the larger racial edifice of Nazism, and its repulsive fantasies of physical and spiritual perfection.

This process is given queasily powerful expression in *Olympia*, Leni Riefenstahl's propagandising hymn to the 1936 Olympics, where scenes featuring a sleekly charming and masterful Hitler sit alongside footage of divers in flight. As with much of Riefenstahl's work, it is difficult to untangle the aesthetic power of the whole from the toxic delusions of national and racial supremacy that animate it. Yet as the perfectly muscled bodies of the divers soar and somersault against the sky, the editing speeds up and the montage becomes more grandiose and surrealistic, until they become part of an extraordinary, transcendent whole, beautiful and deadly.

This racialised idea of the swimmer's body was also present on the other side of the Atlantic, perhaps most obviously in the casting of the swimmer Johnny Weissmuller in the film adaptation of Edgar Rice Burroughs' Tarzan stories. Tarzan's racism was baked in from the start, of course: in Burroughs' original novel his Whiteness is constantly, almost anxiously invoked, and his nobility and intelligence are contrasted not just to that of the apes who raise him, but to the 'savage blacks' he battles. Seen through this lens the casting of Weissmuller, with his long-legged, sleekly muscled physique and Aryan good looks, takes on an even more disquieting charge, quite literally embodying the colonial and racist fantasies underpinning the character.

These elements converge in an extended underwater sequence originally shot for 1934's *Tarzan and His Mate*, but cut before the film's release, in which Tarzan and Jane frolic underwater, Tarzan clad only in a loincloth, Jane nude (Weissmuller's co-star, Maureen O'Sullivan, was replaced by Olympic swimmer Josephine McKim for the underwater shots). The scene is as startling for its beauty as it is for its daring. Against the dark backdrop of the pond Tarzan and Jane move effortlessly, their pale bodies dappled in the light from above. Their nudity is at once liberating and sensuous, their bodies careless in their grace and power. And, as if to underline the scene's intense eroticism, O'Sullivan's low, delighted laugh when she finally surfaces is unmistakably, nakedly sexual. Yet despite the scene's beauty there is also something unsettling about it, a sense in which the Whiteness of their bodies is uncanny, even deathly.

Perhaps ironically, *Tarzan and His Mate* was banned in Nazi Germany, whose censors objected to its depiction of an Aryan man living like a savage. Yet one only has to think about the stereotypical images of blond, sun-kissed swimmers and surfers to see how these associations between Whiteness and the water, and Whiteness and the swimmer, persist in many cultures.

Nowhere is this truer than in Australia, where the idea of the classlessness and generosity of the beach is one of the most-treasured myths of White Australian culture, captured in iconic artworks such as Max Dupain's *Sunbaker* (1937) and Charles Meere's *Australian Beach Pattern* (1940). Certainly it is not coincidental that in the speech he gave after winning the seat of Wentworth at the 2007 election, former Australian prime minster Malcolm Turnbull harked back to visiting Bondi as a boy. He evoked a place where bricklayers and barristers lived and played side by side, characterised by what he dubbed, in a characteristically Turnbullian turn of phrase, 'the democracy of the surf club'. Yet as the riots that erupted two years earlier in the beachside suburb of Cronulla, in Sydney's south, clearly demonstrated, this supposed openness and equality disguises an uglier reality: after simmering tensions between locals and visitors of what was often described as 'Middle Eastern appearance' came to a head in a clash between a group of Lebanese-Australians and a lifeguard, a crowd several thousand strong gathered on the foreshore. Draped in Australian flags and chanting 'We grew here, you flew here,' this mob attacked people of colour, and pelted police with beer bottles.

SIMULTANEOUSLY, however, swimming has also long offered a means to resist and subvert these structures of colonial and racist subjugation. Although the distinction between the aquatic proficiency of many of the enslaved and dispossessed and the relative lack of such skills among those who exploited them was used as a way of policing the hierarchy that treated those with darker skin as less human, swimming also provided a source of shared experience and community for the oppressed. Enslaved Africans passed aquatic skills and understandings from their original cultures down through multiple generations, and these shared aquatic cultures often helped the disparate cultural groups thrown together on plantations to build new forms of community. They imbued the waterways and coastline of the New World with beliefs they had brought with them over the Atlantic, transforming them into new waterscapes. In the words of Kevin Dawson, 'cultural resilience served as a mechanism of resistance'.

The water also offered a more literal space of freedom, providing a place for the enslaved to bathe and relax. And at a more fundamental level, the ability to swim helped many of the enslaved escape their captors.

In Hawaii, surfing served a similar function: the in-between space of the *po'ina nalu*, or surf zone, was not just a space of physical and psychological refuge for Hawaiian men, it also provided a place in which they could upend colonial hierarchies by asserting their superior skills. As surfer and historian Isaiah Helekunihi Walker says, as 'they transgressed *haole* expectations and categories in the waves, Hawaiian surfers simultaneously defined themselves as active and resistant Natives in a colonial history that regularly wrote them as otherwise.'

This process was also visible in Australia. In the early 1800s gangs of sealers arrived on the place that the invaders would come to call Kangaroo Island, off the coast of South Australia. Although only about 14 kilometres separates its eastern tip from the end of the Fleurieu Peninsula, the island was uninhabited, seemingly abandoned sometime after rising sea levels cut it off from the mainland about 10,000 years ago. The Ngarrindjeri, whose Country extends from the southern tip of the Fleurieu Peninsula to the Coorong, called it Karta; for them it was the place where the ancestral being, Ngurundjeri, who made many of the islands and bays along the coast, took the spirits of the dead.

The sealers had not been on the island long before they began carrying out raids on the Ngarrindjeri and on Aboriginal communities in Tasmania and the islands around it, abducting women and bearing them back to their camps on the island.

The grief for the communities left behind is inconceivable. The fate of the women was even more horrific. On Karta they were forced to work for the sealers, and were repeatedly raped and beaten. If they refused, they were punished even more brutally: one of the sealers is said to have tied up one of the women, slashed her buttocks and cut off her ear, while another snapped the arm of a ten-year-old boy over his knee as punishment for running away. Despite the consequences if they failed, many women still sought to escape. One woman freed herself and her companions and made it back to the mainland in a stolen boat, and there are stories of another group who fled in a dinghy.

Others, however, chose to swim. This must have been a terrifying prospect. Backstairs Passage, the strait, is notoriously treacherous, not just rough and cold, but fast-moving and filled with great white sharks. Despite these challenges one woman, who had been snatched from the mainland and raped multiple times, slipped away in the darkness and made it back to her people and her child. Others were not so lucky. At least one woman made it back only to be recaptured and returned to the island. Others drowned: one woman attempted the swim with her child tied to her back: the Aboriginal man who found her body on the beach later said, 'Baby died on way. Child on her back.'

THE IDEA THAT THE ACT of swimming might be shadowed by these histories as well as its more usual associations cannot help but alter our relationship with the water, imbuing it with a sort of double-visioning, revealing the degree to which an activity often regarded as entirely natural is in fact shaped by historical forces. But swimming also gives rise to another sort of duality, a sense in which we are simultaneously contained within our bodies and part of something far larger. For those who inhabit a European tradition, the latter is given precedence: in an echo of the Romantic notion of the swimmer as a solitary figure, somehow separate from society. As free diver Johanna Nordblad, who specialises in a form of dynamic freediving that involves swimming as far as possible under ice on a single breath, says, she loves the blackness of the depths because 'Down below there are no colours, no sounds. There's nothing . . . You are completely alone with yourself.' Likewise my friend the philosopher Damon Young argues that ocean swimming brings us into contact with the sublime, reminding us of our physical fragility and finitude in the face of the water's immensity and mystery.

Others have described swimming as a kind of compulsion. In his *Confessions of an English Opium-Eater*, Thomas De Quincey intuits a connection between swimming and taking drugs, citing his observation that he had consumed so much opium he 'might as well have bathed and swum in it', as well as the way opium inclines the mind towards images of moving water. The writer and surfer Thad Ziolkowski makes a similar connection with surfing: if 'surfing is a . . . parable of addiction', replete with the 'thrill of being gathered up and borne along as if by magic,' he argues, then the passage into the physical and psychological zone of surfing – that transitional zone between the shore and the open ocean, between thought and action – also involves a kind of psychosocial liminality. 'Addiction is liminal too,' he writes, 'an interstitial realm divided from workaday reality by an unseen veil . . . Addicts and surfers . . . are . . . threshold people, who exist on the margins of society in a state of social invisibility and lowliness.'

It is not difficult to see why such ideas might appeal: in its vastness and careless violence the ocean often seems to render ideas of selfhood and individuality irrelevant. Yet they also transform the ocean and the water into an idea rather than a real place, stripping out its particularity and reducing it to a philosophical concept. The ocean is not trackless or unknowable. Instead, as I am reminded every time I enter its waters, it is a living space, where the great cycles of the planet – the movement of the tides and currents, the motion of the wind and cloud, the arrival and disappearance of animals across days and seasons – intertwine with the busy life of the immediate, the suck and pull of the kelp in the surge of the waves, the shimmering movement of the fish, the cries of the gulls and the dive-bombing terns. Understood in this way, water's mutability becomes a way of imagining and even inhabiting different ways of being, of acknowledging the intermingling of our bodies and lives with both planetary systems and the intimate, microscopic life that surrounds and permeates us.

Perhaps it would do better to attend to the experience itself, the way that surrendering to the weightless embrace of the water allows us to be in our bodies in the most intimate way while also connected to these larger cycles. Understood like this, swimming is more than a way of being: it is a way of knowing. When we swim, we become one with water's fluidity. Suspended in it we must learn to follow its eddies and currents, to allow its motion to bear us where we want to go. As anybody who swims in surf will know, it is not possible to fight the waves; instead it is necessary to move with them. Sometimes that is exhilarating, sometimes it is meditative, occasionally it can be terrifying. But it also offers a different way of thinking about our relationship to the world around us. When we swim our bodies become part of the tidal flow and movement of water, the great pulse of the planet's systems, the act of giving ourselves over to their rhythms a form of communion, of embodied connectedness. As the Blue Humanities scholar Steve Mentz has eloquently put it, 'swimming can become an ecological meditation for the Anthropocene'.

This meditation cannot exist outside of the histories embodied in the ocean. Describing the way 'water holds Black memory', the writer Makshya Tolbert quotes the poet Christopher Gilbert, 'Floating brings the Atlantic to light, floating allows me the spaciousness of being "the long black language that reaches all the way back"'. For Tolbert water is a place in which slavery's afterlives reform and recur, but the act of becoming water also offers the possibility of learning to live within and beyond those afterlives, a way to 'transmute violence into healing practice'.

But neither can it be divorced from the unknown space ahead. For as sea levels rise, they bring with them the legacy of centuries of extraction and colonial violence. To survive that future we will all need to be swimmers.

3

MIGRATIONS

'Wake: the track left on the water's surface by a ship; the disturbance caused by a body swimming or moved, in water; it is the air currents behind a body in flight; a region of disturbed flow.'
Christina Sharpe, *In the Wake*

EACH NIGHT, AS THE LINE that separates day from night sweeps across the face of the ocean, a vast wave of life rises from the ocean's depths behind it. Made up of an astonishing diversity of animals – myriad species of minute zooplankton, jellyfish and krill, savage squid and a confusion of fish species ranging from lanternfish to viperfish and eels, as well as stranger creatures such as translucent larvaceans and snotlike salps – this world-spanning tide travels surfaceward to feed in the safety of the dark, before retreating to the depths again at dawn.

Known as the diel vertical migration, this nightly cycle is the single largest movement of life on Earth, with some estimates suggesting the biomass of the animals that make the journey may total 10 billion tonnes or more. So dense is this cloud of bodies, in fact, that in World War II, scientists working on early sonar were perplexed by readings showing a phantom sea floor that rose and fell at dusk and dawn. Now dubbed the deep scattering layer, this phenomenon unsettled many commanders, and later gave rise to research into the possibility submarines might be able to disguise themselves within the obscuring fog.

The animals that make this journey mostly reside in the twilight of the mesopelagic zone. Beginning approximately 200 metres below the surface and extending to approximately a kilometre down, the mesopelagic is the transition between the well-lit upper layer of the ocean and the stygian gloom below. Its upper boundary is defined as the point where a mere 1 per cent of the Sun's light still penetrates; its lower limit lies at the point where all light is finally extinguished.

This lightlessness is both sanctuary and trap. Down in the gloom of the mesopelagic, zooplankton and other animals are safe from the predators that patrol the sunny waters above. But no sunlight also means no photosynthesis, and no food save that which drifts down from above in the slow rain of dead plankton, decaying fish and other organic matter known as marine snow. And so the residents of the mesopelagic wait, safe in the dark, until night falls above, and then rise to feast, some on the plankton, some on their fellow migrants from the deep.

As our understanding of the diel vertical migration has grown more sophisticated, more detail has emerged. In the Arctic, for instance, where the Sun barely rises for months at a time, the migrating animals time their movements to the appearance of the Moon instead; elsewhere, solar eclipses and even heavy cloud can trigger movement upwards. More significantly, this colossal movement of life does not simply provide food for the denizens of the mesopelagic, it plays a vital role in maintaining life on our planet. The eddies created by this movement of trillions of bodies help move nutrients, oxygen and heat through the ocean, with some studies suggesting the diel migration may play a bigger part in ocean circulation than the wind. It even helps regulate the climate. For as the migrating animals consume the photosynthetic phytoplankton, they ingest the carbon the phytoplankton have absorbed from the water and transport it downwards, away from the surface and into the mesopelagic.

IMMENSE AND PROFOUND as it is, the nightly dance of the diel vertical migration is only one of countless migratory journeys undertaken by animals above and below the waves. As the days and seasons cycle past, animals and other organisms shift with them, tracing patterns across the surface of the planet. Some of these journeys cover staggering distances: humpback whales, for instance, travel many thousands of kilometres every year, swimming from Antarctic waters to their breeding grounds in the tropics and back again. Sea turtles also undertake startlingly long voyages: in late 2017 a loggerhead turtle released from an aquarium in Cape Town travelled more than 37,000 kilometres back to her breeding ground on Western Australia's Pilbara coast. And Arctic terns – tiny black-headed birds that weigh a mere 100 grams – leave their breeding grounds in the Arctic as the northern summer ends and travel south, to the waters off Antarctica, where they spend the southern summer on the islands and the edge of the ice before heading north once more, a wandering path that covers an average of 70,000 to 90,000 kilometres a year.

These extraordinary journeys are only the most extreme examples of an ocean's more widespread tendency. Like animals on land, creatures in the ocean are constantly in motion. In some cases their movement is hostage to natural forces, as they drift with the currents or the winds. In other cases it is deliberate, as they propel themselves through the ocean. Like the more passive wanderers, these journeys sometimes take advantage of currents. In the Pacific, mako sharks ride the enormous rotating currents known as gyres, moving with them in search of the prey that gathers on their edges. Likewise, loggerhead turtles use the North Atlantic Gyre as a sort of conveyor belt, slipping into the Gulf Stream in the waters off Florida as hatchlings, then riding the gyre in a great loop across the Atlantic towards Portugal and North Africa before it bears them back to their breeding grounds. Still others move more freely, wandering great distances – such as albatrosses, some species of which will travel the distance to the Moon and back seven times over across the course of their fifty-year lifespan, or the seals that regularly swim back and forth across the icy waters between Antarctica and the Subantarctic islands, and sometimes as far as Australia and New Zealand.

This movement of life is not confined to surface waters and the skies. In 2020 researchers found evidence of what they dubbed 'a great migration' deep beneath the South Atlantic. By analysing seven and a half years of time-lapse photos produced by two observatories positioned 1.4 kilometres down on the continental slope off Angola, they found the numbers of fish fluctuated according to the season. More significantly, these fluctuations were synchronised with changes in the abundance of life in the upper layers of the water, suggesting the fish were following the slow fall of nutrients from the surface, their annual movements tied to the seasons just like those of animals far above.

Human cultures have also long relied upon these vast, interlocking cycles of migration. Coastal Indigenous cultures in Australia possess sophisticated understandings of the seasonal movements of fish and other marine animals. Up and down the east and west coast many greet the arrival of the humpbacks from Antarctica with songs and ceremonies, and incorporate the arrival of other species into the systems of knowledge that shape their relationship with Country. In some Queensland cultures the number of rainbow lorikeets helps predict the size of the mullet run, while the D'harawal People whose Country lies just south of Sydney knew that when the coast myall, or kai'arrewan, flowered, it meant the prawns were schooling in the estuaries. On the opposite side of the Pacific, the Salish people and many other First Nations cultures regard the salmon that course through their coastal waters and rivers as gift-bearing relatives, and the annual migration that sustains not just human communities, but bears and whales and many others, as part of the complex interplay of human and non-human agencies that structure the world. The salmon do not just provide food, they are a reminder of the existence of more-than-human presences, and of the reciprocity that underpins First Nations understandings of the world.

For Europeans these seasonal fluctuations were also an integral part of the natural calendar. But despite this, European science largely rejected the idea of animal migration until well into the eighteenth century. Faced with the annual appearance and disappearance of migratory birds European scholars argued these species somehow hid themselves during the winter months, hibernating in holes, burying themselves, or – as Swedish priest Olaus Magnus suggested of swallows in the sixteenth century – submerging themselves in the mud at the bottom of lakes and streams. Others followed Aristotle in believing some birds underwent a kind of metamorphosis, changing from one species to another as summer verged into winter, before changing back again with the arrival of the spring. Thus the robin and the redstart could never be seen in the same place, because, like avian versions of Clark Kent and Superman, they were actually the same creature. On those rare occasions when migration was considered a possibility the explanations were even more fanciful than the stories of lakes filled with slumbering swallows: in 1684 the physicist and philosopher Charles Morton argued that the storks that nest in Europe in the summer spend their winters on the Moon, declaring that the journey took a month each way, and the birds were aided in their journey by the lack of atmospheric pressure and gravity.

At least in part, these theories reflected the constraints of their historical context: without any means of following birds and other animals, or of easily sharing observations with people in other parts of the world, Europeans had little access to information that would challenge the assumptions of their predecessors. Yet there was another dimension to their resistance to the idea of migration. As European empires expanded outwards they took possession of new territories, violently subjugating their inhabitants and transforming their landscapes.

The horrifying cruelty and violence of this process demanded ways of imagining other peoples that placed them outside the sphere of moral concern. And so Europeans began to develop racialised hierarchies that treated Africans and Indigenous peoples as less than human. As these divisions hardened, the idea of a world in which racial

categories mapped onto levels of social and cultural development –
and therefore human worth – became naturalised, giving rise not just
to the repulsive race science that would find its ultimate expression in the gas chambers of Auschwitz and Treblinka, but even the
organisation of the planet's space: notions of the natural industry
and advancement of those from cool climates and the degeneracy of
those from the tropics were commonplace, and in 1858, when British
zoologist and geographer Philip Sclater proposed the system of zoo-
geographic realms that with minor alterations remains in use today, he
drew explicitly upon these hierarchies.

The notion that birds and other animals might move from place to
place conflicted with these assumptions: after all, if animals were not
confined to one part of the world where did they fit in the hierarchy?
Yet as the European powers continued to expand, it became harder
and harder to deny the reality of animal migration. At first most of
these examples concerned land birds: sailors occasionally reported
great flocks of swallows and other birds skimming across the waves
or nesting in the rigging, and as early as the 1770s, Gilbert White,
English clergyman and author of the luminous *A Natural History of
Selborne*, cited reports of the movement of swallows across Gibraltar
and Andalusia as evidence the birds were travelling south in the
winters, and noted the crowds of birds the changing seasons brought
to parts of England: 'more', he thought, 'than can be bred in any one
district'. But the first unambiguous evidence of long-distance migration did not arrive until 1822, when a hunter in the north of Germany
shot a stork with something embedded in its neck. On examination
it turned out to be a 76-centimetre-long spear of African origin. The
bird was sent to the University of Rostock where it remains on view
today, the spear angled grotesquely upwards through its long neck, its
diamond-shaped head high above the stork's.

Over the next century or so scientific understanding of the annual
movements of birds grew more detailed. But it was not until the

advent of radar during World War II that the true scale of bird migration began to become apparent. In 1941 Britain's Chain Home radar system began to detect strange signals that suggested huge formations of German bombers. Yet when fighters were scrambled to repel the invaders, all they found was empty air.

As sightings of these phantom squadrons multiplied, the mystery of what they were deepened. They seemed to form without warning and to disappear just as quickly, sometimes travelling with the wind, at other times against it. For Chain Home commander Sir Edward Fennessy, the enigmatic signals were a distraction: years later he would declare his men were too busy fighting a war to chase phantoms; in the absence of a credible explanation, British pilots took to saying the blips – or 'angels' as they came to be known – were the ghosts of fallen soldiers returning to defend their country.

The first person to suggest the mysterious angels might be birds was the biologist David Lack. A committed Christian and pacifist who spent much of World War II assigned to a secret radar research group in the Orkneys, Lack was also, in his civilian life, a dedicated birdwatcher, with considerable experience observing the behaviour of flocking birds.

Lack's commanding officers rejected this idea, refusing to believe birds migrated at night, or that they could fly at the sorts of speeds the 'angels' regularly attained. Confirmation of Lack's theory came late in the war, when angels detected over south-eastern England turned out to be huge flocks of starlings startled from their nests by V1 rockets passing overhead.

These discoveries did not just lead to a fuller understanding of how many birds are on the wing at any one time, they make it possible to see that while the land, sky and the ocean are often seen as separate, they are in fact part of a vast interconnected whole that is bound together by the movement of animals. Just as turtles and fish navigate upon the ocean's currents, birds ride the winds above the ocean.

Many of these journeys take place far above the waves: while seabirds tend to fly close to the water over the ocean and rise higher over land, this pattern is reversed in migrating land birds. Nor are birds alone in migrating across the oceans; many insects do as well: the wandering glider, or globe skimmer dragonfly, undertakes an 18,000-kilometre multigenerational migration on the monsoon winds from India to Africa and back, with individual insects sometimes travelling up to 6000 kilometres.

IF THE MIGRATIONS of birds and insects were mysterious in the years before the twentieth century, the seasonal movements of marine animals were even more so. While First Nations communities on the East Coast of North America have long relied upon the annual arrival of species in the rivers and estuaries, European invaders marvelled at the astonishing numbers of fish that filled those waterways in the spring, when species such as striped bass, salmon and halibut migrate inwards to spawn. In his history of human impacts on the ocean, *The Unnatural History of the Sea*, the marine biologist Callum Roberts cites the planter and author William Byrd who in 1728 wrote that when the alewife arrive, 'all streams and waters are completely filled with them, and one might believe, when he sees such terrible amounts of them, that there was as great a supply of herring as there is of water. In a word, it is unbelievable, indeed indescribable, as almost incomprehensible.' Fifty years later Captain George Cartwright wrote that the rivers of Labrador were 'so full of salmon, that a ball could not have been fired into the water without striking some of them'.

On the opposite side of the Atlantic, herring fisheries were similarly bountiful. In the late eighteenth century Oliver Goldsmith described the arrival on the British coast of columns of fish 8 to 10 kilometres long and 5 to 7 kilometres wide, the fish within them so numerous their massing bodies created a bow wave that moved before and around the column, and the water flashed and glittered gold and blue where the Sun struck their bodies. In the wake of this army of fish came pods of sperm and fin whales, basking sharks and birds, all gorging themselves on the racing herring, their annual appearance evidence of cyclical movement within these species as well. The writer George Monbiot describes similar accounts from the early twentieth century of huge schools of massive bluefin tuna, some of which weighed nearly 400 kilograms, that preyed upon the schools of herring and mackerel that massed in the North Sea.

Given the sheer number of herring and other fish, the question of where they came from, and where they went, was a subject of much curiosity during the nineteenth and early twentieth century. But by 1912 the French scientist Marcel Hérubel was able to write that 'Hardly has [the herring] issued from the egg when the fish first migrates. It travels from the west towards the deep bottoms or the open sea, and its movements are as invariable as the migrations of birds of passage.'

These movements of fish drove human migration as well. In Britain the herring fleet followed the herring spawning along the coast, followed on land by the thousands of mostly female gutters and curers who trailed the boats from town to town, sometimes travelling from the north of Scotland to the south of England and back in a single year. Along the west coast of Africa, fishers still follow the seasonal explosions of life created by ocean upwellings in the Gulf of Guinea as far north as Mauritania, and as far south as Cameroon, while each year in India tens of thousands of men travel from the east coast to the west coast and Tamil Nadu in the south to take advantage of the fishing season.

Many of these massive spawning events, and the immense populations of fish and other animals they sustained, are now gone, eradicated by overfishing and the disruption of rivers and estuary environments.

Yet even in their depleted state they offer a reminder of the dynamism of both animal and human populations, the ways in which they ebb and flow in response to changing environmental conditions. More tantalisingly, it is possible that these events provide a glimpse of much larger temporal scales: with each spawning event coral reefs offer a reminder that this process has been going on for tens of millions of years, the reefs rising and falling with sea levels. Likewise, the annual dance of the fifteen or so species of eels that inhabit Europe, the Americas, Australia and elsewhere can be traced back to a common ancestor breeding in the waters of the Tethys Sea more than 100 million years ago.

ONCE IT WAS UNDERSTOOD that birds and other animals were travelling great distances across the planet, scientists were confronted by the mystery of how they found their way. At first it was assumed they followed landmarks, travelling from one to the next like somebody hopping from stone to stone. Yet this notion posed obvious questions. How, for instance, did birds navigate by night, or in cloud or fog? How did they adapt if landmarks changed, or were erased by human activity or natural disasters? And perhaps most troublingly, how did they navigate out of sight of land, or over unfamiliar terrain?

As early as 1873, Charles Darwin theorised that homing pigeons must possess 'some part of the brain [that] is specialised for the function of direction', but the first serious attempts to document the navigational abilities of animals did not begin until the early twentieth century.

We now know that many terrestrial animals possess a wide range of navigational adaptations. Not only do many mammals employ internal maps based on visual and olfactory landmarks, bees and other insects can use the position of the Sun to find their way.

Birds possess a particularly complex suite of navigational abilities. Many rely upon celestial navigation, orienting themselves by reference to the Sun and the stars (the use of the stars also helps explain why so many migratory species travel by night). Others use smell, relying on olfactory cues to find their way from place to place. Still others possess the ability to perceive the magnetic field produced by the Earth's molten iron core.

Exactly how this ability works is still not properly understood: some researchers believe it relies upon small amounts of magnetite in their beaks, others suggest it relies upon special proteins known as cryptochromes found in the eyes of some birds that pick up the flickering cues of the Earth's magnetic fields and lay it over other visual information. Some birds even seem to possess an inbuilt magnetic map of the entire planet, a little like a biological GPS; this allows them to find their way even when thousands of kilometres from their home territory.

The first person to suggest marine animals might also be using some form of magnetic navigation was the American ecologist Archie Carr. During the 1950s Carr became fascinated by the ability of turtles to find their way back across thousands of kilometres of open ocean to the beaches where they were hatched, to nest. Particularly confounding was the behaviour of the green sea turtles that fed off the coast of Brazil yet were somehow able to return to their breeding ground on tiny Ascension Island, more than 2000 kilometres away in the middle of the Atlantic.

Carr entertained various explanations for this ability. Because the adult turtles seemed to swim against the prevailing currents, he wondered whether they might be following some kind of scent trail. He also proposed they might be homing in on particular sounds, such as the snapping of shrimp, or might be using features on the sea floor, or perhaps using the Sun or the stars or even the Coriolis force created by the Earth's rotation to orient themselves. More speculatively, he even wondered whether the journey might be a relic of an earlier age, when the island was larger, and that as the island had slowly sunk into the sea the turtle's homing ability had become progressively refined.

Yet it was his intuition that the homing abilities of turtles might involve magnetoreception that was to prove most important. Although it took almost half a century, scientists eventually established that turtles, like some birds, seem to be able to use the Earth's magnetic fields to find their way across the open ocean.

As our understanding of the abilities of turtles has grown more sophisticated, it has become clear this magnetic sense is not simple, or unidimensional. Turtle hatchlings appear to have some form of inbuilt compass that allows them to sense both the intensity and the inclination of magnetic fields; in combination this allows them to maintain the correct heading as they paddle outwards and into the currents that bear them away, across the ocean, at the beginning of their lives. Once out in the ocean the juveniles use local fluctuations in magnetic fields as signposts, or markers, ensuring they do not drift too far off the correct migratory route. As they mature, turtles appear to develop magnetic 'maps', learning the magnetic topography of the ocean and developing the ability to use this topography to remember the location of feeding and nesting sites. But while these maps are effective, they are not always terribly detailed: the hawksbill turtles that breed in the Chagos Archipelago in the middle of the Indian Ocean seem to use their magnetic sense to guide them in the general direction of the islands, before turning to visual and olfactory cues for the final stage of the journey. Similarly, leatherback and loggerhead turtles seem to augment their magnetic maps with some form of solar or stellar navigation like that used by birds.

Whales also appear to use the Earth's magnetic fields to navigate. In early 2016, forty-five sperm whales stranded themselves on beaches in the North Sea. Autopsies showed the whales were in good health and revealed no sign of hunger or injury. Faced with the mystery of what might have caused the whales to strand themselves, researchers in Germany found the event had coincided with a sequence of massive solar storms that had disrupted the Earth's magnetic fields potentially playing havoc with the whale's navigation. Sperm whales are not alone in being able to sense the contours of Earth's magnetic fields: strandings of gray whales seem to correspond with the incidence of sunspots, which rise and fall across an eleven-year cycle, driving fluctuations in the Earth's magnetic fields, and humpbacks often congregate around the submarine mountains known as seamounts, many of which create magnetic anomalies.

This ability to sense our planet's magnetic fields seems to be made possible by the presence of minute amounts of magnetite, the metal used to make lodestones, in the brains of whales. But it is also only one part of a much more complex suite of navigational abilities possessed by cetaceans. Humpbacks, which often swim in straight lines for many thousands of kilometres through the open ocean, seem to orient themselves by combining their magnetic awareness with visual observations of the position of the Sun and the stars. Some whales even seem to be able to perceive subtle permutations in the Earth's gravity, allowing them to pass an understanding of the bumps and ripples of the planet's gravitational topography from one generation to the next.

Other marine animals rely upon chemical cues. Salmon imprint on the odour of the stream or lake in which they hatch; when they try to return to breed as adults, they seek out its particular chemical signature. But while salmon are the most famous example of this sort of behaviour, they are far from alone. Lampreys – eel-like jawless fish possessed of round, sucking mouths, many of which feed on other fish by attaching themselves to them like leeches – use smell to find their way to suitable spawning grounds. In the case of the sea lamprey, which has moved upriver into America's rivers and the Great Lakes, researchers have weaponised this ability against the species: sea lampreys are so disturbed by the scent of dead lampreys they will immediately flee an area upon detecting it, allowing them to be steered into traps or kept away from areas of high ecological value.

Smell also plays an important part in helping many seabirds find their way across the ocean. Shearwaters that are removed from their nesting sites and transported up to 800 kilometres away will fly straight back, but if their olfactory receptors are removed the birds can no longer find their way home. Similarly analysis of the flight patterns of petrels and albatross shows they rely on scent to find their way to and from feeding and nesting sites. Exactly how they do this remains a mystery, but it seems likely that at least part of the answer lies in the production of dimethyl sulphate by phytoplankton, and that across the course of their often very long lives – albatross and related species such as storm petrels and shearwaters can live forty to sixty years or more – the birds learn to recognise seasonal changes in the concentration of these chemical cues, transforming the seemingly featureless expanse of the open ocean into a dynamic olfactory landscape shaped by the action of the winds and the movement of life beneath the surface.

IN RECENT YEARS our understanding of animal movement has been transformed once more by the advent of new and increasingly sophisticated tracking technologies. One of the most significant of these exists under the umbrella of Australia's Integrated Marine Observing System (IMOS). Created in 2006, IMOS relies upon a complex network of floating sensors, deep-water moorings, ocean radar and other technologies to collect information about ocean processes and systems, with a particular emphasis upon the effects of climate change.

However, IMOS depends upon more than just buoys and moorings. It also uses tracking devices that are attached to seals, penguins, turtles and other species. As the animals move through the ocean, sensors in these devices collect information about location, depth, salinity and temperature.

The brainchild of Rob Harcourt, professor of marine ecology and head of the Marine Predator Research Group at Sydney's Macquarie University, the Animal Tracking Facility has revolutionised our understanding of the behaviour of a number of species, in particular elephant seals, which spend up to ten months of every year at sea, often far from land. During these months the seals spend long periods underwater, often only surfacing for a few minutes at a time. Until the 1980s it was assumed they spent most of this time in the top 200 metres of the ocean, but once scientists began affixing trackers to them, something surprising emerged: 'When researchers first started putting time-depth recorders on elephant seals they kept getting crunched,' says Harcourt. 'At first people assumed that was because the seals were rolling on them, but then they realised they were being crushed by the pressure because the seals were going much deeper than anybody expected.'

We now know elephant seals regularly dive 400 to 800 metres, and often much deeper: the deepest dive on record is 2388 metres, and dives of more than a kilometre are common. Because the seals descend to these depths to feed upon the squid and fish that lurk in the darkness of the mesopelagic, their movements allow researchers to infer changes in the distribution of life at those depths, improving our understanding of the web of life and energy that connects the different layers of the ocean.

As the example of the elephant seals suggests, the trackers do more than simply extend our understanding of the behaviour of the tagged animals. They transform the animals into remote sensors, allowing them to be used to gather information about water temperature and salinity, and the movement of other species. As scientist Mark Hindell has observed of one study involving elephant seals, 'It's almost as if we had 287 mini-submarines exploring the Southern Ocean.'

The volume of the data collected in this way is significant: in Hindell's study information was collated from over a million dives in more than 568,000 locations. This is particularly important in the Antarctic and Subantarctic, where research is difficult, expensive and often dangerous. 'The Southern Ocean is probably the most critical area of the world when it comes to climate change,' says Harcourt. 'As the circumpolar current sweeps around the globe it effectively regulates the amount of heat in the oceans. By diving under the ice through the winter the seals collect a huge amount of data we wouldn't otherwise be able to get. This data has improved the accuracy of the Intergovernmental Panel on Climate Change's (IPCC) ocean state models by about 15 per cent, which is a huge change in our under-standing of how much energy is being pumped into the ocean, and has transformed the IPCC's predictions.'

IMOS is only one of a growing number of animal teleme-try projects. In Germany, researchers at the Max Planck Institute of Animal Behaviour have developed an ambitious system known as International Cooperation for Animal Research Using Space, or ICARUS, which is placing tiny solar-powered tracking devices on thousands of individual animals from a growing array of species, ranging from rhinoceroses, giraffes and hyenas to birds and bats and seals. Weighing a mere 5 grams, these devices incorporate a GPS module capable of pinpointing an animal's location to within a few metres, and a radio transmitter that beams the information to a receiver placed on the International Space Station (ISS). The small size and weight of the transmitters is vital, because it allows the devices to be attached to birds and other small animals. But future generations of transmitter are in development that will weigh no more than a gram and could potentially be attached to insects.

The compactness of the ICARUS transmitters also allows them to be deployed at scale, so many thousands of animals can be tracked simultaneously, providing information about collective behaviour, social networks, predator–prey relations and other interspecies interactions, allowing researchers to observe and monitor the movement and behaviour of animals at both an individual and ecosystem level.

Yet like Harcourt's work with IMOS, ICARUS will do far more than track animals. Despite their minuscule size the transmitters are equipped with a range of sensors capable of gauging speed, magnetic intensity, temperature, humidity and air pressure, transforming any animal fitted with one into a mobile sensor array, and providing incredibly fine-grained information about a range of environmental conditions in real-time. And – no less relevantly – they transform the animals themselves into data, something reflected in the characterisation of the project by ICARUS's founder, German biologist Martin Wikelski, as an 'internet of animals'. The potential applications of this are startling: as well as making it possible to observe changes in the distribution and behaviour of animals due to climate change and other forces, improving our ability to protect them in a rapidly changing world, Wikelski anticipates ICARUS will allow scientists to better predict outbreaks of pathogens and future pandemics by tracking the movement of species such as bats and birds that act as vectors for the transmission of disease. ICARUS may even be able to provide an early warning system for earthquakes and tsunamis by harnessing the tendency of some animal species to become agitated or to behave abnormally prior to seismic activity.

ALTHOUGH THE TECHNOLOGIES deployed by IMOS, ICARUS and similar projects are directed at important scientific goals, they are also emblematic of a world in which surveillance grows more pervasive and panoptic every day. Technologies such as social media and artificial intelligence rely upon the uncontrolled harvesting of human knowledge and labour for private profit, but they also allow the creation of new architectures of control. These are designed to anticipate our needs and desires, allowing them to be monetised, but increasingly they also shape our behaviour, our politics and – as the growth in the reach and power of conspiracy theories and alternative facts makes clear – even our reality.

Nor is it coincidental so many of the technologies that have enabled the development of scientific understanding of the movement of animals were originally military technologies. Just as the development of many of the technologies underpinning tracking projects like ICARUS were jumpstarted by the rise of the security state in the aftermath of 9/11, sonar, radar, GPS, even early satellite tracking harnesses are all, to a greater or lesser extent, the product of military research. Similarly, the military has often sought to harness knowledge about animal movement in the ocean and elsewhere: during the Cold War the American Navy devoted considerable resources to finding ways the deep scattering layer could be used to disguise submarines from enemy sonar, while others have suggested integrating animals into military surveillance networks by using satellites to track their movements. As the writer and naturalist Helen Macdonald has observed, these ideas prefigure contemporary visions of the battlefield as a digitised world in which 'military superiority is built on knowing where everything is, coupled with the ability to intervene in real time', a place where drones strike from the sky on the basis of decisions reached in offices continents away. In her study of the Nazi official and architect of the Holocaust Adolf Eichmann, the philosopher Hannah Arendt wrote of 'desk murderers', the bureaucratic enablers of state violence, and of the way their actions depend upon abstracting those who are its victims. Remote warfare is the logical endpoint of this phenomenon, a process in which the effects of violence are kept separate from those who authorise and benefit from them, and are instead inflicted upon the victims and those who must carry out the orders.

This matters because as the climate crisis accelerates, new migrations are beginning. Animals are already in motion: as the ocean heats up, free-swimming species at the equator are declining in number, and elsewhere marine animals are shifting polewards at an average of 6 kilometres a year. As they move, ecosystems are being disrupted, reducing diversity and simplifying foodwebs. This process is already visible in the destruction of kelp forests in Australia and North America by sea urchins and fish species arriving from warmer waters. But it is also rapidly altering marine ecosystems elsewhere: off Sydney, where water temperatures have warmed 1.5 degrees over the past century, coral is beginning to appear, and tropical fish that were once only vagrants, washed south in summer, have taken up permanent residence. And further south, in Antarctica, warming waters have allowed king crabs – previously confined to the deep waters of the ocean floor – to ascend the continental slope around the continent, threatening the unique and fragile ecosystems that flourish deep beneath its waters.

Elsewhere migratory behaviours are being disrupted as breeding grounds grow too hot or disappear beneath the waves. On the Great Barrier Reef's Raine Island, the largest green turtle breeding ground in Australia, hatching rates of eggs have dropped by as much as 30 per cent because the eggs cannot tolerate inundation by seawater for extended periods. As waters rise further the island will eventually disappear, as will other turtle hatcheries around the world. Similarly in recent years the arrival of migrating shearwaters in Tasmania and elsewhere on the Australian coast has become increasingly unpredictable, their numbers depleted by mass deaths because warming waters and Arctic heatwaves have disrupted the krill on which the birds rely. Without the krill the birds eat whatever they can find, gorging themselves on plastic or, in 2013, the floating pumice from the eruption of Havre, an underwater volcano in the Kermadec Islands between New Zealand and Tonga that carpeted Australian beaches for the next two years. Even blue whales – the largest animal ever to have existed on Earth – are being affected: as krill populations shift, the whales, which rely upon their prodigious memories to calculate where they should feed, are finding it more difficult to locate food.

Nor is it just animal populations that are on the move. According to the United Nations High Commissioner for Refugees the number of people displaced globally reached a new record of almost 110 million in 2022, ten million more than the year before, and triple the number only a decade earlier. Meanwhile the Internal Displacement Monitoring Centre estimates that in 2022 nearly 61 million people were displaced internally, with weather-related dis-asters accounting for well over half of that number. Many of these were the result of tropical storms and flooding, but the changing climate is destabilising communities elsewhere: in the Arctic subsidence and hastening erosion caused by melting permafrost are forcing Indigenous communities from their homes. These displacements do not just place people at risk and push them into lives of impermanence and economic and social vulnerability, they also rupture bonds of community and culture, shattering the web of connections that shape people's identities.

There is no question global heating is also accelerating conflicts in many parts of the world by placing pressure on food, water and other resources, or promoting instability by affecting economic stability and damaging infrastructure. In the Sahel region, between the Sahara and Sudan, rapidly declining food production due to climate-change-induced desertification, has already led to horrify-ing violence and driven more than two million people to flee their homes. Meanwhile on the Horn of Africa six consecutive years of climate change-fuelled drought has displaced more than 2.5 million people, and pushed at least 20 million into acute food insecurity. Syria has also seen the impact of slow-onset climate disruption, with civil war precipitated in part by drought causing almost six million people to flee the country.

Many of those displaced by these disasters have attempted to find new lives in Europe and elsewhere. Since 2014, more than two million people have attempted to cross the Mediterranean into Europe, mostly from Africa and the Middle East. Across this same period more than 25,000 have died. Most of the dead have drowned; others have died of asphyxiation or been crushed on overcrowded boats, their bodies pressed together in the reeking holds. It is 'just like a slavery boat – the same,' said one aid worker. Thousands more have died attempting the crossing from Africa to the Canary Islands and fleeing across the Gulf of Aden towards Yemen.

The bodies of most of those who have died have never been recovered; of those that have been brought ashore, most lie unidentified in unmarked graves. Yet each of them is a human being who leaves behind parents and children, wives and husbands, brothers and sisters and lovers, who will never know what happened to their loved ones. Those who work with refugees speak of parents who have lost children and cannot stop looking for them or grieve because their fate remains uncertain. 'I can't sleep. I cry every time I see a picture of him. I can't stop thinking about him,' 'I see the other children here, and I think, where is my child? Only God knows,' said one woman who spoke to journalist Molly Blackall.

These tragedies are inseparable from the fossil fuel economy, and not just because it is greenhouse gases that are sending temperatures soaring. In many places the convulsions that are driving illegal migration have been intensified by war and violence associated with securing oil supplies, as well as by the longer histories of colonial extraction that continue to impede development in the Global South.

Nowhere is this process more sharply delineated than in the

recent history of the Middle East, a region that has been shaped
by the attempts of the US and the British to control oil supplies,
a process that culminated in the catastrophic intervention in Iraq. The
writer and activist Naomi Klein cites the work of Israeli writer Eyal
Weizman, who has observed that if one traces what is known as the
aridity line, which delineates the limit at which it is no longer possible
to grow cereal crops without extensive irrigation, and currently runs
in a long arc through the Sahel, Ethiopia and Somalia, and across
North Africa through Gaza, Syria and Iraq to Afghanistan and
Pakistan, one finds both the sites of many of the bloodiest conflicts
of recent years and, more recently, growing numbers of drone strikes.
Or in Klein's formulation, 'just as bombs follow oil, and drones follow
drought, so boats follow both: boats filled with refugees fleeing homes
on the aridity line ravaged by war and drought'. And as Klein also
recognises, 'the same capacity for dehumanizing that justified the
bombs and drones is now being trained on these migrants, casting
their need for security as a threat to ours, their desperate flight as
some sort of invading army'.

In wealthier nations this dehumanising rhetoric is being used
to justify increasingly draconian border controls and the systems of
surveillance and detention that underpin them. The human cost
of this can be counted in the experiences of children separated
from their parents at the US border, or in the accounts of despair,
sexual violence and suicide in the detention camps operated by the
Australian Government in Nauru and elsewhere. But it is also visible
in the growing influence of the far right in developed nations, and
its conflation of hastening environmental breakdown with anxieties
about mass migration from poorer countries and toxic racist fantasies
about infiltration and replacement.

Part of what makes these ideas so unsettling is the way they invert
reality, erasing the history of extraction and colonial violence that
has given rise to both these imbalances of wealth and the hastening

environmental breakdown of the planet, and turning the blame for the world's growing instability onto the victims of that instability. No less importantly, these fantasies ignore the fact that the climatic and environmental upheaval that has been unleashed is displacing us all by sweeping away the certainties of the past, and propelling the world into a new and radically unstable future.

4

ECHO

'Sound is sea: pattern lapping pattern.'
Ronald Johnson, *Ark*

ON A GLOOMY WINTER AFTERNOON I step off the back of a boat moored near the outer edge of the Great Barrier Reef. The water is cooler than I expect, and I gasp a little as I kick away from the boat, anxious to separate myself from the rest of the group. Although I can still hear the other divers talking and laughing in the distance, I am mostly aware of the sound of my flippers splashing behind me and the rattle of my breath in my snorkel. As I reach the first outcrop of coral I take a breath and slip below the surface.

Almost at once the soundscape from above fades and is replaced by a murmur of muffled noise. As I stop moving and pay attention, distinct sounds begin to emerge out of the confusion. First comes the creak and clank of the boat moving on its mooring, and the distant murmur of voices, but beneath that I can also make out an assortment of cracks and creaks, and behind it all, a static of clicks and pops.

My inability to make immediate sense of this submarine sound-scape is not surprising. Although hearing evolved in the ocean, my ears are adapted for the gaseous medium of the atmosphere, not the aqueous space beneath the waves. For the creatures who live here, though, this constant play of sound forms a complex and dynamic aural environment they must navigate and exploit in order to survive.

The significance of sound in the ocean is partly a result of the lim-itations of sight: although many marine species use eyesight to hunt and navigate, vision declines in effectiveness as one leaves behind the light-filled upper layers and is easily overwhelmed in murky or turbid conditions. Sound, by contrast, travels more than four times as fast through the water than through the air, transmitting information through oceanic spaces with remarkable efficiency, and over great distances. A specially designed hydrophone placed at the deepest point of the Mariana Trench in 2016 picked up what sounds like the soundtrack of a peculiarly unsettling dream: earthquakes rumbling off Guam, the whirr of propellers from ships almost 11 kilometres above, and the resonant moans and alien whoops of whales, as well as the thump of seismic survey air guns, low-frequency sonar and the white noise of a typhoon. Sound can even travel between oceans: in 1991 scientists placed loudspeakers several hundred metres below the surface off Heard Island, a remote speck of land midway between Africa and Australia in the Southern Ocean. The low-frequency sound produced by the speakers was detectable 17,000 kilometres away in waters off Nova Scotia, on the east coast of Canada.

The substrate of this sonic environment is the sound of the planet itself, perhaps most obviously the crash and surge of the waves, together with the wind and rain and other disturbances at the surface. But these are only one part of a much more complex soundscape. The tides and currents produce turbulence that results in extremely low frequency infrasound, as do underwater volcanoes, ocean vents, and seismic activity such as earthquakes and submarine landslides. In polar regions the rumble and crash of melting glaciers and the grinding of sea ice creates a cacophony that can be heard over hundreds of kilometres. Even the Earth hums – a constant, low oscillation that seems to be caused by ocean waves colliding in the shallow waters along the continental shelf creating a standing wave that vibrates the entire planet.

Over and around this clamour is a panoply of biophonic sound produced by marine animals, and in particular fish. The capacity of fish to make and hear sound is not a recent discovery: many human cultures have been aware of fish song for millennia. While aboard a boat with Ghanaian fishers, the writer and musician Bennett Konesni was shown how to listen to the fish by placing his ear against a wooden paddle. 'I heard the high whistle of one type, *moi*. Two other varieties also whistle, another two make low grunting noises, and one makes a soft "kwa kwa, kwa kwa." That fish is named after its shout – *kwa kwa.*'

Fishers working the waters west of the Japanese island of Kyushu have also long relied upon the noise produced by the fish known as *koichi* and *shiroguchi* – species of drummer and croaker, respectively – to guide them to their prey. The volume of these cries rises and falls across the year but is at its loudest during the spring and summer, when the fish gather to spawn. As late as the 1970s the clamour of the fish at this time of the year was loud enough to wake fishers sleeping in their boats. And the waters of the Batticaloa Lagoon, in Sri Lanka's Eastern Province, have long been known for the songs of the fish that live in them, songs that are especially audible at night, under the full moon. The local Tamil fishers call the singing fish *oorie coolooroo cradoo*, or the crying shell; interviewed in 2014, one elderly man described hearing them for the first time while crossing the Kallady Bridge after an evening movie show in 1946: 'I heard a musical sound. I thought it was the noise made by the telegraphic wires.' The teacher and politician Prince Casinader described it as 'like a man idly playing on the keys of a piano. Bass notes and treble notes and various things. Many people who heard it said it sounded like a Jewish harp. Some people said it was akin to somebody with a wet finger rubbing on the rim of a wine glass.'

Fish are not the only animals contributing to this underwater symphony: many invertebrates also produce sound. Sea urchins are particularly noisy, the rasp of their teeth scraping against the rocks amplified by their calcareous shells (or tests) and the fluid within their bodies to produce a kind of crackling, a little like that of a fire, that will be familiar to many divers and snorkellers. Various crustaceans also make a significant contribution to the submarine soundscape, none more so than snapping shrimp, which use a special mechanism on their pincers to produce a shockwave that is powerful enough to stun or kill worms, other crustaceans and even smaller fish that have the misfortune to be in proximity. The constant detonation of these tiny shockwaves creates a persistent background crackle that sounds like an underwater Geiger counter. Present in oceans around the world, it reaches such a crescendo in some tropical waters that it can drown out the sound of many other activities. When the Japanese Navy began installing sophisticated networks of hydrophones to monitor underwater activity in harbours and coastal waters in the 1930s, American submariners quickly learned to use the racket of the shrimp to enter Japanese harbours undetected. The massed roar of snapping shrimp also has a history of interfering with military sonar, and even today poses challenges for underwater surveillance and research. Even humble plankton make noise, the release of oxygen as they photosynthesise generating a fizzing within the audible range of humans. This sound rises through the day when the Sun is shining, and fades in the night, when its rays are not present. And there is speculation that the discovery in corals of genes associated with the reception and production of sound in anemones makes it possible adult corals communicate with sound, and that coral larvae may use sound to steer themselves towards reefs.

Once, many of these noises were assumed to be random. Now we know a high proportion of them are connected to particular places and activities. Like the shifting choruses and patterns of birdsong above the water, the choruses of fish and other marine species rise in intensity at dawn and dusk, and at the full Moon, as well as during breeding seasons, when many fish cry all day. And like animals everywhere, fish use sound to attract mates and warn off rivals, to alert each other to approaching threats and to strengthen bonds between parents and young.

LAUREN HAWKINS is a researcher at Curtin University in Western Australia. She became fascinated with soundscape ecology as an undergraduate: on a trip in the Amazon, her guide turned off the light and told her to just listen. Astonished by the complexity and wonder that act of listening revealed, she returned to Australia and began to work on fish vocalisations. Her most recent project is an attempt to map fish choruses around the entire Australian continent.

As we speak Hawkins plays me a recording made off the North West Shelf of Western Australia. At first it is difficult to make sense of, but gradually patterns begin to reveal themselves. One fish calls with a sound like somebody sawing, a movement back and forth that rises in volume for several seconds before falling away over and over again; in the background another releases volleys of loud pops against a static of growls and crackles.

Hawkins often spends hours at a time listening to recordings, noting down the sounds she hears. 'It's very immersive: each time you flip open your laptop it's like you're descending into a completely different world. And every time you're encountering things nobody has ever heard before.'

Her work has revealed that one of the dominant elements in the marine soundscapes off Western Australia is the crescendo that occurs each night as fish and other species rise upwards during the diel vertical migration, the sound of which is often so loud it can be picked up 200 kilometres away. 'It's an immense biological event, but it's also a huge acoustic event. It's an amazing thing to be able to listen in on something of that scale.'

Much of the din created by the diel vertical migration occurs as the fish ascend. Hawkins thinks this is probably because these sounds are associated with social behaviours: basically the fish are calling out to each other as they head upwards. But some noises also seem to be connected to feeding. 'These guys make a particular pop that's quite percussive, and one theory is that they're using that sound to stun the plankton they're feeding on.'

Exactly which fish are producing this clamour is difficult to know for sure, although Hawkins has detected the cry of one in particular – possibly a species of trumpetfish or one of the family of deep-sea fish known as *berycidea* – at locations from Bremer Bay in southwest Western Australia to Bass Strait. 'That means this biological activity is happening along the continental shelf from one end of the continent to the other. It's ridiculous!' she says, laughing.

Hawkins has also established that the composition and timing of fish choruses varies considerably between the tropical waters off northern Western Australia and the waters in the temperate south. This reflects differences in the distribution of species but is also related to environmental factors.

'In the warmer waters up north the feeding patterns of planktivorous fish are connected to the phases of the Moon. Down south they're driven by the tides and upwellings where you get large amounts of plankton. So in temperate waters their cries can be heard every morning and every evening without exception, except for a period each year that seems to correlate with the breeding season, when they call all day. But in tropical waters they rise and fall in intensity across the lunar cycle, so you'll have a period each month, either at the New Moon or the Full Moon, when they're super-intense, and then they'll peter off for a fortnight or so before they begin to get louder again.'

These patterns offer a window into a world largely invisible to humans, revealing a degree of complexity and sociality that many will find surprising. But this sort of acoustic ecology is also a powerful tool for understanding the health of fish populations. 'In the Perth Canyon, we can see that the times when the chorus is super-intense correlate with periods of really high productivity,' Hawkins continues. 'That opens up the possibility of using the loudness of the fish as an indicator for productivity. That's especially useful if what you want to do is keep tabs on populations of commercial fish species, because it means that in some circumstances you can estimate abundance using hydrophones and echo sounders instead of having to rely upon bycatch counts or intrusive sampling methods.'

While the possibilities for such work are significant, much of Hawkins' work continues to be focused on listening to recordings to identify calls. 'Of the 34,000 or so species of fish in the world, there are only about a thousand known to produce particular sounds. Some of those are things like drummers that are well known, but a lot of the fish calls I record aren't ones anybody recognises.' In order to help overcome this gap in our knowledge, researchers recently created an open-access website that allows scientists to upload recordings of fish calls to aid with identification. For Hawkins the motivation for her work goes beyond simply identifying fish species: she sees it as a privilege, a glimpse of another way of being that has altered her perspective on the world. 'I haven't eaten fish since I started this project because when you hear them talking day after day, they become a bit more human in a way. Because they can communicate.'

SOUND ALSO HELPS SHAPE the human experience of the ocean, connecting us across space and through time. In the Gulf of Guinea, Ga and Ewe fishers sing as they work, the music helping synchronise and ease the burden of repetitive activities like rowing and hauling nets, as well as providing a private language that allows them to speak freely in the presence of outsiders.

These songs can have deeper purposes as well. The Mukkuvar fisherpeople who live along the shores of Kerala and Tamil Nadu in southern India and in parts of Sri Lanka also use song to structure their work, singing and swaying rhythmically as they pull their nets along the shore. Known as *eilamidal*, this ritual incorporates the recitation of myths and stories, and helps the fishers absorb themselves in their work. There is a Mukkuvar saying: '*Mukkuvanu meen pattāl muthappayil kaṇṇu*' (While fishing is in progress, a fisherman sees only the float). On Yorke Peninsula in South Australia, the Narungga people sang to the dolphins and sharks they used to herd fish into the traps they constructed on the beaches. Similar practices were recorded further west on Eyre Peninsula and in Moreton Bay on the east coast, while in Eden, on the southern coast of New South Wales, where the Yuin cooperated with killer whales to kill migrating hump-backs, it is likely people addressed songs to their orca kin. And the people of the Marshall Islands and other Pacific cultures use chants that contain complex understandings of the winds, the currents, and other environmental and cultural knowledge to assist in navigation. In the words of the scholar Karin Ingersoll, 'Merging the body with this rhythmic sea enables a reading of the seascape's complex habits, as well as all the memories created and knowledges learned within this oceanic time and space.'

Sound and song also took on deeper resonances on the slave ships as they bore their human cargo westwards across the Atlantic. Aboard these vessels the newly enslaved were kept in conditions of unspeakable cruelty, jammed into the hold, their bodies bound in irons and pressed against each other in fetid darkness. Sailors who travelled on these ships spoke of the sounds of unfathomable sadness and pain that came from below deck, 'the long groan', and 'shrieks of woe and howlings of despair', while the emancipated slave Ottobah Cugoano recalled 'the rattling of chains, smacking of whips, and the groans and cries of our fellow-men', as he and other captives were taken on board.

These cries of loss and suffering were not the only sounds on the slave ships. 'They frequently sing,' said one sailor who completed four voyages to Africa in the late 1760s and early 1770s. 'The men and the women answering one another, but what is the subject of their songs I cannot say.'

We do not know what these refrains were. There are no record-ings, and songs, like speech, are only a movement of air. Some were clearly songs of sorrow: a sailor on board the British slave ship *The Syren* wrote of the singing of the enslaved that he 'had never found it anything joyous, but lamentations . . . for having been taken away from their friends and relations'. Others seem to have been a way of preserving and sharing memories of their lost homes: William Butterworth, who served as a seaman on board the slave ship *Hudibras* in the 1780s, describes the female captives gathering around a woman who was both a songstress and some form of orator; kneeling in the centre of the circle she would sing 'slow airs, of a pathetic nature' that Butterworth believed to be of 'friends far distant, and of homes now no more'.

Still other songs must have been a way of entreating the spirit realm for relief from suffering or for freedom: conditions on the ships were so cruel that until the beginning of the eighteenth century one in five imprisoned died before they reached the New World; a century later the figure was still one in eighteen. The Martinique poet Édouard Glissant called the ships womb-abysses, 'pregnant with as many dead as living under the sentence of death', and described the balls and chains used to weigh down the bodies of those cast overboard as 'underwater signposts mark[ing] the course between the Gold Coast and the Leeward Islands'. Similarly the Saint Lucian poet and Nobel laureate Derek Walcott wrote of the way the space of the Atlantic had become the repository of that grief, the silences that fill its vast cathedral a palpable absence:

> Where are your monuments, your battles, martyrs?
> Where is your tribal memory? . . .
> in that grey vault. The sea. The sea
> has locked them up. The sea is history.

Sound bound the ships together in other ways as well. Butterworth describes an incident in which a surgeon taunted an enslaved man by telling him he would soon be setting off on another journey just as long as that which he had just suffered through. The news moved through the ship 'like a train of gunpowder, ignited at one end'. When the misunderstanding was sorted out, 'songs of joy superseded the vociferations of discontent; and the powerful chorus bespoke the concord of souls now pacified: they continued to sing nearly the whole of the night'. Others speak of the enslaved using their own bodies as instruments, slapping their thighs and clapping, while in some instances the ship itself became an instrument and a telegraph, as slaves beat out rhythms on the timbers or tapped messages on the bulkheads. Sometimes these messages included plans for revolt – on board the *Hudibras* boys carried messages back and forth between the sections containing the men and the women, passing them through the bulkheads.

It is not easy to find language commensurate with the truth of what happened aboard the slave ships. In the cramped and reeking holds abducted Africans were subjected to a regime of constant psychological assault and systematic violence designed to strip away any sense of individual identity or agency and transform them into something less than human.

Some of the enslaved found ways to fight this process – sometimes directly, through insurrection, or, most brutally of all, by choosing to take their own lives and thereby deny their captors the use of their bodies. Songs and music were also a form of resistance, offering a way to remember the places and people the enslaved had lost, as well as creating a medium in which they could communicate without their captors overhearing. But perhaps most importantly, they offered the newly enslaved a way to assert their humanity and survival in the face of unimaginable grief and cruelty. That the crews understood this is clear from testimony like that of Dr James Arnold, who served as surgeon aboard the Guineaman *Hannibal* in the late seventeenth century. Arnold describes his captain's fury when the women on the

ship sang songs that told 'the History of their Lives, and their sepa-
ration from their Friends and Country'. To punish their assertion of
connection to their pasts, the captain 'flogged the women . . . in so
terrible a manner, that the witness has been a fortnight healing the
incisions'.

Yet even as their identity was stripped away from them the enslaved
were beginning to forge new identities capable of transcending the
often disparate or even antagonistic cultures and ethnic groups that
were thrown together in the holds of the ships. Once again song and
music played an important part in this process, offering a way not just
to speak across those lines, but of binding beliefs and cultural practices from the world left behind into new configurations.

These new identities – and the music that helped shape and express
them – followed the enslaved from the ships and onto the plantations
and farms of the New World. In Brazil they evolved into traditions
of music and dance such as jongo, maracatu and the samba; in Cuba,
where slavery continued until 1886, they can be heard in the complex
rhythms of rumba and other forms; and in the United States they
fed into the blues (which bears a remarkable resemblance to musical
structures employed in Senegambian music) and the extraordinarily
rich and powerfully expressive work songs and spirituals of African
Americans.

These musical traditions offered a link to the world the enslaved
had left behind, but they also helped bind communities together
in ways that made them stronger and more resilient. In the United
States in particular music played an important role in helping African
American communities survive the dehumanising brutality of slavery
and segregation. Spirituals did not just offer a way of transcending
the pain and grief of daily life: like the songs on the ships they helped
spread information in ways that could not be overheard by overseers
and others; similarly the work songs that helped make labour bearable
often also hummed with fury.

This connection to the past continues to reverberate through the contemporary world, audible not just in rock and roll, gospel, soul and R&B, but also in the thriving musical traditions of Brazil, Cuba and contemporary African music, which has drawn into itself music from the opposite side of the Atlantic. These interwoven traditions form part of what the historian Toby Green describes as 'an intellectual production that embodied many of the pathways – human and other – criss-crossing the ocean'. But these traditions can also be seen as a shared hymn of survival that transcends the rage and grief that suffuses the past and present of the Black experience. Or as Zadie Smith has said, they enact an 'alchemy of pain' that transforms sorrow into beauty and celebrates joy, sensuality and faith in the face of oppression.

NOT ALL SONGS in the ocean are human. In the years after World War II engineers using hydrophones to scan for the sound of Russian submarines at a secret US Navy listening station in Bermuda found themselves picking up a babel of strange, eerie sounds: skittering blips, long cries, whoops and basso moans.

One of the engineers, Frank Watlington, realised the noises must be being produced by the humpback whales that gather off the Bermuda coast to breed each winter, and began to make recordings of the sounds. Concerned his discovery might be used by whalers to find and kill already highly threatened whale populations, Watlington kept his recordings to himself until 1967, when he was introduced to the biologist Roger Payne and his first wife, zoologist Katy Payne, who had come to Bermuda in the hope of studying the humpbacks. Watlington invited the pair aboard his yacht and played them his recordings.

Both Roger and Katy Payne immediately understood that what they were hearing was something profound. 'I had never heard anything like it,' Katy Payne said later. 'Oh my God, tears flowed from our cheeks. We were just completely transfixed and amazed because the sounds are so beautiful, so powerful, so variable.' Roger Payne was, if anything, even more powerfully affected. 'Few experiences have had a deeper effect on me. It changed my life.'

After they returned home, Katy Payne found herself listening to the tapes over and over again. Eventually she used a sonograph machine to create visual representations of the sounds, which revealed they weren't random; instead, they had structure.

Meanwhile Roger Payne had shared the tapes with his friend Scott McVay. A poet and naturalist, McVay had worked with controversial dolphin researcher John Lilly, whose vision of communicating with cetaceans had led him to dose himself and captive dolphins with LSD. Together with his mathematician wife, Hella, McVay had come to a similar conclusion: the sounds the whales were making not only had a regular pattern and shape, but they repeated. As Roger Payne and Scott McVay write in their groundbreaking 1971 paper 'Songs of Humpback Whales', 'Humpback whales . . . produce a series of beautiful and varied sounds . . . and then repeat the series with considerable precision.'

Roger Payne's next step was to release the album *Songs of the Humpback Whale*, which he hoped would help raise awareness of the plight of the whales. It succeeded beyond his wildest dreams: within a year it had sold more than 45,000 copies, and inspired songs by Pete Seeger and Judy Collins, and a symphony by American composer Alan Hovhaness. *Rolling Stone* said the record stretched the mind 'to encompass alien artforms'. Bob Dylan even played excerpts from it at concerts. 'There was a burst of realisation that the world could change its relation to wildlife,' Katy Payne said later. 'The reaction to hearing these sounds made whaling obsolete!'

The structure the Paynes and McVay had found in the humpbacks' songs was remarkably complex. The whales employ a variety of sounds – whistles, whups, low keening moans and liquid blips and blops – that range from 0.15 seconds to about eight seconds in length. These individual sounds are arranged into phrases that usually last somewhere between five and thirty seconds. These phrases are then repeated with minor variations to create sequences known as themes that contain up to twenty phrases and take several minutes to complete. The themes are then combined to form songs. The songs usually contain four or five themes and last somewhere between seven and twenty minutes, although more complex examples contain as many as eight themes and last up to half an hour. The humpbacks sing these songs over and over again in long sessions that can last many hours.

Although at the time Roger Payne and Scott McVay published their paper it was not clear whether it is all humpbacks that sing or just some of them, we now know that while female humpbacks also vocalise, only the males sing. Usually they do this while hanging motionless in the water with their heads pointed downward. The photographer Jem Cresswell, who has spent many years photographing whales in Tonga and elsewhere, describes seeing one humpback position himself in front of a curving coral reef in one of the main channels of Vava'u. Cresswell heard the whale from his boat, and after dropping into the water found the whale about 15 metres down, 'floating vertically, upside down, with its head almost on the steep sloping reef that descended into the abyss from the outer wall of the pass'. He watched the whale for about forty-five minutes, during which time it would surface occasionally to take a breath, then descend once more and begin singing again, but 'always with its head right near the sloped reef wall, as if it was using it to amplify the sound', until finally a mother and a calf turned up, and the male left with them.

More intriguingly, it has also become apparent that the songs are specific to particular populations, so that at any given time all the whales in particular areas will be singing the same song. But while the overall pattern of the songs remains constant within populations, individual humpbacks improvise within the songs by dropping themes, swapping phrases around, or repeating phrases within themes to create new rhythmic patterns. They also vary the frequency, volume and duration of particular sounds, as well as altering the rhythmic emphasis within the phrases in the same way human singers emphasise or extend particular words to produce particular effects.

Nor is the structure of the songs random. Instead the humpbacks tend to alternate particular themes (ABAB) or use particular patterns of doublets and triplets. These patterns seem to help the whales learn and remember the songs. Like humans, the whales seem to rely upon rhyme, especially in longer and more complex songs, where adjacent themes frequently begin and end in similar ways. The songs also have internal structure, especially within phrases, which, rather like a human singer singing an old song with new lyrics, exhibit characteristic shifts in frequency, amplitude and duration that remain the same even as the sounds themselves change.

This process of innovation means the songs change and develop over time, as new variations evolve and are adopted or discarded. This process makes clear the whales are learning the songs from each other, rather than simply producing them from some kind of genetic script. Even more remarkably new songs can jump from one population to another, supplanting old songs like musical fads spreading from one human community to another. In the South Pacific the songs travel eastwards, leaping from the humpbacks that winter on the east coast of Australia to the whales that congregate around New Caledonia, and on through American Samoa to the Cook Islands, finally reaching the whales that winter in French Polynesia. And it is likely this is only one part of a much larger picture: recent research shows the songs continue to travel eastwards from French Polynesia to the coast of South America, implying a system of cultural transmission that spans the entire Pacific, and possibly the entire southern half of the planet.

Precisely how the new songs are transmitted from population to population is not clear: perhaps males join different populations in different seasons, or perhaps the whales overhear each other during the summer months in the Antarctic or while travelling along shared migration routes. There is no question the songs carry huge distances through the water: as Roger Payne has written, 'when you swim up next to a singing whale through the cool, blue water, the song is so loud, so thundering in your chest and head, you feel as if someone is pressing you to a wall with their open palms, shaking you until your teeth rattle.'

Exactly why the humpbacks sing also remains a mystery, yet the fact that it is only the males that sing, and most often during the mating season, suggests it is related to breeding. This means it is almost certainly being used by the males to attract females, to compete with other males, or both. Nonetheless the lack of a direct relationship between singing or the complexity of the songs and sexual success has led some researchers to speculate that the males may also use their songs as a sort of aural code or secret handshake among themselves, an idea that finds some support in observations of humpbacks off Hawaii, where singing males seem to seek each other out before heading off in a group to search for females.

There is of course a violence contained in this sort of reductive analysis. We can analyse human culture in purely functional terms, but to do so risks missing the most important qualities of that culture. Art and music have a cultural purpose, but the success or otherwise of an artform is not reducible to that purpose: we love songs and other art because they speak to us in some way, or because they are irresistibly catchy, or joyous or gorgeous. Is it not possible the whales find beauty in their songs in the same way? Or that the idea of beauty is not confined to humans?

Humpbacks are not the only whales that sing. The bowhead whales that live in the Arctic and Subarctic sing almost constantly across the winter months, often engaging in sessions that last twenty-four hours or more. Their songs resemble those of humpbacks and evolve and change in a similar way. But unlike humpbacks, bowhead communities often sing several pieces at once, and cycle through them with startling rapidity, producing new songs weekly or even daily: between 2010 and 2014 one population of about 300 bowheads off Greenland produced at least 184 distinct songs. 'When we looked through four winters of acoustic data, not only were there never any song types repeated between years, but each season had a new set of songs,' said oceanographer Dr Kate Stafford.

Blue whales also sing, producing long, mournful tones pitched so low humans cannot hear them that are arranged into single themes the whales repeat over and over. Like humpbacks, blue whales in different parts of the ocean have different songs, suggesting the whales are divided into separate populations. But where the songs of humpbacks are constantly evolving and being passed from population to population, the songs of the various blue whale communities show little variation over time. In the words of Australian marine ecologist Professor Tracey Rogers, 'Humpbacks are like jazz singers. They change their songs all the time. Blue whales . . . are more traditional. They sing very structured, simple songs.'

Blue whale songs are also enormously powerful, their low-frequency pulses audible over hundreds of kilometres. This allows the whales to find each other even when separated by huge distances, and has also led to the discovery of previously unknown populations of the whales in remote parts of the ocean: in 2018, French bioacoustician Dr Emmanuelle Leroy was examining data from the network of hydrophones used by the Comprehensive Nuclear Test-Ban Treaty Organization to monitor the ocean for possible nuclear bomb blasts when she noticed an unusually energetic signal. Realising it was too powerful to be a single whale, she examined the acoustic signature of the song and discovered it was being produced by a previously unknown population of pygmy blue whales that gathers near the Chagos Archipelago. Leroy also discovered that seasonal changes in the pitch of the whales' songs are caused by singing louder in the summers so they can be heard over the rumble of melting ice.

Fin whales and minke whales sing as well. The songs of fin whales, like blue whales, are relatively simple, but travel huge distances. Minke whales, on the other hand, produce a range of blips and blops and pulses that sound startlingly like the audio effects from 1970s *Doctor Who* (one call has been dubbed 'the *Star Wars* call') – as do Bryde's, Omura's and sei whales. And while the data is sparse – partly because numbers have never recovered from the devastation of whaling – there is evidence that at least some right-whale populations sing. Indeed of all the baleen whales, it seems only gray whales do not sing.

WHILE SONG REMAINS the preserve of baleen whales, toothed whales also use sound in extraordinary ways. Dolphins in particular are fantastically vocal, chattering and whistling and firing bursts of sound back and forth at each other almost constantly.

The question of what these sounds might mean has fascinated scientists for decades. It is broadly accepted dolphins are highly intelligent: they are self-aware and compulsively social, they possess culture, can solve complex problems, play and hunt collaboratively and have been observed using tools. Some populations even enjoy getting high by dosing themselves with pufferfish venom.

There is also no real doubt dolphins use their vocalisations to communicate. They vocalise when cooperating and when collaborating on problems, and seem to actually converse: pairs of dolphins have been observed firing bursts of sound back and forth, apparently listening and replying to one another, and in an experiment in which a mother and her calf were placed in separate tanks connected by an audio link the pair immediately began chattering with each other over the intercom. They also have signature whistles: unique calls that are taught to infant dolphins by their mothers in the same way humans give each other names. These whistles are used by dolphins to identify themselves to close social partners, to call to each other when they are separated, or in some instances, combined with the signature whistles of close associates, to form a shared, compound whistle.

Given all this, the assumption dolphins are talking to one another seems irresistible, yet evidence that dolphins possess language remains elusive. At one level this is surprising: multiple studies show dolphins are capable of *learning* language. They are able to associate particular sounds with particular objects, to arrange symbols and sounds into meaningful sequences, and even to understand a question symbol that was designed to elicit a yes or no answer. Wild dolphins taught to associate a particular whistle with a specific species of seaweed were later observed using the same whistle among themselves when they encountered that species. They are even capable of learning to communicate with other species: in the Firth of Clyde on Scotland's west coast a solitary dolphin that regularly associates with harbour porpoises has been observed using the porpoises' system of high-pitched clicks rather than the pulses, clicks and whistles of her own species.

Part of the problem is the question of what we mean by 'language'. Animals, plants and fungi communicate in all sorts of ways – with sound, but also using chemical and visual cues, touch, even electrical impulses. But while these systems of communication can be surprisingly complex, they are also different in both type and degree from the sort of language humans use.

The exact nature of these differences is widely debated, but there are two key distinctions. The units of language humans use have meaning, or semanticity, and syntactical rules that govern the ways those meanings change when arranged in specific ways. The communication systems of some animals have elements of the former – certain birds and monkeys use specific cries to warn of particular predators, and African elephants, which hate bees, use a special alarm call to signal the presence of their apian enemies – and parrots, dolphins and apes all seem able to learn some syntax.

But while none of these communications exhibits the complexity of human language, tantalising evidence of unrecognised complexity within dolphin vocalisations is beginning to emerge. Dr Julie Oswald is a researcher at the University of St Andrews in Scotland. Oswald has been using artificial intelligence to identify the sounds produced by dolphins, a process that has revealed the dolphins are using a repertoire of several hundred distinct whistles that appear to remain stable over time. While it is tempting to assume these sounds are a vocabulary, Oswald is wary of making that leap: 'Dolphins seem to have a lot to communicate about, and they produce a lot of whistles, and I'm confident they're not doing that for no reason and the whistles mean something to them. But whether that's a language or a vocabulary in the way we understand it or something different, I really don't know.'

Despite evidence suggesting the dolphins are using different whistles for different things, Oswald argues 'The line between language and not language isn't necessarily cut and dried . . . If you could understand a dolphin, would you really understand it? Because their brains work so differently from our own, and they have such different things to communicate about.'

An even more ambitious initiative is underway in Dominica, where an international consortium of biologists, linguists, cryptographers and computer scientists are hoping to unlock the mysteries of sperm whale communication. Named Project CETI in a nod to SETI, the long-running search for extraterrestrial intelligence, the

Cetacean Translation Initiative does not just aspire to identify the structure in the whales' vocalisations, but to unravel its meaning.

The largest of the toothed cetaceans, sperm whales mostly make their home in the open ocean. Females and immature individuals form family groups in temperate and subtropical waters, while the males live solitary lives, often roaming into polar waters. They have the largest brains of any animal on Earth and, like their much smaller cousins the dolphins, are capable of a wide variety of vocalisations. These range from the slow clicks of the solitary males, to the accelerating creaks they produce when they close in on their prey deep underwater, to what are known as codas – short patterns of clicks lasting a second or so that sperm whales produce while socialising near the surface.

Codas seem to help reinforce social bonds in sperm whale groups, and may well aid in identification in much the same way as the signature whistles of dolphins. Individual whales occasionally produce codas on their own, but more often the whales call back and forth to one another, creating overlapping patterns. Sometimes they will do this while floating close to each other, or across larger spaces of water, but they have also been observed swimming and playing alongside each other, clicking back and forth for as long as forty minutes at a stretch.

Young sperm whales learn their codas from their mothers and other family members (like dolphin calves and human babies, sperm whales babble before they talk). Each social unit of sperm whales also possesses a distinctive repertoire of codas, which in turn share family resemblances with the codas of certain other social units. These larger groups of whales possessing similar codas are known as vocal clans, and can consist of thousands of whales spread over huge expanses of ocean. There is no evidence suggesting these different clans are genetically separate: instead, like human languages splintering into dialects as communities diverge, these differences in coda repertoire map onto behavioural and geographical differences between different sperm whale clans, suggesting an ongoing process of cultural evolution.

CETI relies upon a complex network of dedicated hydrophones in combination with specialised underwater video cameras, submersible and floating drones, and even cameras that attach to the whales themselves. Operating in concert, these are in the process of amassing a library of millions of sperm-whale codas, together with information about social behaviours associated with these vocalisations.

This library of videos and audio files will be fed into artificial intelligence systems, which will try to map the underlying structure of the codas in the hope of isolating discrete sounds and patterns that act as basic units in the way words and phonemes do in human language. If these can be found, the next challenge is to try to establish whether there are rules governing the use of these units. And if there are, is there some way of establishing the meaning of the sounds?

Dr Shane Gero is a Canadian biologist and one of the founders of CETI. He has spent a decade and a half observing and recording the sperm whales of Dominica. He says it is important to recognise the challenges of the undertaking: 'There are some fundamental similarities about being a terrestrial mammal and a marine mammal, but there are a lot of things about their experience of the world that will be hard for us to understand.' Nonetheless he believes the project is already yielding results, especially when it comes to mapping the internal structure of the whales' vocalisations. 'What's really interesting is that on the face of it the codas seem to be a very simple communication system. But what we're already finding is that what looks like a simple, Morse code kind of structure seems to contain nested levels of complexity that we haven't been looking at before now.'

For Gero the importance of the project goes beyond the question of what the whales are saying to each other. Instead it contains within it a recognition of the right of the whales to inhabit the world on their own terms. 'All we're really doing is paying attention and listening. But perhaps the most important thing we need to do right now is to learn to listen to beings that are different from ourselves.'

THESE DIFFERENCES ARE PROFOUND. The ancestors of today's whales were wolflike mammals that turned to an aquatic lifestyle around 50 million years ago. As they spread across the planet's oceans the structure of their bodies changed, their limbs transforming into flippers and their nostrils migrating upwards onto the tops of their heads. Sometime around 35 million years ago the whale lineage diverged again, as what would become the baleen whales split from the ancestors of today's toothed whales.

This split seems to coincide with the period of significant cooling that followed the opening of the Drake Passage between South America and Antarctica, and the Tasmanian Seaway between Australia and Antarctica, which changed the distribution of food within the oceans. While the ancestors of the baleen whales began to feed on smaller animals and gradually developed the baleen they now use to filter krill and other organisms from the water, the ancestors of the toothed whales began hunting schools of fish collaboratively, resulting in a shift to a social lifestyle, a rapid increase in the size of their brains and – perhaps most importantly – the development of echolocation.

Thirty-five million years later, and dolphins and the other toothed whales are, quite literally, wired for sound. Not only have their brains been reorganised to allow for lightning-quick processing of sensory information, their auditory and visual systems are interconnected, allowing what scientists describe as 'highly developed cross-modal sensory processing', or the ability to 'see' with sound. Their bodies are also superbly adapted to produce and receive sound. The massive, blunt heads of sperm whales, for instance, are filled with huge sacs of the waxy oil known as spermaceti that led to their slaughter by whalers. These structures can hold almost 2000 litres of spermaceti and allow the whales to produce and amplify complex patterns of sound. Similarly dolphins and other toothed whales possess a mass of fatty tissue called a melon that sits in the bump on their foreheads that they use to focus the sound they produce into bursts. These systems are surprisingly powerful: dolphins use bursts of sound to harry sharks and fish and to discipline their young, while the sound of sperm whales firing their sonar at the hulls of ships produces a sound like hammering. Nineteenth-century whalers sometimes called them carpenter fish as a result, and the scientists Hal Whitehead and Luke Rendell have speculated that inquisitive sperm whales may be the origin of the legend of Davy Jones, the malevolent fiend that sailors believed haunted the ocean's depths, knocking on the hull.

It is difficult for us to imagine the world the toothed whales' sonar reveals to them. For a start it is etched out in ultra-high definition: experiments with dolphins show they can 'see' objects the size of golf balls from 100 metres away, and detect differences of a few tenths of a millimetre from 10 metres away. But because sound waves pass back and forth through objects, sonar also renders the world around dolphins transparent, allowing them to 'look' through other animals and each other and granting them the ability to discern different materials.

A world encountered in this way is startlingly different from our own. Dolphins are not just able to see through objects, they have access to a detailed three-dimensional understanding of the internal structure of the animals and objects around them. The philosopher Thomas White compares this ability to having instantaneous access to the sort of information about the internal structures provided by x-rays and MRIs and other forms of medical imaging. Other animals resemble living versions of anatomical models, in which the organs float in glass, and the beat of the heart and the pulse of blood are visible. As a result, they may well be able to tell when other individuals are ill or injured or pregnant: they tend to take particular interest in people with surgical steel or other artificial objects in their bodies, and clearly understand when humans are pregnant (one researcher was even alerted to her own pregnancy by the dolphins she worked with "buzzing" her belly with a form of very intense echolocation). Similarly, there are instances in which they seem to have been able to recognise illness or death. The dolphin researcher Denise Herzing has told of an incident in which a group of dolphins she knew well behaved unusually when her research vessel encountered them, refusing to approach the boat or ride the bow wave, and instead flanking the boat at a distance, like 'an aquatic escort'. In the middle of this perplexing encounter a crew member discovered one of the passengers had suffered a heart attack, and was lying dead in his bunk below decks. After the body of the passenger and his grieving wife and daughter had been returned to shore, Herzing and her crew headed out once more; this time the dolphins behaved exactly as they usually did, greeting the boat and frolicking alongside it. 'Twenty-five years later I have never again observed the dolphins escort the boat in the same manner,' writes Herzing.

It is even possible that as boundaries become less distinct so too does the idea of where one dolphin ends and another begins. Dolphins can hear the reflected clicks of the dolphins around them, meaning they have access to the sensory experience of other dolphins. Shane Gero likens this ability to running through a darkened forest with a group of people all wielding torches: although your attention is focused on what is revealed by your own torch, you are also able to glance sideways and see what others are seeing. The cetacean expert Ken Norris has suggested this ability to process the sensory input of an entire pod creates what he dubs a sensory integration system, and that this explains the tendency of pods of dolphins to act as a single entity, especially in situations where the group is under threat. As Norris has observed, when spinner dolphins are 'under attack . . . individuality is reduced close to zero.' In such situations the dolphins seem unwilling or unable to take action that might benefit them or even save their lives unless the whole group can do the same. The same phenomenon may be a factor in mass whale strandings: even if moved back into deeper water, stranded whales often just re-beach themselves unless most or all of the other whales are also moved offshore. Perhaps significantly, mass strandings also mostly involve matrilineal toothed whales, such as pilot whales, with extremely tight social structures.

This perceptual entanglement is not restricted to small or tight groups. Instead it may be experienced over long distances: sperm whales are able to listen in on the vocalisations of whales many kilometres away. Being together for whales may not require physical proximity; it is possible they feel connected to one another even when dispersed over tens or even hundreds of kilometres of ocean. But it also raises the possibility dolphins and other toothed whales do not experience selfhood in the way we do: rather they inhabit a world in which they are simultaneously themselves and part of a larger, social or collective self, bound together by the lightning-quick exchange of reflected sound.

WONDROUS AS THEY ARE, these ways of being are under assault by human activity. Since the introduction of powerful low- and mid-frequency naval sonar seventy years ago the frequency of whale strandings has increased dramatically. Indeed one recent study concluded that of 136 mass strandings of beaked whales reported between 1876 and 2004, 126 took place after the introduction of mid-frequency sonar in 1950, and strandings in Britain, Greece, the Canary Islands and the Pacific have been linked to the use of naval sonar.

Many of these strandings involve Cuvier's whale, a deep-diving species of beaked whale that lives in tropical and temperate waters around the world. Exactly why Cuvier's whales are so vulnerable isn't clear, but when sixteen beached themselves after a US Navy sonar exercise in the Bahamas in 2000, autopsies showed they had suffered haemorrhages in their ears and brains. The cetologist Ken Balcomb, who found several of the whales, points out that the resonance frequency of the whales' airspaces almost exactly matches the frequency of the sonar used in the exercises. 'Envision a football squeezed to the size of a ping-pong ball by air pressure alone. Now envision this ping-pong ball compressing and decompressing hundreds of times per second. Imagine this ping-pong ball located in your head, between your two ears. This is what the . . . whales experienced as a result of the Navy's sonar.'

The effects of sonar on whales are only one example of a much larger phenomenon. Over the past fifty years human activity has transformed the soundscape of the ocean. Shipping noise has increased dramatically: the din produced by large ships has been doubling decade on decade since the 1960s, resulting in a more than 32-fold rise. Although the roar of propellers is loudest along shipping lanes, it is definitely not confined to them: in the open ocean the low-frequency rumble of ships is audible for tens of kilometres in every direction, while in shallower waters it bounces back and forth off the ocean floor to produce a constant roar of reflected noise. In many places this din competes with the clang and grind caused by the construction of oil and gas platforms, the roar of drilling, and the deafening thunder of seismic surveys. These surveys, carried out by fossil fuel companies searching for new reserves of oil and gas, use ships equipped with arrays of air guns on the bottom of their hulls that direct bone-shattering pulses of sound at the ocean floor. The guns fire every few seconds around the clock for weeks and months at a time, filling the environment around them with a constant barrage of noise so loud it can be heard up to 4000 kilometres away.

The impact of this cacophony on marine animals is significant. Whales exposed to the sonic assault of seismic testing suffer damage to their ears and hearing loss, and even shipping noise causes distress and interferes with communication and other behaviours. Humpback and bowhead whales exposed to the roar of ships will move away or dive, reducing their ability to feed and socialise normally, and the faeces of right whales, gray whales and orcas exposed to shipping noise contain higher levels of cortisol and other chemicals associated with stress. Measurements of surfacing intervals suggest that even many dolphins find the noise of ships stressful.

Fish also avoid noisy ships, and when drowned out by the acoustic smog that now suffuses ocean environments, tend to raise their voices like humans trying to be heard over the din of traffic. This affects social behaviours, and in extreme cases can lead to fish simply abandoning their habitat and moving elsewhere. And the problems don't stop there: fish exhibit elevated levels of stress hormones in noisy environments, and continued exposure to noise can induce neurophysiological changes and interfere with normal reproduction by killing larvae or affecting the ability of fish to care for their young. Reproductive success in the spiny chromis, a species of damselfish found on the Great Barrier Reef that spends months of every year guarding its eggs and young, has been shown to fall by more than a third when the fish are exposed to the sound of dive boats, and individual chromis exposed to noise show signs of agitation and distraction.

Invertebrates are suffering from the rising volume underwater as well. Seismic surveys kill lobsters and crabs; those that survive can suffer hearing loss and damage to their balance and coordination. Hermit crabs exposed to the vibrations produced by pile drivers have been observed leaving their shells and inspecting them before

climbing back in – behaviour that increases their vulnerability to predators. Soft-bodied animals such as cephalopods are even more severely
affected: seismic surveys kill giant squid and other species, but even
the low-intensity sound that now permeates marine environments can
cause significant trauma to the delicate system of tiny sacs and filaments called statocysts that cephalopods use to control their balance,
disrupting the ability of a wide range of squid, octopus and cuttlefish
species to orient themselves. Noise affects shellfish as well. Sometimes
these impacts are direct: in 2010 seismic surveys in Bass Strait killed
at least 24,000 tonnes of scallops. But noise also has more insidious
effects on shellfish: not only do scallop larvae exposed to sonic pulses
develop abnormally, but blue mussels exposed to ship noise produce
a stress hormone that alters their DNA, leading to reproductive
anomalies. The noise from sonic blasts also kills zooplankton up to
1.2 kilometres away.

Human activity is altering the sonic environment beneath the
waves in less direct ways as well. As the oceans heat up, the soundscape of the oceans is changing. In the tropics, increased storm and
wave activity is adding to levels of noise produced by rain and water.
Meanwhile in the Antarctic the thunder of collapsing ice shelfs grows
louder with each passing year, a quickening metronome that counts
out the heartbeat of a warming world. And in the Arctic, where the
groaning of the sea ice in the polar night is replaced by the susurrus of
the wind on the open water and the rumble of shipping and air guns
in search of fossil fuels, the cries of species such as killer whales are
being heard further north than ever before.

Elsewhere, as currents shift and disruption to seasonal cycles
accelerates, the calls and choruses that once rose and fell with the
passage of the seasons are also changing. The soundscapes of coral
reefs, sponge gardens and kelp and seagrass beds are being simplified
or silenced as bleaching and warming water reduce their diversity, or
species migrate to cooler waters. Because sound travels faster through

warmer and more acidic water, global heating is even changing the acoustic quality of the water itself: as scientists Jérôme Sueur, Bernie Krause and Almo Farina have put it, climate change is 'altering the acoustic fabric of the planet . . . detuning natural sounds' and breaking the Earth's beat.

IN THE EARLY MONTHS OF 2020, as commercial shipping and industrial activity slowed due to the spread of COVID-19, noise levels in the ocean fell dramatically. Along shipping lanes, and in regions such as the Pacific Northwest and the North Sea, the oceans were quieter than they had been in many years. Around the world, marine animals began to reappear: in the Gulf of Oman and the Persian Gulf, sharks and rays appeared in silent marinas and huge pods of dolphins swam close to shore, and off Sri Lanka a superpod of more than 300 sperm whales was sighted.

This brief window of quiet, part of a larger cessation of human activity that some have dubbed the anthropause, provided a glimpse of what an ocean unburdened by human sound might be like for its inhabitants. But it also offered a powerful reminder that the world is full of other presences, and other beings, and of what might be gained by caring enough to listen to them. For as the work of Shane Gero and others reminds us, listening is not a passive state; instead it is a matter of showing concern, an attitude of care.

Similarly, as Amitav Ghosh has observed, it is not coincidental that so much of the Western intellectual tradition is about silencing and rendering mute. The project of ocean-borne imperial expansion and colonial violence out of which it grows is, at its core, a process of denying the worth of other ways of being in the world. Writing of the 'war of the soul' that took place on the slave ships, the historian Sowande' M. Mustakeem compares the process to that of actual war, and the need to disown the injuries inflicted upon both enemy and self.

Seen like this the 'deafening secrecy' of the historical record about conditions on the ships becomes a way of denying the truth of what took place upon them. To acknowledge that reality, to listen to the gaps in the record and attempt to imagine the bodily conditions on the ships, the sounds and sights and smells and sensations, thus becomes a way of beginning to reckon with not just the human cost of slavery, but the degree to which its wounds – and the silencing and denial of those wounds – continues to haunt our world. The same is true of the violent silencing of Indigenous peoples, who were placed outside the circle of human concern, their homes destroyed and their stories, songs and knowledge ignored and suppressed. Any reckoning with the truth of history demands attention to that suffering, and an attempt to hear the voices that were silenced.

Perhaps it also demands a recognition not just of how much has been silenced and suppressed, but that there are other voices in the world, non-human ones. The sounds of the ocean's voices are many and varied – reminders of the complexity and infinite variety of life. The ocean of sound that surrounds us is in fact a babel, filled not just with human voices, but those of other species. To hear them demands letting go of the idea that only our way of being in the world contains meaning. What could listening to those voices teach us, what different ways of being in the world might it teach us? A blue whale's heart beats just two times a minute when it dives – to listen to recordings of it is to find yourself wondering about the meanings contained in its songs, what it must be to inhabit those rumbling pulses and longing cries. The songs of whales, of dolphins, of fish are all reminders of the presence of other ways of being, ways of being that hum with their own purpose and structure and value. The ocean is alive with meaning; to listen to it is to be made aware of the immensity, complexity and impossible beauty of the world. But it is also to recognise that we are not separate from that world, and to think that we are is to do violence not just to it, but to ourselves.

As Roger Payne writes in his often strikingly beautiful memoir, 'The first time I ever recorded the songs of humpback whales at night was off Bermuda. It was also the first time I had ever heard the abyss. Normally you don't hear the size of the ocean when you are listening, but I heard it that night. It was a bit like walking into a dark cave, dropping your flashlight, and hearing wave after wave of echoes cascading back from the darkness beyond, realizing for the first time you are standing at the entrance to an enormous room. The cave has spoken to you. That's what whales do; they give the ocean its voice, and the voice they give it is ethereal and unearthly.'

5

BEINGS

'The mind evolved in the sea.'
Peter Godfrey-Smith, *Other Minds*

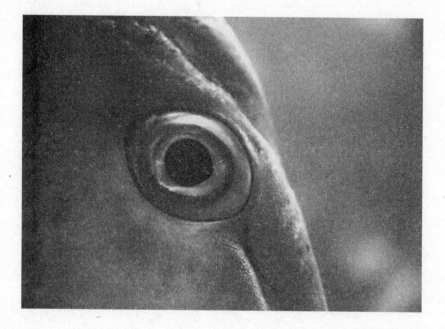

IN EARLY 2020 while on a field trip in the Cocos Islands, I took some time out and went snorkelling. I was in a shallow channel between two islets and the tide was running, so my only real option was to let the current carry me, which it did, and quickly, sending me shooting over an expanse of broken coral and sand. Although I had seen fish earlier in the day, there weren't that many about in the water I was moving through, but after a few minutes a trevally came angling in towards me. It was a striking creature with a silvery, streamlined body perhaps 80 centimetres long and a vivid blue stripe that ran down its spine and back along its midline. It swam towards me, moving fast and without fear, until at the last moment it shot off to one side and swooped behind me. I turned, watching as it turned back to circle me again, and then again, occasionally swinging in close, then out once more.

It kept up this process for the next ten minutes or so, shadowing my path as the tide carried me through the channel and into the lagoon beyond. At first I was concerned it might be thinking of attacking me – certainly there was an edge of aggression to its constant motion – but what most struck me about its behaviour was its air of purpose, the sense I was being observed and assessed. There was no question that this was a creature with its own intentions and agenda, or that when I met its gaze it was looking back.

Anybody who spends time in the water will have had similar encounters. My brother sometimes feeds the blue groupers that live in Sydney's rocky coves, cracking open sea urchins and holding the meat with his teeth, so the groupers must come close as if they are kissing him, their movements flighty, yet purposeful. I'm not sure I want to get that close to a grouper's mouth, but I have handfed huge stingrays, delighting in their puppylike playfulness as their strange bottom-facing mouths suck and pull at my fingers like an affectionate sea squirt. Although their behaviour is quite different to that of the trevally, it is no less obvious that they are thinking and acting in the same way as the terrestrial animals most of us are more familiar with.

Despite such encounters, fish are often dismissed as little more than slimy automatons and – at least outside of Indigenous cultures – almost never considered as subjects of moral concern. When most humans do think of them it is as food or as pets of a purely ornamental kind. Pescatarians who would regard slaughtering a cow or a pig or a chicken as unutterably cruel and unethical happily consume fish as if they are little different to vegetables. Even our language often erases their particularity. In English we speak of one fish or many fish, as if they are so interchangeable it is not worth according them the dignity of a plural form. Likewise scientists and economists talk blithely of fish stocks, transforming the lives of millions of individual beings into a resource, little different to metal or timber.

These assumptions elide a world of astonishing complexity. Not only have fish inhabited our planet far longer than air-breathing animals such as ourselves, having first evolved over half a billion years ago, but they are also extraordinarily diverse. The more than 34,000 species of fish known to science make up 60 per cent of all vertebrate species, or more than mammals, birds and reptiles combined. Fish range in size from the minute *Paedocypris progenetica*, which is found in peat swamps and blackwater streams of Sumatra and Bintan and is less than 8 millimetres in length, to the immense whale shark, *Rhincodon typus*, which grows to almost 13 metres in length and can weigh well over 20 tonnes. I once swam with one whose tail was taller than I am; I have never forgotten the casual way it shot away with a leisurely flick of that tail.

Fish are also found in almost every watery environment on Earth. They make their homes below the ice of the polar oceans and the blood-warm rivers and seas of the tropics, on the mudflats and intertidal zones of mangroves and more than 8 kilometres below the surface in the darkness and bone-liquefying pressure of the ocean trenches. Some species lay eggs; others bear live young. More than 400 species are known to be hermaphroditic, and capable of changing sex. *Eviota sigillata*, a tiny reef fish found in the Indo-Pacific, lives a mere eight weeks, the shortest lifespan of any vertebrate. By contrast analysis of radioactive deposits in the eyes of the mysterious Greenland shark – which lives in the icy darkness of the North Atlantic and Arctic Sea and feeds on fish and the carcasses of seals, whales and even polar bears – has revealed the species lives for hundreds of years. One specimen was shown to be between 272 and 512 years of age, making it the longest-lived vertebrate in existence. The species lives so long that females are not believed to reach sexual maturity until they are a staggering 150 years old.

And this remarkable diversity is only the beginning. A growing body of evidence makes it clear that fish not only think and feel, but exhibit complex social behaviours and sophisticated cognitive abilities, are capable of learning, problem solving and tool use, and possess culture and even the sort of self-awareness previously assumed to be restricted to primates, dolphins, elephants and a few species of birds. These discoveries do not just demand a rethink of assumptions about the cognitive capacities of our finned cousins, they challenge ideas about how to identify and measure intelligence in species so different from ourselves.

ONE OF THE LEADING FIGURES in this emerging space is Culum Brown, professor in Biological Sciences and Vertebrate Evolution at Macquarie University in Sydney. Brown's early research was on rainbowfish, small river fish native to Australia, New Guinea and Indonesia. A popular aquarium species, rainbowfish possess the curious mixture of nervousness and glassy regard that tends to lead humans to dismiss the idea fish might be intelligent. Yet Brown's work reveals they recognise each other, have complex social lives and hierarchies, and are capable not just of learning to avoid dangers such as predators and traps, but of passing these techniques on to other rainbowfish. Furthermore, these abilities are not rudimentary: experiments show rainbowfish learn to associate signals with food three times as fast as rats and twice as fast as dogs.

Brown's work with rainbowfish demonstrates that contrary to the old joke about goldfish remembering nothing for longer than thirty seconds, many fish have excellent long-term memories. When tested again almost a year later, the rainbowfish immediately recalled the required responses and responded as if no time had passed. Similar results have been observed in many other species of fish: tilapia taught to associate a signal with netting, for instance, remembered the signal and responded accordingly seventy-five days later, while gobies, which form highly detailed mental maps of the tidal pools in which they live, can still remember the location of neighbouring pools for at least forty days after being removed from their own pools.

Even more importantly, Brown's research makes it clear that not only are fish capable of remembering, they are capable of learning from each other through observation and interaction, meaning behaviours can be passed between individuals and, even more significantly, between generations. This capacity to acquire and transmit stable behaviours is the foundation of culture.

At one level these discoveries should not come as a surprise. As Brown points out, 'We've known about social learning and cultural transmission in animals for sixty years. Everybody looked for it in chimps first because they're so much like us. But since then the search has cascaded through nearly all the animal taxa, to the point where I think it would be fair to say social learning and cultural traditions are present in almost all animals.'

Much of this social learning relates to foraging and food acquisition. Archerfish, for instance, develop their ability to shoot insects from the air by firing water from their mouths by watching other fish, while sea bass can learn to press a lever to obtain food by observing other individuals doing the same. But fish are also capable of learning more complex information about migratory routes and where to find food from each other. Guppies pass on information about where to go to find food to other fish as they join the group.

Evidence of this sort of cultural transmission in fish is widespread, especially in long-lived species. French grunts, a yellow fish native to the West Atlantic, spend their days sheltering from predators among the spines of sea urchins; when they emerge at sunset they follow paths to their feeding grounds that are learned from older fish. Similarly on reefs off Panama young female bluehead wrasse learn the route to the best breeding spots from older females. As with all forms of cultural transmission this process is also highly vulnerable to disruption, with studies showing cultural traditions are quickly lost if the individuals that possess them are removed from the group.

As their various vocalisations suggest, fish also possess complex social lives. Although humans often struggle to tell them apart, fish have no trouble recognising and remembering each other. And, like other social animals, they often form attachments, preferring the company of shoalmates and fish they are familiar with over other members of their species. Brown and his team recently discovered that Port Jackson sharks – a gentle, bottom-dwelling species of bullhead shark whose heavy head and elegant brown markings will be familiar to many in southern Australia – have well-established social networks and prefer the company of individuals of the same age and sex. In other words they socialise with their peers, in much the way humans do (the same study revealed the sharks also perform remarkable feats of navigation, migrating thousands of kilometres from their homes on

Australia's east coast to Bass Strait before returning to the same reef they left from).

These abilities are not restricted to recognition and attachment. Fish often employ cooperative behaviours, especially when hunting: yellowtail amberjacks off the Californian coast have been observed using U-shaped formations to separate out and trap smaller fish, behaviour that closely resembles the sophisticated hunting techniques of mammals such as wolves and dolphins. Some species of cichlids – a diverse family of fish native to lakes and rivers in Africa and the southern Americas – also rely upon cooperative techniques to protect and raise young. Conversely fish are also quite capable of excluding and rejecting individuals that do not behave appropriately. When approaching potential predators, sticklebacks use a distinctive stop-start swimming motion to share risk by taking turns at the front. If an individual is reluctant to take the lead, or cheats by hanging back, its shoalmates will refuse to cooperate with it in future, meaning the sticklebacks recall the identity of malingering individuals and remember they are not to be relied upon.

It seems likely these behaviours are only isolated examples of a much more widespread phenomenon – as Brown observes, the practical obstacles to detailed observation of the behaviour of fish means our perspective on their lives is extremely limited. Yet there is no question that as in other species this social complexity requires high levels of cognition, and suggests the tendency to regard fish as primitive, emotionless automata is fundamentally mistaken, and that fish inhabit social worlds at least as complex as those of many mammals and birds.

No less tantalisingly, there is evidence a number of fish species are capable of using tools. At one level this is surprising – unlike mammals and birds, fish lack grasping appendages, and the aqueous environment makes manipulating and striking difficult. But wrasse and tuskfish have been observed using rocks as anvils to break open shellfish, and it is tempting to regard the archerfish's use of water to shoot prey as a form of tool use. Fish are also capable of developing innovative ways of manipulating their environment, as in the case of the group of Atlantic cod in an aquaculture facility in Norway that took to stealing food from a feeding device after they discovered tags attached to their bodies could be used to activate it.

Some scientists have dismissed the idea such behaviours should be regarded as tool use because they do not involve the use of one object to manipulate another. Yet Brown believes these objections are based on a deliberately restrictive notion of what constitutes a tool. 'What's interesting is that when Jane Goodall and others came up with a definition of tools it had nothing to do with appendages: it was all about using an object to achieve a goal. But then the whole concept got hijacked by the primatologists, who substituted a primate-centric definition.' Instead Brown argues we should return to Goodall's original definition, and its emphasis upon the intention of the animal to achieve a goal it could not otherwise realise.

This emphasis upon intentionality excludes many examples of seemingly innate behaviours resembling tool use that have been observed among insects. More intriguingly, however, it allows behaviours such as nest-building to be a form of tool use, permitting a broader approach to our understanding of the evolutionary origins of tool use. 'There's clearly a lot of overlap between building nests and using tools,' says Brown. 'Both are clearly about manipulating the environment in a way that enhances your fitness, so you're either more reproductively successful, you're getting more food, or you're safer from predators and other environmental stresses.'

This is particularly relevant to fish, of which more than 9000 species build nests. Sometimes these structures are relatively simple: certain species of wrasse, for instance, create mucous cocoons in which to shelter while they sleep, and eels and other species produce bubble nests to protect their eggs. But many also construct surprisingly complex structures out of stones and coral, often on a daily basis. 'It isn't simple or boring behaviour. It can be seriously sophisticated,' says Brown. 'If you're building an igloo out of coral rubble it has to be structurally complex enough that it won't fall down around your ears.'

Nests also provoke fascinating questions about the inner lives of the creatures that construct them. As Helen Macdonald has observed in the context of birds, nests may be part of the phenotype of the animals that make them, yet they are also highly responsive to local conditions, suggesting the animals that build them are capable of innovation and adaptation. And, more deeply, they unsettle our assumptions about the difference between skill and instinct. When we consider the elegant concentric rings, rays and doodles that pufferfish create in the sand it is difficult not to wonder, as Macdonald does of birds, whether the animals making these structures are merely following a sequence of cues hard-wired in their biology, or whether, like us, they begin with some kind of mental image of the nest or symbols they create, which they then bring into being, step by step, refining as they go. As human beings we are familiar with the interplay between plan and form, that sense that altering one detail or rearranging an element will result in a greater sense of unity or balance: is it possible fish feel something similar? And what might it mean for our sense of who and what they are if they do?

ANOTHER EXAMPLE of unexpectedly sophisticated behaviour in fish emerged in 2006, when Professor Redouan Bshary, a specialist in behavioural ecology, noticed groupers in the Red Sea interacting with morays in a highly novel manner. The process began with the groupers approaching the morays and performing a distinctive head-shake or shimmying motion to attract their attention. That done, the tropical odd couple would head off together in search of prey.

As fast-moving predators groupers excel at catching fish in open water, but often lose prey when it takes shelter in holes or under rocks. The long, sinuous bodies of morays, on the other hand, are perfectly adapted to catching fish in cracks and crevices – but the morays tend to lose them if they escape into open water. Thanks to their complementary skills, groupers and morays in combination were able to cut off all routes of escape for their prey, making them far more effective hunters than either is in isolation. Bshary later observed the groupers teaming up with humphead wrasse as well, using the much larger humpheads' ability to suck prey from crevices with their powerful jaws to complement the groupers' open-water hunting skills.

This form of interspecies cooperation is incredibly rare. Honey badgers and honeyguide birds work together to find hives and honey, and orcas have been known to help human whalers catch humpbacks in return for the prized tongues of the humpbacks; but other examples are few and far between.

Yet there was more to come. In cases where the fish the groupers were chasing managed to conceal themselves somewhere inaccessible, the groupers would stop and perform a headstand over its location.

To human eyes it seems obvious the grouper are pointing at the prey hiding beneath them. But as anybody who has ever tried to direct the attention of a dog or a cat by pointing at an object will know, referential gestures that seem natural to us are nothing of the sort for other animals. Although Australian ethologist Gisela Kaplan has also observed what appears to be pointing in magpies, and some dogs seem to be able to learn to understand what we mean when we point, the only species known for certain to understand referential gestures are humans, chimpanzees and corvids. Even human children only become capable of using and understanding referential gestures towards the end of their first year of life.

As a result, biologists have developed a fivefold test behaviours must pass before they are accepted as referential. To be truly referential a gesture must be directed towards an object; it must be communicative rather than mechanically effective; it must be directed at a recipient or potential recipient, whose response must be voluntary; and, perhaps most importantly, it must exhibit the hallmarks of intentionality.

With this in mind, Australian scientist Dr Alex Vail set out to investigate cooperative and signalling behaviours. For his study Vail chose to work with coral trout on the Great Barrier Reef, which are not only closely related to the groupers Bshary studied, but also engage in very similar cooperative behaviours with morays and several other species. Remarkably, Vail found that the behaviour of the trout meets all five of the criteria required to qualify as a form of referential gesture. Or, in simpler terms, the trout don't just look like they're pointing, they *are* pointing.

In mammals and birds, the ability to use referential gestures is associated with high intelligence. So what does this tell us about the cognitive capabilities of coral trout? To try to clarify this question, Vail devised a series of experiments modelled on tests used to measure the collaborative abilities of chimpanzees. These showed that the ability of the trout to assess whether a situation requires a collaborator and which individual would make the most effective collaborator in that situation is almost identical to that of chimpanzees – and actually exceeds that of chimpanzees in some circumstances.

Vail, who now works as a cameraman on major natural history projects such as *Blue Planet 2* and Netflix's *Our Planet*, admits he was not hugely surprised by the results: having grown up at the Australian Museum's Lizard Island Research Station on the northern Great Barrier Reef, he has been diving with coral trout all his life and has no doubt they are highly intelligent. Describing an experiment that involved spending time in a tank with them, he laughs. 'It's amazing how quickly they notice your intent. I mean you can spend ten days with them sitting right beside you, watching everything you do, and then the moment you produce a net, even before you've attempted to use it, they're off to the other side of the tank.'

Similar behaviour has been observed in the Chagos Archipelago, where groupers have learned to team up with octopuses. As with the trout Vail studied, the groupers and octopuses combine their very different hunting specialisations to increase their overall success at capturing prey. But in the Chagos the groupers and octopuses have developed a complex system of visual signals they use to communicate with each other while hunting, using changes in colour and gestures with their tentacles to coordinate their behaviour. Some of the signs were so obvious that the researchers studying the behaviour became able to recognise particular signals.

Another, even more impressive example of interspecies cooperation involves several species of cleaner wrasse. Small, highly social fish that inhabit coral reefs from the Red Sea to the Pacific, cleaner wrasse maintain 'cleaning stations', which other fish visit so the wrasse can remove dead skin and parasites from them.

These stations are extremely popular, attracting large numbers of regular clients as well as more casual visitors, and the wrasse are capable not just of recognising and remembering more than 100 clients, but also of remembering *when* they interacted with different clients. Yet while the wrasse are skilled at their job, they are not entirely trustworthy, and will occasionally seek to supplement their diet of dead skin and parasites with a mouthful of live skin and mucus.

Not surprisingly, their clients tend to be unimpressed by being nipped in this way. But the wrasse choose their victims carefully. To begin with they only ever bite non-predatory species unlikely to retaliate by eating or attacking them. They are also able to distinguish between new clients and regulars: faced with a choice between a fish that has been to their station before and a new client, the wrasse will usually clean the newcomer first, and will almost never nip them, thereby increasing the chance of repeat business.

The sneakiness of the cleaner wrasse doesn't end there. Not only are they less likely to bite if they know they are being observed, they will chase after victims who respond badly to being nipped, dancing around them and rubbing their backs and pelvic fins in an attempt to mollify them. They even seem to understand the relationships *between* different client species, and if pursued by an irritated client will swim towards a predator of that species to prevent them chasing them further. Nor is their craftiness always directed at other species: cleaner wrasse have been observed visiting the cleaning stations of other wrasse and biting the fish at them, seemingly in the hope of disrupting their competitor's business by confusing and upsetting their customers.

This Machiavellian behaviour implies cleaner wrasse are capable of attributing mental states to other fish and responding accordingly, suggesting they possess what scientists and philosophers refer to as a 'theory of mind'.

More evidence cleaner wrasse possess higher-order mental states emerged in 2019, when a team led by Japanese scientist Masanori Kohda published evidence showing the wrasse are capable of recognising themselves in a mirror – a feat previously confined to humans and the other great apes, elephants, dolphins and a handful of birds, and usually regarded as a sign an animal is self-aware.

Kohda's team began by placing the wrasse in a tank with a mirror. Confronted with their reflection the wrasse initially reacted with the sort of aggression they would usually direct at a rival. They subsequently abandoned these behaviours in favour of highly unusual activities such as swimming towards the mirror upside-down, before finally spending time in non-aggressive postures in front of the mirror, gazing at their reflections and seeming to visually examine their own bodies. Although these responses were spaced over a number of days, this sequence of behaviours is not dissimilar to the responses of species such as chimpanzees, dolphins and corvids – all of which are also capable of passing the mirror test.

Kohda's team then removed the wrasse from the tank, anesthetised them and placed a mark on their faces or necks before returning them to the tank. This time, when the wrasse encountered their reflection, they assumed postures that allowed them to examine the mark, and then attempted to remove it by scraping the affected area against the wall of the tank or a rock.

Kohda's study was not the first to suggest some fish are capable of self-recognition: a 2017 study by Csilla Ari and Dominic D'Agostino showed that manta rays placed in a tank with a mirror did not attempt to interact with it in the way they would if they thought it was another individual. Instead they circled in front of the mirror, waving their fins, as well as blowing bubbles, a behaviour they had not previously displayed.

Although highly suggestive Ari and D'Agostino's results were not conclusive; as critics pointed out, it is possible the mantas were simply curious about the reflection or engaging in exploratory behaviour of some sort. The cleaner wrasse study was much more definitive. Although the wrasse took longer to stop acting aggressively than species such as elephants and dolphins, their behaviour once they did was consistent with that of other animals that have passed the test.

Despite this, Kohda and his team faced considerable difficulty getting their results published, and widespread criticism from evolutionary psychologists and animal behaviourists who disputed their conclusions. One of the most vocal critics of the study is the man who first devised the mirror test, evolutionary psychologist Gordon Gallup. Gallup argues the behaviour of the wrasse is ambiguous and cannot be taken as evidence the fish recognise themselves or possess self-awareness. Diana Reiss, whose work has shown dolphins are capable of passing the mirror test, is similarly sceptical, suggesting that the fact the mark resembled a parasite – which the wrasse are hyper-evolved to detect – may have skewed the results.

Culum Brown is contemptuous of such responses. 'The mirror test has been the gold standard for individual recognition for fifty years. When primates passed, people said, "Oh yes, of course." Then when a dolphin passed they said, "I guess that makes sense." And when some corvids passed they said, "Okay, maybe; corvids are pretty smart." But when a *fish* passes, suddenly the test must be broken. It's kind of mind-boggling, but it just goes to show that even scientists have these massive biases.'

Alex Jordan, an evolutionary biologist and leader of the Behavioural Evolution Research Group at the Max Planck Institute of Animal Behaviour in Konstanz, Germany, and senior author of the Kohda paper, is even less diplomatic. 'In one of the reviews of the paper somebody wrote "This is a paper about fish – why are we even discussing it?" But if you say that, you shouldn't call yourself a scientist. Science is a process of updating your opinions on the basis of data, not a faith-based system in which you refuse to even discuss things that violate your view of the world.'

Jordan argues that the real question may not be about the cognitive abilities of the fish, but the validity of the test itself. 'If passing the mirror test is evidence of self-awareness in chimps and elephants and dolphins and all the other animals you pay $10 to see in the zoo, then if a fish passes you either have to accept it's self-aware or – pardon my French – your test is fucked.'

Interestingly Jordan takes the latter view, arguing scientists who see behaviours suggesting self-recognition as clear evidence of self-awareness are over-interpreting the results of a flawed test. 'The mirror test isn't unequivocally telling us that an animal has theory of mind, and is abstracting visual images to cognitive representations of self.' Instead he argues it is impossible to discount the possibility the wrasse learn by a process of association that the mirror reflects their world and thus how to use it as a tool. 'That would be a very interesting and amazing finding, and extremely cognitively complex. But it wouldn't involve self-awareness – and all the other behaviours we observed could flow from that.'

In response primatologists such as Frans de Waal have suggested self-awareness might exist on a spectrum – what biologists Marc Bekoff and Paul Sherman call a continuum of self-cognisance, 'ranging from self-referencing to self-awareness to self-consciousness'. Intuitively it's an appealing idea, but it's possible it doesn't go far enough. Professor Danielle Celermajer, a philosopher at the University of Sydney, believes the real problem with the mirror test is that it is built around a very human idea of selfhood. She argues we would do better to recognise that consciousness is radically heterogenous, and takes forms that are not necessarily defined by human parameters.

So are the wrasse self-aware? When I ask Culum Brown he laughs and says that as a fishy person he has no doubt the wrasse passed the test: 'The social complexity of the cleaner and client relationship clearly illustrates the wrasse are extremely sophisticated, so is it really surprising that they're capable of self-recognition as well as recognising all those other different species? Probably not. Probably it's part of a generalised social intelligence thing that includes self-recognition.'

Jordan is more circumspect, although his reservations are methodological rather than cognitive. 'I remain open to the idea as a scientific hypothesis, but while we have some evidence that suggests it might be the case I haven't yet been able to design an experiment that would satisfy me,' he says and then laughs. 'If I had I would have done it, and I'd be back battling with those old monkey-fucker primatologists to get it published.'

WHAT WOULD IT *mean* for a fish to be self-aware? What might it be like to be a fish? A century ago the German biologist Jakob von Uexküll coined the term *umwelt* – literally 'surrounding world', or 'self-world' – as a way of capturing the way the sensory world of an animal shapes its reality. For Uexküll the *umwelt* arises out of the interplay between an animal's sensory world and the way it interacts with its environment: in combination this perceptual world, or *merkwelt*, and the effector world created by an animal's physical abilities, or *wirkwelt*, create the closed unit of the *umwelt*. Thus the world of a dolphin, coursing through a fluid world given shape by reflected sound, is quite different to that of a turtle following the contours of the Earth's magnetic field, or an octopus tasting its surroundings through the receptors on its tentacles as it moves across the sea floor. No less importantly, these different subjectivities are not arranged in a hierarchy, with humans at the top and other species spread out below; rather each creates a different way of being in the world, a different kind of mind shaped by the constraints and possibilities of its sensory apparatus.

The world of a fish – its *umwelt* – is, therefore, radically different from our own, and not just because fish inhabit a weightless world of tides and currents and eddies in which they can move upwards and downwards as easily as they can horizontally. Their way of being is unlike our own because their sensory worlds differ from those of humans in often profound ways.

If we want to begin to understand the nature of these differences a good place to begin is with vision. While the basic physiology of the fish eye is similar to our own, many fish possess extra receptors granting them tetrachromatic vision, allowing them to see wavelengths and colours we cannot. In some – especially freshwater fish – this allows them to see into the infrared, making it easier for them to see in the muddy, red-shifted waters of rivers and lakes. Others are able to see ultraviolet light, an ability many species use to recognise each other and when selecting mates: damselfish, for instance, have intricate patterns on their faces that are only visible in ultraviolet light, allowing them to identify familiar individuals and to alert companions of danger (the ability to see ultraviolet light is mostly confined to species that do not live long enough for the damage it causes to become a problem, whereas longer-lived species tend to have ultraviolet filters in their eyes).

Some fish also seem to be able to detect polarised light, which affords migratory species the ability to orient themselves by the Sun and makes it easier to judge distances. Others, especially those adapted to life in caves or in the lightless world of the deep, are functionally blind, existing in a world given dimension not by light and movement, but sensation and sound. And some even have visual worlds that change across the course of their lives: the eyes of salmon, for instance, are calibrated to see blues better when they are in the ocean and reds better when in freshwater, while trout can see in the ultraviolet when young, thus improving their ability to catch the zooplankton on which they depend at that stage of their existence, but lose this ability when they migrate to deeper waters later in life.

The visual worlds of some fish differ from our own in more radical ways: seahorses (which despite their appearance are a variety of fish), gobies and flounders possess eyes capable of independent movement, presumably allowing them to process two separate visual fields simultaneously. Four-eyed fish have retinas that are divided in half; when aligned with the plane of the water they are able to see above and below the surface at the same time. And certain species of deep-sea fish have tubular eyes in which the rods are aligned upwards to catch the faint light from above, perhaps most strikingly in the barreleye, whose eyes are encased in a transparent dome of soft tissue resembling a helicopter's cockpit.

The way fish experience sound differs from our own as well. As their complex vocalisations suggest, not only do most fish have excellent hearing, many have auditory abilities that exceed our own. Sharks, for instance, are believed to be able to hear prey from up to 250 metres away, and the American shad can hear sounds up to 180,000 Hz, well above the range of human hearing, presumably so they can detect the whistles and cries of the dolphins that prey upon them. Some species of fish also appear to rely upon infrasound, employing the subterranean rumble of the currents and the tides as they break and flow against underwater terrain and the shore to find their way. This allows them to construct an auditory map of their environment, one that does not represent it as a homogenous three-dimensional space, but what researchers describe as a 'complex acoustic and hydrodynamic landscape, with distinct landmarks and information about distant structures'.

Alongside their hearing and vision, fish also possess a form of chemoreception to detect chemical traces in the water around them. In most cases this is done using cells in the nostrils, although some species, like catfish, have chemoreceptors spread across their entire bodies, allowing them to taste anything they touch.

These chemoreceptors are often extraordinarily acute. Some sharks, for instance, can detect blood in amounts as low as one part in a million, allowing them to scent prey over long distances. In great whites an astonishing 14 per cent of their total brain mass is devoted to olfaction. Many species find their way through the ocean using these chemical traces. Salmon, for instance, learn the smell of the river in which they are born as smolts; later they follow the threads of that scent back to the same stretch of river. Eels do the same thing in reverse when they abandon their rivers and lakes and return to their breeding grounds in the ocean.

Fish also use chemical signals to communicate. Many are able to release a pheromone in the presence of danger. Known as *schreckstoff*, or 'scary stuff' (the name was coined by the substance's discoverer, the Nobel prize-winning ethologist Karl von Frisch), the compound is transmitted extremely quickly, and triggers defensive and evasive behaviours in a wide range of species. Others use these signals in courtship and to signal fitness or sexual availability, as well as to help cement social bonds.

And these abilities are just the beginning. Many fish have sensory dimensions quite alien to our human experience of the world. Some can sense minute electrical currents, allowing them to identify other fish by their electric signature. As embryos and pups, many sharks and rays use this ability to avoid danger by falling still when they detect the presence of predators. As adults that process is reversed, and their electrical sense is deployed to detect prey: the distinctive shape of the hammerhead shark's head and the frightening-looking toothed snouts of sawfish both evolved to house complex arrays of the ampullae they use to detect electrical signal. Some sharks also rely upon the electrical fields generated by the movement of ocean currents to navigate. A number of freshwater species take this ability a step further by generating their own electrical fields. Electric catfish use these fields to identify one another and communicate, while knifefish, which can communicate in a system of electric 'chirps', also use electrical signals to organise complex hierarchies.

Still other species possess magnetoreceptors that allow them to follow the lines of force generated by Earth's magnetic fields: juvenile Doederlein's cardinalfish rely on the Sun to guide them towards their reefs during the day, but at night they use magnetism to orient themselves. Similarly American eels seem to possess an inherited magnetic map of the path they take as they return to the Sargasso Sea to breed and die, which they use in combination with a compass sense.

Even more significantly, fish are extremely sensitive to changes in pressure and water movement. As anybody who has ever examined a fish is likely to have noticed, most possess a faint line running from their gills to their tails. Known as the lateral line, this is not just decoration – it is a sensory organ containing specialised cells that resemble tiny hairs encased in gel. These cells, known as neuromasts, register minute shifts in water pressure. This finely tuned system plays an important role in schooling behaviours, allowing fish to respond almost instantaneously to the movements of other fish, even in the dark. But just as the sonar of bats allows them to 'see' at night, neuromasts enable fish to use hydrodynamic imaging to extend the perceptual field of their bodies and 'feel' objects and animals around them, and in some cases to create mental maps that can allow them to remember their surroundings.

IS IT POSSIBLE to comprehend such a radically different way of being? Many of us struggle to imagine our way across lines of gender or culture. What must it be like to exist in a world where magnetic fields have dimension? Or to see the direction of light, be able to form mental maps from the sound of the movement of the tides, or extend the boundaries of your body's perception by registering changes in pressure or tiny movements of water? What is it like to be part of a school, moving in unison, one body among many?

Faced with the problem of such radical subjectivity, some philosophers have argued against the possibility of understanding other ways of being in the world. The Austrian philosopher Ludwig Wittgenstein famously suggested that 'If a lion could talk we could not understand him.' Likewise American philosopher Thomas Nagel, in his seminal paper 'What is it like to be a bat?', argued consciousness is essentially subjective, meaning that not only is it impossible to reduce the experience of consciousness to a purely material explanation, but – in an echo of Uexküll – that there is no such thing as objective experience.

Yet even if we accept we can never truly know, perhaps there are ways to at least begin to intuit these other ways of being. Fish fall for many of the same visual illusions as humans, suggesting similarities in our visual processing, and open-source software now exists allowing ecologists to simulate the visual experience of other organisms. We also share behaviours: sociality, curiosity, memory, even culture, as well as emotional lives. There is even evidence fish dream, just as we do, suggesting they possess inner worlds of subjective experience: in the words of philosopher David M. Peña-Guzmán, the fact of animal dreaming makes it inescapable that 'There exist, alongside ours, endless other worlds – utterly "Other" inhuman worlds . . . Worlds without human contours. Worlds with nonhuman centres.'

Perhaps the real problem is that we are asking the wrong question. Maybe a better place to begin is not to wonder whether it is possible for us to imagine such different ways of being, but what it might mean for us to try. How might that change the way we imagine fish? How might that change *us*?

The scholar Donna Haraway writes of the importance of making kin with other species, of recognising our connectedness with the non-human world. Making kin is not necessarily about recognising similarity, but about acknowledging difference, strangeness: kin are, in Haraway's words, 'unfamiliar . . . uncanny, haunting', and making kin defamiliarises not just them, but us. Many Indigenous cultures are also founded on a recognition of ways of being outside the human, and of the complex webs of reciprocity and care that bind different species together. These conceptual frameworks are often understood as being somehow in opposition to a scientific understanding of the lives of animals, but the two can also complement and enrich one another. As Robin Wall Kimmerer writes of the study of the salmon that course upriver along the Pacific coast of North America, 'Science can be a way of forming intimacy and respect with other species that is rivalled only by the observations of traditional knowledge holders. It can be a path to kinship.'

Seen through this prism it is possible to glimpse the paucity of many human frameworks for imagining not just fish, but all other organisms on this planet. Ours is not the only way of being in the world, nor can sentience or self-awareness necessarily be defined in human terms, or as something that can be measured on a simple scale.

Instead, as Culum Brown suggests, mightn't consciousness be better understood as something multivariate, a multidimensional space shaped by the sensory and cognitive abilities of different species? A function, in other words, of its *umwelt*? Or as Alex Jordan puts it, 'There are many, many paths to get to many, many different places. To argue there's just one developmental scale of cognition and that's the way a fish or a bacterium or a plant or any other thing interacts with the world is pure narcissism.'

RECOGNISING FISH for what and who they are could shift our perspective in other ways as well. The systems of knowledge that prevail in Western culture emphasise an anthropocentric viewpoint that strips animals of agency. Trapped within that viewpoint we tend to conceive of intelligence and culture as relatively recent developments, ways of being in the world that only arrived with the advent of hominids like ourselves. Yet the jawed fish that fill the ocean today first evolved more than 400 million years ago, meaning they possessed minds and emotions for eons before our ancestors came down out of the trees. Imagining the presence and persistence of so many species, so many ways of being in the world across such an inconceivable expanse of time, cannot help but alter the way we think about our own significance in the larger story of life on our planet.

But it also demands we think again about our attitudes to fish, and why so many of us find it difficult to think of them as creatures with their own minds and ways of being in the world. Although the architecture of their brains differs from that of our own, fish possess pain receptors similar to other vertebrates, produce the same pain-suppressing chemicals as mammals and birds, and experience flurries of neural activity in sensory areas when exposed to unpleasant stimuli. They also respond in observable ways to such stimuli, retreating and behaving as if distressed, and are quick to learn to avoid such experiences. Crucially they even seem to experience pain consciously: trout given injections of acetic acid become less cautious, suggesting they are distracted by the discomfort.

Armed with this knowledge, why would we continue to regard their suffering with scepticism? Perhaps it is partly a function of their unfamiliarity and otherness: the blank faces and glassy eyes of fish are difficult for our mammalian brains to relate to. But might it not also be because thinking of them as conscious is just too confronting?

We cannot know for certain how many fish humans kill every year. Studies based on officially reported catches estimate that between 2007 and 2016 somewhere between 790 billion and 2.3 trillion wild fish were caught and killed annually. But since these figures do not include illegal and unreported captures, or fish that were killed accidentally or escaped only to die afterwards, it is likely the true number is far higher. And this is only wild-caught fish: between 2017 and 2019 as many as 167 billion fish were killed each year in aquaculture.

Almost all of these fish die slowly and painfully. Most commercially caught fish die of asphyxiation, remaining conscious as they slowly drown in the air. This process usually takes somewhere between one and four hours, although some fish can survive considerably longer. Others are live-gutted, although even then they frequently survive half an hour to an hour, and as the term suggests, the relative brevity of their distress comes at a price. Fish caught in nets suffer horribly as well, sometimes remaining trapped for many hours or even days, with studies showing that by the time they are landed they are severely exhausted and extremely stressed, and in cases where they have been brought up from deep waters, suffering the agonising effects of the bends in the same way a human diver would. These processes are largely unregulated: despite international agreement about the use of humane slaughter techniques on farmed fish, fish are specifically excluded from animal welfare legislation in many countries, and even when they are included, non-compliance is rife.

In this context, arguments that fish cannot feel pain seem not just misguided and anachronistic, but grotesquely self-serving. If studies make it clear we routinely downplay the mental abilities and capacity for suffering of the land-based animals we eat, is it so unlikely the same psychological strategy is at work with fish?

Perhaps the time has come to think more deeply about the view of philosophers such as Christine Korsgaard who argue there is no morally justifiable way to ignore the suffering of other beings, whether human or non-human, and that our tendency to emphasise intelligence as the basis of moral concern ignores the reality of animal suffering and emotion. As Korsgaard writes, 'It is a pain to be in pain. And that is not a trivial fact.'

We also need to recognise the urgency of this challenge. Alex Vail chose to work on coral trout on the Great Barrier Reef instead of the groupers in which the pointing behaviour was first observed because in the six years between Redouan Bshary's initial observations and Vail's study the grouper population in the Red Sea had almost entirely disappeared. And this collapse is only a footnote to a catastrophe of planetary proportions: over the past fifty years humans have wiped out at least half the fish on the planet, a process that will only accelerate as rising ocean temperatures kill coral reefs and radically reshape habitats and migratory patterns.

Some of this damage is now irreversible. Many migratory fish find their way across the ocean by relying upon their collective memory, which is held by the oldest and most experienced members of the population. Removing these older individuals before they have the ability to pass it onto younger fish can result in the loss of migratory behaviours that have existed for thousands of generations. 'Effectively you're destroying animal cultures which are unlikely to be ever the same again,' says Culum Brown, who describes this process as 'cultural genocide' and believes its consequences for the future fitness of these populations have probably been underestimated.

Perhaps we will never truly understand the lives of fish; maybe they are too different, too strange, too *fishy*. But maybe by recognising the wonder of their particularity, their individuality, their capacity to think and feel and learn, it might be possible to begin to see not just them, but ourselves, differently. Perhaps we might, as Haraway says, begin to make kin with them. Perhaps by saving them we might save ourselves.

6

BEACHES

'This all come back now.'
Evelyn Araluen

IT IS A WARM afternoon in late November, and I am standing in the damp sand on the edge of Minjerribah, or North Stradbroke Island. Overhead, pearl-grey clouds streak the sky; behind me the mangroves glow radioactive-green in the diffuse glare of the afternoon light. Across the expanse of Moreton Bay the towers of Brisbane are just visible.

I am here with the Quandamooka artist Megan Cope and her assistant, Dhana Merritt, to carry out maintenance on one of Cope's most recent – and most significant – works, a living sculpture known as *Kinyingarra Guwinyanba*, or 'place of the oyster rocks'. Simultaneously a work of art and an act of ecological and cultural recovery, *Kinyingarra Guwinyanba* is made up of a series of wooden poles erected along the shoreline. Arranged in circles about 3 metres in diameter and standing a bit over a metre tall, each pole has several hundred oyster shells threaded around its top half. At the moment the poles are exposed but they are positioned so that the tide covers them twice a day, allowing the shells to be colonised by larval oysters – a process that is gradually transforming the circles of poles from inanimate objects into living reefs.

It is quiet as we cross the wide expanse of the intertidal zone, the only sound that of the waves and the susurrus of the armies of soldier crabs that scuttle in great sheets across the sand, their massed bodies shifting here and there like grass moving in the wind. Although the tide is out the sand is pitted with round stingray holes filled with water, in which toadfish startle and jump as we pass; elsewhere lines of upturned sand are traced across the beach like giant Keith Haring doodles.

When we reach the third circle Cope stops to assess its condition. Although most of the poles are doing well, on several of the others the strings of shells have come loose or fallen to the ground. Handing me a bucket, she tells me to begin digging the fallen shells from the sand, then scoops up a handful of shells for herself and sits down on an upturned milk crate beside one of the poles on the opposite side of the circle.

While Cope and Merritt tighten the wires that hold the shells to the poles and rethread those that have come loose, I finish filling my bucket with shells and carry them towards the water to wash them. Although these poles have only been here for a few months, finger-nail-sized juvenile oysters are already growing on a number, their presence a reminder of the capacity of living systems to restore themselves and, no less importantly, of the process of recovery and repair at the heart of the entire project.

Oysters and oyster reefs have long held immense significance for Indigenous Australians. As the middens that are still to be found at various places around the continent's perimeter attest, oyster meat was a vital food source for many coastal communities, while the shells were used to create cutting tools, fishhooks and items of adornment.

Early European explorers and invaders were staggered by the abundance of shellfish along the coast. When Cook landed in Botany Bay in 1770, he noted 'vast heaps of the largest Oyster Shells I ever saw', later observing that 'On the sand and Mud banks are Oysters, Muscles, Cockles, etc., which I believe are the Chief support of the inhabitants, who go into Shoald Water with their little Canoes and peck them out of the sand and Mud with their hands, and some-times roast and Eat them in the Canoe, having often a fire for that purpose'. Similarly, in Port Jackson in 1788 Lieutenant William Bradley described the 'great quantity' of oysters of 'amazing size'; while here in Moreton Bay, the reefs were said to be so abundant that a man might '[tie] his boat to a stake, then [commence] to dredge for six months'.

For invading Europeans these oyster beds were not evidence of Indigenous care and cultivation of their Country; they were simply another resource to be exploited. Within a decade of the establishment of the first colony in Moreton Bay in 1824 Europeans were harvesting oysters and burning their shells for lime; by the 1860s large-scale exploitation had begun, alongside the mining of middens – a process that reached its peak in the early 1890s, when at least 35 million oysters were hauled in from the waters of the bay.

The sheer scale of this catch was not sustainable: within a decade of this record haul, production had dropped almost 40 per cent, and by the 1920s it had dropped as much again. Reefs that were metres thick and spread over many kilometres of coastline were destroyed by dredging, the oysters that remained assailed by disease and parasites. This was a process that was repeated around the continent: although it is difficult to know for sure how many reefs there were prior to the arrival of Europeans, estimates suggest that close to 90 per cent of rock-oyster reefs and more than 99 per cent of flat-oyster beds have been wiped out. So total was this devastation that the existence of the reefs was almost entirely forgotten by scientists and historians until comparatively recently.

'They dredged the beds and then they took the rocks away and used them in colonial gardens as decoration, until it was completely gone,' says Cope. 'And then once they'd wiped out the oysters that were already here and torn up the reefs, the Moreton Bay Oyster Company came and established their own industry in their place.'

Yet, as elsewhere, on Minjerribah this was only part of a larger process of destruction. While the oyster beds were being pillaged, Europeans were also butchering the dugongs – cousins of the manatee native to Australia, southern Asia and east Africa – that made their home in the shallow waters offshore so their bodies could be boiled down for oil (this slaughter inspired one of Cope's most powerful works, a mound of dugong bones crafted from glass and lit from within so they glow with the spectral luminescence of a sea creature). When the dugong population collapsed, the invaders moved on to whaling, harpooning the humpbacks that gathered in the bay each winter to raise their young. Meanwhile, others had begun transforming the island itself: first by digging up the middens of oyster shells created by the Quandamooka to use as lime in mortar, and later by sand mining. 'They sell the island to tourists as somewhere pristine and natural,' says Cope. 'But it's not. It's a profoundly traumatised landscape.'

Cope began exploring the idea oysters might offer a way to give shape to this culture of extraction a bit over a decade ago. Initially she sculpted oyster shells emerging out of little concrete bricks. 'I wanted to connect the material to the object so it would help tell the story of the way our middens were erased from the physical landscape and repurposed to help build the foundations of colonial buildings.'

As time went on, Cope's oyster sculptures increased in scale, culminating in a series of artificial middens incorporating tens of thousands of artificial oyster shells. Constructed out of concrete and copper slag, these creations echo the architectural structure and scale of the real middens that once stood on the shores of Moreton Bay: a haunting power inheres in their sculpted lines, the way they speak simultaneously of erasure and survival.

For Cope these works were a way of pushing back against White Australia's ongoing denial of Aboriginal culture and ingenuity. 'There was really sophisticated aquaculture, where rocks were placed in the intertidal zone to create reefs. There are stories about living in the mission, which is right near the poles. That whole area was covered in oyster reefs and mangroves, and back then you could pick the oysters off the mangroves nice and easily and get a feed.'

But Cope's early sculptures were also an attempt to capture something of the significance of middens, not just as symbols of occupation stretching back many thousands of years, but as living reminders of Indigenous connection to Country. 'We've spent generations being told our people are uncivilised because we didn't have buildings. And that makes me angry, because we did, they just weren't limited to four walls in that way. I wanted to make that connection, to show that the middens were architecture, that they were structures and forms located in place. And to think about what it means when those forms are destroyed. Because that connects to the fundamental mechanism of colonialism, the way it renders things invisible by extracting and destroying what's already there and replacing it with its own structures.' She shakes her head and laughs. 'The surf club by Main Beach is built on a midden, and there's an article in the local paper from 1971 that says they must have been formed by the wind and the waves pushing the shells up the beach, because blackfellas are too lazy to make shell piles.'

Yet even as Cope was crafting these extraordinary works, she was growing increasingly uncomfortable about the way her use of concrete and similar materials was recapitulating the extractive logic she was critiquing, and of the contradictions of creating work that was supposed to be about speaking to Country but was confined to galleries and museums. 'Connecting with those sites of trauma, being in that negative zone where you're pushing up against the colony, it gets hard after a while. So I decided to step back a bit and try to think about the problem a bit differently.'

Cope began by talking to relatives who work as oystermen on the island. 'It's so hard for them, because the industry's been set up in a way that's designed to eliminate us, and to wipe out traditional aquacultural practices.' That led her to Uncle Ricky Perry, whose family holds the oyster lease on the stretch of beach where the poles now stand. 'That lease has been in Uncle Ricky's family since the 1940s, or even longer. And when I asked him what he would like to see there, he said all he wants is for the reef to come back, so the oysters can grow naturally.'

Cope's original plan was to place rocks in the intertidal zone, thus recreating the physical structure of the reefs that existed before the rocks were removed. But that plan was scuttled when she discovered that the terms of Uncle Ricky's lease forbade him to place any stones or oyster shells on the sand. 'There's never any leeway for Aboriginal people who break the rules. And I couldn't have lived with myself if this project meant Uncle Ricky lost his lease.'

This prohibition looked set to sink the project until Cope discovered that while it is illegal to place rocks on the sand, there is nothing forbidding the use of wood. With the idea of the poles in place, she set about finding shells. She laughs, remembering. 'We got a lot of the shells from bougie restaurants in Byron Bay.'

The first installation took place on Australia Day 2022. Officially a day of celebration that marks the first raising of the British flag at Sydney Cove on 26 January 1788, the date is also known as Invasion Day by many Indigenous (and non-Indigenous) Australians, who regard it as a day of sadness on which the only thing to celebrate is Indigenous survival in the face of more than 230 years of colonial violence. Cope insists the timing was accidental and had more to do with the availability of volunteers and the timing of the tides, but the symmetry pleases her.

Cope says that being on Country and creating work that does something in the world has been important, but acknowledges it can only achieve so much. 'Even when our mob are working with scientists and marine parks and national parks, we're still being extracted a lot of the time. We're not given agency to do anything. Everything is still micromanaged and there's always intense power plays going on. All we want to do is restore our Country and live here and preserve it for future generations.'

The act of constant repair embodies the relationship of care Cope describes. 'I planned for a thousand poles, but we're not there yet,' she says, and laughs. 'I suspect I'll be doing this for the rest of my life.'

As sunset approaches, Cope says we have to leave: the tide is coming up fast, and we need to get back through the mangroves. Before we depart I stop to look around. The water is already lapping about my feet; in an hour this entire expanse of sand will be submerged.

In the fading light the poles seem at once ancient and impermanent, a reminder not just of the transience of the physical, but also that here, in this space between land and ocean, past, present and future are all present simultaneously.

THIS IMPERMANENCE IS FUNDAMENTAL to the nature of the beach. Beaches are sites of encounter, where sea meets land and land meets sea, each altering the other as the energy of the ocean is released. That process is frequently destructive: storms erode beaches and eat away at the land. But it is also creative, sustaining biological communities by releasing food and energy, and supporting breeding, or both, as the clamour of birds and sharks that gather to devour hatching turtles testify. As much a process as a place, the natural state of the beach is one of constant change, each moment as fleeting and unrepeatable as Heraclitus's river. With each rising tide the sand is reshaped, the evidence of the last swept away. Driftwood and sea wrack are deposited and removed again. Fish and worms and crustaceans emerge to feed as the water washes in, while birds and crabs hurry across the newly exposed sand as it washes out again.

These daily changes are contained within larger, cyclic variations. Growing up in Adelaide I always loved the storms that blew in each winter, transforming the waters of St Vincent Gulf into a roiling mass of waves. In their wake the beach would be scoured and remade, its waterline heaped with banks of dead seagrass so thick it was possible to sink up to one's thighs in them. Oftentimes these same storms would strip so much sand away that they exposed the layer of thick, black, sulphurous mud that lay beneath it. Presumably this was the rotted remains of the mangroves and seagrass beds that lined the coast prior to European invasion, but in my mind it was always the vestige of some primeval swamp populated by dinosaurs and other ancient creatures.

This ceaseless flux is inextricably entwined with the rise and fall of the tides, their daily movement up and down the sand. Some form of this cycle has existed since the Earth first formed: even in its molten, moonless primordial state, the Sun's gravity must have pulled the magma of the Hadean planet towards itself.

Today it is the Moon that is principally responsible for the tides. As it swings around in its orbit it is held in place by the Earth's gravity. But while the Earth's gravitational pull is far more powerful than that of the Moon, the mass of the Moon pulls on the Earth as well, dragging it upwards towards the Moon.

This force is so powerful it deforms the entire planet, creating a bulge that flows not just through the ocean and the atmosphere, but the rock, stone and magma of the Earth's crust. This crustal tide is not detectable without special instruments, but in some places the crust rises and falls by more than 50 centimetres every twelve hours or so. The same process of tidal flexing, albeit on a much larger scale, helps contribute to the tidal heating that results in the warm, salty, possibly life-rich oceans that lie beneath the surface of ice moons such as Europa and Enceladus.

The friction of the tides is also gradually slowing down the Earth's rotation. Although the effect is so small as to be almost imperceptible, adding a mere 2.3 milliseconds to the day each century, across the impossible space of geological time it has added many hours to the day. Physical evidence of this is able to be discerned in the growth bands of ancient shells and corals, which show that 400 million years ago a day lasted almost three hours less than it does today, and 4.5 billion years ago, while some mathematical models suggest that in the aftermath of the Moon's cataclysmic creation, the Earth may have spun so fast that a day was as short as six hours. This gradual slowing may even have paved the way for complex life: as the days grew longer the cyanobacteria that filled the ocean more than two billion years ago began to produce more and more oxygen, thus triggering the Great Oxygenation Event.

While human cultures have long recognised a connection between the rise and fall of the ocean's waters and the cycles of the Moon, the precise nature of this relationship is surprisingly complex. The tides are not regular, except in the most general sense; instead they rise at different times and to different heights every day, peaking with the spring tides that coincide with the full and new moons, and moderating at the neap tide when the Moon is at the first and third quarters. Nor are tides the same in different parts of the world. In the eastern Mediterranean and the Sea of Japan, for instance, the water rises and falls only a centimetre or two each day – a stark contrast to places such as the Kimberley region of northern Australia, where tides can exceed 12 metres, or the Bay of Fundy, on Canada's east coast, where the average difference between high and low tide is almost 12 metres, and the difference between the highest and lowest tides is several metres higher. There are also variations in the number of tides each day in different places: although in most places there are two tides every twenty-four hours and fifty minutes (because the Moon orbits in the same direction as the Earth rotates, the time it takes to reappear in the same spot overhead is slightly longer than twenty-four hours), in certain locations in the western Pacific and Indonesia there is only one tide a day, while in other places, such as Tahiti, the tides follow the movement of the Sun rather than that of the Moon.

These complexities are so mysterious and confounding that Aristotle is sometimes said to have drowned himself in frustration when his attempts to divine the mechanism underlying them foundered, and as late as the beginning of the seventeenth century no less a figure than Galileo rejected the notion the Moon was responsible for the tides, arguing instead they were the result of the rotation of the Earth.

It was not until the end of the seventeenth century that Isaac Newton sketched out the beginnings of our modern, scientific understanding of the tides. Newton wasn't the first to intuit that the Moon was somehow drawing the waters of the ocean towards itself – the Muslim astrologer Abu Ma'shar al-Balkhi wrote in the eighth century that it 'pulled' upon the Earth. But unlike earlier thinkers Newton realised that the force acting upon the oceans is the same force that draws planetary bodies towards one another: gravity.

Newton's theory – now known as the equilibrium theory – radically oversimplified the problem by assuming a planet entirely covered by water of a uniform depth. And while it was completely ineffectual at predicting the timing or the height of the tides in the real world, it did manage to explain several of the most perplexing questions regarding the tides – in particular the way they rose and fell twice a day, instead of once, as might have been expected from the movement of the Moon, and the connection between the phases of the Moon and the size of the tides.

Over the next 250 years, various mathematicians and scientists contributed further pieces to the puzzle, but the next major advance emerged out of the work of the polymathic French mathematician, physicist and astronomer Pierre-Simon, Marquis de Laplace. Unlike Newton, Laplace took into account aspects of the real world, such as the effects of the Earth's rotation on the oceans, the behaviour of fluid in motion, and the interaction between the water and the land.

This insight allowed Laplace to derive a series of equations capable of describing the interaction of these forces. These equations made sense of many oddities that had previously eluded explanation, as well as revealing new wrinkles. No less importantly, though, they showed it was possible to understand the tides not as a single phenomenon, but as the interaction of several different tides caused by different forces.

While Laplace's formulas are relatively simple, the calculations they involved were so complex that they were of little use without powerful computers. But what they did show is that it is possible to understand the tides as a long-period wave rolling around the world twice a day. This opened the door to William Thomson, the 1st Baron Kelvin's realisation that the tides might be able to be understood and predicted by using the mathematical procedure known as harmonic analysis.

At its simplest, harmonic analysis relies upon the fact that any wave, no matter how complex, can be understood by breaking it down into a collection of simpler, 'pure' waves, each of which is able to be described mathematically. Applied to the tides this allowed them to be understood as the interaction of constituents, ranging from the gravity of the Moon to the depth of the water to variations in the curvature of the Earth.

Kelvin's ideas were extended in the late nineteenth century by Charles Darwin's scientist son, George, and later by the oceanographer Arthur Doodson. By drawing upon the lunar theory of the astronomer Ernest William Brown, Doodson identified 388 tidal frequencies and a system of equations that allowed the frequency of any given tide to be calculated. These advances also allowed the construction of tide-predicting machines. Essentially mechanical computers, these devices used combinations of wheels and pulleys to perform the equations in physical form, each component contributing an element. By adjusting the components it was possible to generate predictions for different locations for months or even years in advance. Doodson used the machine that is still on view at the Liverpool Maritime Museum to pinpoint the best time for the Allied landings at Normandy on D-day in 1944.

Tide machines were eventually replaced by computers capable of handling the complex mathematics of tidal prediction. Brown's calculations have likewise been superseded by extremely accurate measurements of the Moon's orbit obtained by directing a laser at a reflector left on the surface of the Moon during the Apollo missions. Nonetheless there is something remarkably poetic about the idea of the tides as a wave flowing around the Earth, breaking and dividing and feeding back into itself, day after day, year after year, century after century, backwards and forwards through time.

THE DAILY EBB AND FLOW of the tides also offers a reminder that our experience of time is inextricably bound up in the physical world – the passage from day to night and back again, the long cycle of the seasons and the associated transformations and movements of plants and animals.

Once, this was how all humans experienced time – not as discrete units, or minutes and hours on a clock, but as something more fluid, an interwoven and overlapping system of natural patterns and cycles. For hundreds of thousands of years our lives were shaped by these cycles: we relied upon the stars and the Moon and the movements of migratory species to structure social and cultural life, marking out our lives not just by changes in the seasons, but by the longer cycles of human life. More than 30,000 years ago people in Southern Europe etched detailed lunar calendars into mammoth tusks and antlers, while stones at Wurdi Youang, outside Melbourne, which indicate the position of the Sun at the equinox and solstice, may date back more than 10,000 years. Similarly the ocean crossings that allowed early humans to spread out across the globe would have been impossible without an understanding of the annual cycle of the winds, or the ability to use the movement of the heavens for navigation.

The idea of time as something external only really began to displace this older way of inhabiting time in the first half of the last millennium, with the development of mechanical clocks and their spread through northwestern Europe. Although the notion of the hour was already established, and was sometimes measured with sundials, the division of time into units was mostly fairly fluid, with day and night often divided into twelve hours that varied in length depending on the season. But as clocks displaced these more organic divisions, time became something external and separate – what the historian Moishe Postone refers to as 'abstract' time, a measure of activity, a way of regulating and valuing work.

This process accelerated with the spread of industrialisation, and the growing need to coordinate large numbers of workers. As the task-oriented time of rural life was replaced with a system regulated by the rhythms of factory life, with its shifts and set hours, time became a commodity, able to be bought and sold. This gave birth to new ways of organising economic and social activity. Simultaneously the increasing technological sophistication of industrial society allowed it to decouple from natural rhythms. In cities, the abundant energy of the fossil fuels that drove the metronomic motion of the machines made it possible to banish darkness, while in rural areas, where people still lived closer to the land, the arrival of fertilisers and machinery allowed the production of crops to be accelerated. This engendered a shift in the tempo of the human relationship with the planet, speeding up rates of exploitation and consumption, and helping lay the foundations for the catastrophe that now surrounds us.

The development of abstract time also played an important part in the expansion of European power across the world. The linear notions of temporality contained in ideas of progress and modernity provided the justification for the invasion and genocidal subjugation and exploitation of other cultures. Similarly, the coordination of empires across space and time demanded systems of measurement and regulation that depended upon standardised time systems. Even the solution to the problem of measuring longitude – a significant barrier to accurate navigation, and therefore the maintenance of lines of trade and borders – was solved by the development of a special chronometer capable of measuring time accurately at sea.

As time was further standardised by the intercontinental telegraph network of cables across the ocean floor, these systems of economic gain grew ever further divorced from natural systems. And, as more and more of the world fell under the control of European empires, these systems of abstract, standardised time were applied to the task of suppressing and controlling their new subjects, both by enmeshing them in economic systems that severed them from traditional ways of life, and by suppressing Indigenous cultures and systems of knowledge.

This process is starkly evident in Australia. Today Australians imagine the seasons through the prism of European agriculture, speaking of spring, summer, autumn and winter. Yet Aboriginal cultures understand the year's cycle quite differently. In what is now Sydney's south the D'harawal People had six seasons, each timed to annual changes in the environment such as the migration of the eels outwards into the ocean to answer the call of the great eel, Parra'dowee, and the gathering of the flying foxes. Like the heedlessness and violence inherent in the destruction of the oyster beds, this imposition of an alien temporal lens is part of the same colonial logic that Cope describes, a way of relating to the world that erases the webs of meaning and relationality that bind it together and replaces them with something else entirely. As the sociologist and theorist of time Barbara Adam has observed, abstract time is 'a central part of the deep structure of environmental damage wrought by the industrial way of life'.

The world most of us inhabit today is entirely enmeshed in this abstract time. Around the world devices connected to the internet are synchronised to atomic clocks accurate to millionths of a second, creating the architecture that sustains the economy and much of our personal lives. Meanwhile we organise every aspect of our lives by the clock, seeking to optimise our work, our exercise, even our sleep. Even the global positioning system that increasingly structures our conception of space depends upon a standardised time.

ON THE BEACH it is possible to feel this way of being slip away. Suspended in its liminal space abstract time is replaced by the rhythm of the waves, regular as breath, the movement of the clouds overhead, the rising and falling of the Sun and the Moon. This release offers a reminder that there are many ways of relating to time and the planet. Even the idea of time as something linear and unidirectional is cultural, an invention of the peculiar eschatologies embedded in Christian thought, and its often fevered imaginings of an end time. For many other cultures time was – and is – a much more fluid concept. The Mayans understood existence as simultaneously linear and cyclical; likewise in Hindu philosophy, Buddhist ideas of impermanence are situated in an understanding of the universe where time is both cyclical and repeatable.

But as Cope's poles suggest, there is another, even older notion of time. For Indigenous Australians time is understood through the lens of the complex web of relationships and obligations that structure all life. Within this web, past, present and future are not separate, and neither are Country and culture: they are all inseparable parts of a living, interconnected whole. Dubbed the everywhen by the anthropologist W.E.H. Stanner, this way of relating to and understanding the world is an integral element of a tapestry of incredibly rich cosmogonies spread across the Australian continent and the islands around it.

Within such a conception, time is not something external. It is a way of being in which time is relational, the past inhabits the present and human life, culture and even language emerge from the landscape. Thinking about time in this way has profound implications. Historians Ann McGrath, Laura Rademaker and Jakelin Troy observe that 'the concept of everywhen unsettles the way historians and archaeologists have conventionally treated time – as a linear narrative that moves towards increasing progress and complexity'. But more than that, it places the people of today in a relationship of responsibility to the past and future, emphasising connection and obligation between generations.

These relationships are a reminder that beaches are not simply places of encounter between land and sea, but between cultures and temporalities. Seen in this way beaches also suggest another kind of longer view. Twenty-one thousand years ago, during the coldest part of the last ice age, sea levels were somewhere between 120 and 130 metres lower than they are today. Around the world, shorelines were many kilometres further out, and land bridges connected landmasses that are now separate. Yet as the world began to warm, and the ice sheets started to melt, sea levels began to rise.

This process was not gradual; nor did it unfold at a consistent rate. At first the water rose slowly, adding a few metres a millennium, but then, just under 15,000 years ago, the first and largest of a series of meltwater pulses pushed sea levels as much as 25 metres higher in less than 500 years. Over the next 8000 years or so they continued to rise, occasionally spiking upwards by several metres a century, before eventually stabilising between 6000 and 7000 years ago.

The rising ocean redrew the outlines of the terrestrial world, inundating coastlines and drowning low-lying areas. Perhaps the most famous of these lost worlds lies beneath the surface of the North Sea. Dubbed Doggerland, after the Dogger Bank, a huge sandbank in the middle of the North Sea, it stretched from what is now the west coast of England to the Netherlands, and was home to animals such as mammoth and reindeer, cave lions, wolves and hyena, as well as communities of modern humans and Neanderthals, who lived and hunted across its grasslands and wide river plains for tens of thousands of years.

As sea levels began to tick upwards Doggerland was slowly eaten away, its woods and marshes lost to the rising water and its hills transformed into islands until finally, about 8200 years ago, an earthquake off Norway triggered a tsunami that swept away most of what remained. Today all that is left of this once-bountiful landscape are relics buried in the mud on the sea floor: mammoth tusks, fragments of tooth and bone and wood and pollen, and artefacts made by human hands, such as elegantly knapped flints and carved arrows and spearheads.

Over the past century or so, archaeologists have used these remnants of this lost world to draw its forgotten landscape back into the light. Yet elsewhere the memory of the world that existed before the oceans rose is not merely encoded in physical objects but survives as part of the living body of Indigenous knowledge.

On a sunny winter's morning I meet Uncle Bunna Lawrie at Bondi. Uncle Bunna is an elder of the Yinyila Nation of the Mirning People, whose Country lies along the edge of the Great Australian Bight, which the Mirning call Billia Mocalba. Now in his fifties, Uncle Bunna has had a long career as a songwriter and musician. A founding member of the groundbreaking Indigenous band Coloured Stone, which had hits such as 'Black Boy' in the 1980s, he has also recorded with Midnight Oil, played with the Whaledreamers and contributed to many other projects and recordings – achievements that were recognised in 2022 when he received the Wisdom Treasure Award from international First Nations organisation Seeds of Wisdom. On the day we speak he is on his way home from performing with Coloured Stone at the Barunga Festival in the Northern Territory, and has come to Bondi to watch the humpback whales on their annual migration north.

Uncle Bunna's desire to see the whales is more than just curiosity: it reflects a far deeper connection. The Bight is a nursery for the southern right whale. The species gathers in the shallow waters beneath its limestone cliffs every winter to give birth and raise their young. The Mirning's Creation stories tell of the ancestral whale being, Jeedara, who created the land, sea and sky. Jeedara conceived a passion for the Seven Sisters, the group of stars we know as the Pleiades, and that the Mirning call Yargaryilya. The Mirning are the children of that union, and so are the whales. 'We are the children of the Seven Sisters. The whales are the children of the Seven Sisters. They are our family, our brothers and sisters,' says Uncle Bunna. He tells me the Mirning and the whales share the same yalgu, or blood, and goonminyerra, or friendly law and customs.

Within the web of sacred duties that structure the Mirning's belonging and interconnection with Country, Uncle Bunna is a marban bai and whaleman, a senior elder who sings to the whales to welcome them home. 'When they hear you, they come close,' he says.

The landscape along the coast of the Bight is densely patterned with meaning and significance, a complex fabric of sacred places where Jeedara and other ancestral beings shaped the cliffs, bays and the ocean's waters. Mirning woman Iris Burgoyne has written of growing up on the Bight during the 1940s. She describes caves hidden beneath the cliffs decorated with rock art or filled with carefully packed whalebones, as well as streams of deep, pure springwater that rise up through the limestone. Uncle Bunna also speaks of the caves, and the part they played in Mirning life before Europeans arrived: 'People would dive off the cliffs, swim into the caves beneath them and follow the channels up to little secret openings up above.' These caves and tunnels connect to the cave systems under the Nullarbor, and in the past the Mirning used to take canoes into them at night to catch fish. 'When the tide came in they'd block the mouth,' says Uncle Bunna. 'So when the tide went out it was easy to spear fish in the shallow water.'

The Mirning's memories of visitors to their Country stretch back hundreds of years, to the arrival of Dutch sailors in the seventeenth century. Mirning elders speak of their ancestors meeting these new arrivals and sharing food and water with them. But their memories also extend much further back, to what Uncle Bunna describes as 'the time before the sea', when the coast lay many kilometres further out than it does today. 'In the karraja, a long time ago when Jeedara came, and the world and the sea were young, the water was shallow. If you look down there on the sea floor with sonar or whatever, you'll see there were streams and rivers and creeks that used to run down into the depths of the sea. Our people used to hunt and live on that.'

The Mirning's traditions describe the transformation of this coastline by the rising sea, and the effects upon the animals and other creatures that made their home on the continental shelf, Yetarngarin. Some, like the wallaby they call koogurda, almost disappeared. Others were transformed, becoming seals and other sea creatures. 'I've heard stories from my mother and my grandmother,' says Uncle Bunna. 'One day two men were hunting out there. They'd caught all kinds of things, they had goanna, snake, fish, wallaby and kangaroo – they even had ochre in their bags. And they heard this big noise, like a huge dust dog coming. They didn't know what it was, but they looked up, and there was a sea eagle above them, and the sea eagle said, "You gotta get back to the coast. Go now! Hurry!" So they looked around and they saw what looked like a big hill coming towards them. But it wasn't a hill, it was a huge wave, a tsunami. So they turned and ran. They ran real fast, until they got right to the edge of where the sea is now, and the water got up to their waists, but finally they got out and the water stopped, but it never went out again.'

Although it is now submerged, the land that once extended south from the cliffs and beaches that today mark the edge of the continent remains part of the Mirning's Country, its contours woven into the songs, rituals and belonging of their cosmogony. Known to them as Billiaum, or Deep-Sea Country home, they regard it as a living place, and hold stories about sacred sites far off the coast that it is their duty and responsibility to protect. Perhaps the most significant of these is the place they call Bingarning, which is rich in the red ochre created by Jeedara's blood, and which plays an important part in Mirning ceremonies. Bingarning is protected by the Dreaming owl, Bingarroo, a messenger and protector. Uncle Bunna explains that before the sea came in, his ancestors used to gather ochre from Bingarning; for a time after sea levels rose they continued to collect it by walking out or diving down, but today they must rely on ochre that washes up on the shore.

The Mirning are not the only Indigenous people who remain connected to a time before sea levels rose. Similar traditions have been identified in more than thirty Indigenous cultures around Australia. Some of the most striking of these are to be found in Palawa cultures in Tasmania. Archaeological evidence suggests humans reached Tasmania at least 40,000 years ago, at which time it was connected to the mainland by dry land. That land would have begun to disappear around 18,000 years ago as rising sea levels encroached on lower-lying areas. By somewhere around 13,000 years ago the only place where it would have been possible to cross would have been a narrow path running from Flinders Island and Wilson's Promontory, which would have been swept away by 11,000 years ago. Palawa stories transcribed in the 1830s seem to recall this process, declaring that 'this Island was settled by emigrants from a far country, that they came here on land, [and] that the sea was subsequently formed'. Others of less certain provenance describe a time when icebergs were common off the coast. This time ended when Vena, the wife of Parnuen, the Sun-man, drowned after an iceberg she was sitting on melted, and Parnuen, in his fury, melted the ice. In the aftermath of Parnuen's anger the land called Moo-Ai, which once lay to the north of Tasmania, was swamped by the Teeunna Niripa Leea Tarighter, 'Big Sea Water Flow', which forced the Vetea Parlever, or Moon People, to flee south in a canoe.

On the opposite side of Bass Strait, Gunaikurnai traditions tell of a time when the land still stretched southwards, which was flooded after some children found a sacred artefact that was only for men's use and took it back to camp to show to the women. 'Immediately . . . the earth crumbled away, and it was all water, and the Kurnai were drowned.' Kulin and Bunurong traditions recorded in the nineteenth century similarly speak of a time when it was possible to walk across Port Phillip Bay, and their ancestors hunted there. According to the Kulin, that changed when children accidentally knocked over a magic trough while playing with spears, which spilled so much water it filled up the bay.

It is perhaps unsurprising that people would remember what must have been a convulsive period. In Mirning Country, on the western edge of the Great Australian Bight, the continental shelf slopes so gradually that shorelines may have retreated as much as 500 metres a decade for centuries at a time, while in other places there may have been episodes of sudden change as barriers were breached by rising waters. Historian Dr Billy Griffiths is a senior research fellow at Deakin University. He says that the collapse of the land bridge across Bass Strait may have been extremely abrupt. 'The final sundering of the Bassian Plain wouldn't have happened incrementally, it would have happened within a generation. Even now there are places where the currents surge between the islands; in a storm that force of water could easily have broken through and washed a sandbar or a finger of land away. It's not difficult to imagine that happening so quickly that it separates a family or a community, and people have to make a decision about whether to go north or south and whether to brave the water that's running through the break.'

Traditions elsewhere describe periods of social conflict and disorder as the ocean encroached. Today the Spencer Gulf in South

Australia extends 300 kilometres inland and is 80 kilometres wide at its mouth, but despite its size is less than 50 metres deep, meaning that until around 9000 years ago it would have been solid ground. The Narungga People of the Yorke Peninsula appear to have preserved memories of this lost landscape: accounts recorded in the 1930s describe a time when the Gulf was not water but 'only marshy country reaching into the interior'. That changed when the birds barred the other animals from drinking from the lagoons. This act of arrogance caused great suffering and triggered a long conflict that left many dead, until eventually a senior kangaroo took the thigh bone of one of his ancestors and dragged it from the ocean inland, causing the sea to rush in and creating the Gulf.

There is something vertiginous about the idea of information being transmitted over many thousands of years in this way. These stories are not myths or tales handed down over hundreds of generations – they are part of incredibly ancient and sophisticated systems of knowledge that govern and shape the worlds of the Mirning and other Indigenous peoples, connecting them and their cultures to the world around them across time. But more than that, they suggest that on the beach we glimpse the outer limit of human memory, the deepest time our species knows. Although this contradicts the assumption that true knowledge lies in the written word, in libraries and science perhaps it should not be a surprise that such depth of memory is encoded in the songs and stories of Indigenous cultures. For in such cultures time inheres in the landscape, inseparable from ideas of care for Country and the cycles of the natural world. This stands in stark contrast to capitalist and colonial cultures: decoupled from ecological rhythms and unable to comprehend the world as anything other than a resource, they demand constant growth, consuming everything they encounter.

*

THE EFFECTS of this relentless consumption are increasingly apparent on beaches in other ways as well. Off Australia's northern tip, in the gulf between Cape York and Papua New Guinea, lies the Torres Strait, or Zenadh Kes. Only 150 kilometres across, the waters of the Strait are dotted with islands, and home to the remarkably rich and diverse cultures of the Saibailgal, Maluilgal, Kaurareg, Kulkalgal and Meriam peoples. Skilled mariners who have made their home on these islands for thousands of years, these cultures rely on fishing and hunting dugongs and turtles, as well as growing yams, coconuts and other crops.

In recent decades sea levels in the Strait have been rising at twice the global average, and as tides and storms rise ever higher, erosion and flooding are increasingly affecting Islander communities. In 2022 a king tide on Muralag Island exceeded the previous record for the highest tide by 10 centimetres, and on the islands of Saibai and Boigu high tides are already spilling over seawalls that are barely a year old. And with sea levels now projected to rise almost a metre or more by the end of the century, there is no sign of this process slowing down.

Alarmed by the threat to their homes and way of life, representatives of Torres Strait communities approached the Australian Government demanding they take immediate action to reduce emissions. But the government ignored them: former prime minister Scott Morrison declined an invitation to visit the Strait, and when several of the Islanders visited their local member of parliament to discuss the matter he refused to meet with them. Finally, in 2019, they decided to take their case to the United Nations. Together with traditional owners from Boigu, Poruma and Warraber, a group of Islanders dubbed the Torres Strait Eight lodged a complaint with the United Nations Human Rights Council (UNHRC), alleging the Australian Government's failure to take meaningful action on climate change constituted an infringement of their right to practise their culture.

Yessie Mosby is one of the Torres Strait Eight. A Zenadh Kes Masig man, and a traditional owner from the Kulkalgal nation, Mosby is also an artist whose sculptures and carvings have been exhibited in galleries in Cairns and elsewhere. On the day we meet he is in Sydney for the Biennale, which features an exhibition of the work of Mosby and others that celebrates his and his people's culture and their fight to preserve it; while we talk Mosby shows me the totems he has carved for the installation, carefully explaining their significance.

Mosby's home, Masig Island, is situated on the eastern side of the Strait, about two thirds of the distance from Cape York to Papua New Guinea. A teardrop-shaped strip of green just over 2.5 kilometres long that is home to around 250 people, Masig is, like many of the 274 islands in the Torres Strait, a coral cay, and little more than 3 metres above sea level at its highest point. As the rate of sea level rise has accelerated over recent years its shoreline has been disappearing at an alarming rate, retreating several metres a year in many places.

Mosby realised in 2019 something had to be done. He was working in his garden when his uncle ran past. Mosby called after him, but the older man only shouted out the name of Mosby's wife's great-great-great grandmother. Following the older man to the beach, Mosby found the rising tide had swept away his wife's ancestor's grave, and her bones were rolling around in the waves. 'I grabbed a bucket and started picking up all I could. My brother-in-law grabbed the skull and placed it in a bucket. My kids were picking up forearm and leg bones. It was like shells off the beach.' He pauses. 'Nobody should have to do this. Nobody should have to be walking along picking up their ancestor's remains. Especially not children.' That night it happened again. The high tide carried away the bones of another of Mosby's wife's ancestors. 'It took her. And we still can't find her today.'

Mosby says that day was a turning point for him. 'We've tried our best to stop the erosion, but there's only so much we can do. We don't want to be refugees in our own country. I hear some people say we were stupid to bury our loved ones near the ocean. But they weren't near the ocean. They were the length of a football field inland. That's how much has been eaten away already.' He shakes his head. 'When scientists studied our home in the 1980s, we were losing a metre in five years. Now it's three metres a year.'

For Mosby the fight is not just about losing his people's home. It is about preserving a far deeper connection. 'We cannot go to some other island. It's not ours. We're not in tune with that place.' When he describes this connection his voice has a quiet authority. 'We cannot live without the island. We are helpers and guardians for our animals as well. When this particular bird passes through on its way to Papua New Guinea every year, we get the island ready for them and welcome them to our shore. And when they come, they sing beautiful melodies. You only hear that melody at that particular time, but when you do it lights up the forest, it lights up the island, it lights up us, it's a part of us. And when they fly from Papua back to Australia just before the monsoon season, we farewell them.'

This movement of birds is part of a larger, interconnected network of natural cycles that shape Mosby and his people's lives. 'We know when these birds come because we observe the constellations, and when a certain star sits on the horizon we make ready because we know that the birds will be here either tomorrow or the day after. You see flocks of them, flying to our island along the water's surface.'

It isn't just erosion that threatens the Islanders' way of life. Across the Strait, coral bleaching and dieback of seagrass beds has led to a rapid decline in many species of fish and lobster, eliminating vital sources of income and increasing the Islanders' dependency upon imported food. On Masig and elsewhere, seawater is seeping into the water table beneath the island, poisoning trees and polluting the supply of fresh water. And extreme heat and changing weather patterns are interfering with traditional agriculture. 'The stars tell us when we need to backburn and get our gardens ready for the monsoon. But now the dry season lasts much longer, and when the monsoon does arrive the rain is so heavy it practically drowns our gardens.'

For Mosby, these disruptions are part of a more widespread breakdown of the cycles and systems that structure the natural world and his people's culture. 'The scariest thing is that birdlife and the marine life are suffering. That's how we first knew that there was something happening. It wasn't the erosion. The birds told us. The turtles told us.' He pauses. 'The southeastern trade winds used to stop in September – they blow right through to December now. That interferes with the turtle mating. And down on the ocean bed where there used to be life there's no life. It's like a desert. This particular coral doesn't have the particular algae that grows on it, which means the fish that survive on that algae are not there. And if that's happening down at the bottom of the sea, what's happening on top? If these islands are not here, then that particular bird which uses them as stepping stones on its migrational journey from Papua New Guinea to Australia and vice versa will die of starvation. And if that bird dies, then that particular snake that is living in the forest on the tip of Cape York will die, because that particular snake eats that particular bird. And if those birds die, then these particular trees will die as well because that bird is the only bird which eats that seed and spreads it, which activates that particular seed.'

The endpoint of this process is already visible elsewhere in the Strait. In 2016 researchers from the University of Queensland declared that the Bramble Cay melomys, a species of rodent endemic to a tiny island on the eastern edge of the Strait, was officially extinct. When Europeans first landed on the cay in the 1840s, the melomys were so abundant the sailors shot them for sport, and as late as the 1970s there were hundreds of them, but as sea levels rose storm surges began to wash across the island and poison the trees and shrubs the melomys depended upon for food and shelter, so that by 2014, the melomys were gone.

THE TORRES STRAIT EIGHT won their case. In a landmark decision delivered in September 2022, the UNHRC found the Australian Government's failure to adequately protect Indigenous Torres Strait Islanders against the effects of climate change was an infringement of their human right to enjoy their culture without interference. The newly elected Labor government committed to responding, although given its ongoing support for the expansion of Australian coal and gas exports it is difficult to see what a meaningful response might look like.

What victory will mean for the people of Masig is difficult to know. Mosby sees it as a step forward but does not believe it is the end of the struggle. 'Life is different now. It's harder,' he says. 'Our ancestors were survivors. If they weren't, we wouldn't be here today. We would've been wiped off the face of the Earth. Sickness almost wiped our people out when the Europeans arrived, but we fought through and our bodies became immune. Then World War II came and we fought through that. And we're still fighting to save our race now.'

The peoples of the Torres Strait are not the only ones whose homes are threatened by rising seas. In North America and the Arctic, rising waters and melting permafrost are forcing Indigenous communities to abandon their homes. And across the Pacific, island communities are suffering. In Tonga, Samoa, Fiji and elsewhere, rising water is eroding shorelines and flooding villages, while on Tuvalu, where two of the tiny nation's nine islands have already almost disappeared, the groundwater is already undrinkable, and the effects of heat and pollution are damaging marine ecosystems and affecting the fruit trees and crops the Tuvaluans rely upon. Faced with the almost certain loss of their island home by the end of this century, members of the Tuvaluan Government have floated the idea of creating a virtual copy of their nation, a memorial that could be kept in perpetuity to remind people that they were there.

In the face of this existential threat, Pacific Islanders have led a global movement demanding the developed world act decisively to curb emissions before their homes are lost forever. This movement played an important part in the Paris Agreement's commitment to strive to limit temperature rises to 1.5 degrees above pre-industrial conditions, and the creation of a loss and damage fund to assist poorer nations at the United Nations Climate Change Conference (COP27) in late 2022. Speaking at the Glasgow climate summit in 2021, the Fijian president, Frank Bainimarama, declared that Pacific Island nations 'refuse to be the proverbial canaries in the world's coal mine, as we are so often called. We want more for ourselves than to be helpless songbirds whose demise serves as a warning to others'.

But these successes do not diminish the challenges island communities face, or the power and influence of the forces arrayed against them. As Bainimarama observed in Glasgow, representatives of fossil fuel companies at the summit outnumbered Pacific Island negotiators twelve to one. And when former Australian prime minster Kevin Rudd suggested in 2019 that the citizens of Tuvalu, Kirabati and Nauru could be offered Australian citizenship in exchange for control of the waters and fishing rights, the former Tuvaluan prime minister Enele Sopoaga responded angrily. 'Moving outside of Tuvalu will not solve any climate-change issues . . . I am very worried about this very self-defeatist approach to suggest that people from low-lying, at-risk countries could be relocated. Because it fails to understand the implications of this issue for the entire world.' Or in the words of Marshallese poet Kathy Jetñil-Kijiner:

. . . you tell them
we don't want to leave
we've never wanted to leave

and that we

are nothing

without our islands

The environmental scholar Rob Nixon writes of what he calls 'slow violence'. In contrast to the more immediate and explosive phenomena we normally associate with the idea of violence, slow violence 'occurs gradually and out of sight'. It is 'a violence of delayed destruction that is dispersed across time and space, an attritional violence that is typically not viewed as violence at all'. The ongoing destruction of Indigenous cultures and ecosystems by extractive industry, poverty and rising sea levels and other climate impacts is a form of slow violence. But so is the unequal way in which the impacts of climate change are experienced: the rising waters that threaten Mosby's people's home are the direct result of the burning of fossil fuels, yet Masig didn't even have electricity until the 1970s.

This also makes clear something that has long been obvious to many Indigenous people around the world, which is that what is taking place in the Torres Strait, and in Indigenous communities around the world, is not something new. It is merely the latest stage in a far longer process of cultural and ecological destruction. 'The shore doesn't sing songs of love by the fish any more,' says Mosby. 'The sea cries on an empty stomach because what beautifies her isn't there any more. The island has lost its brightness. It's like a dimming light.' Or as Megan Cope puts it, 'The endpoint of colonialism is the destruction of all life.'

*

THE HISTORIAN Greg Dening saw the beach as a metaphor for cultural encounter, a place 'where we see ourselves reflected in somebody else's otherness'. What those of us who are the beneficiaries of colonial violence and dispossession see reflected is the ruinous cost of the system we inhabit – the long wave of environmental collapse that rolls outwards from the first encounters between Europeans and the cultures of the Americas, Asia and the Pacific. Yet perhaps it is also possible to see the outline of a better way of being. The late Samoan poet and scholar Teresia Teaiwa wrote that 'island' did not have to be a noun: instead it could be turned outwards and made a verb. 'The islanded must understand that to live long and well, they need to take care. Care for other humans, care for plants, animals; care for soil, care for water. Once islanded, humans are awakened from the stupor of continental fantasies. The islanded can choose to understand that there is nothing but more islands to look forward to . . . Let us turn the energy of the island inside out. Let us "island" the world! Let us teach the inhabitants of planet Earth how to behave as if we were all living on islands!'

Or as Mosby puts it, 'I have to do this. Because I don't want my children or my grandchildren walking on the beach picking up my remains. It's an ancient land we're fighting for. It's not just an island. It's our home. The whole island is a burial ground. Our ancestors are scattered right through it.'

7

DEEP

'The only other place comparable to these marvellous nether regions must surely be naked space itself, out far beyond atmosphere, between the stars, where sunlight has no grip upon the dust and rubbish of planetary air, where the blackness of space, the shining planets, comets, suns, and stars must really be closely akin to the world of life as it appears to the eyes of an awed human being, in the open ocean, one half mile down.'
William Beebe, *Half Mile Down*

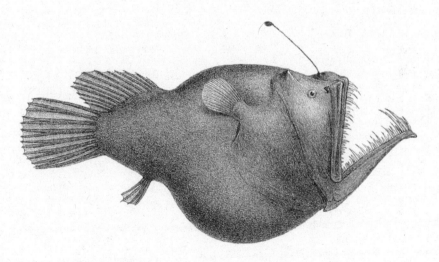

PICTURE YOURSELF inside a glass sphere floating in the ocean. Above the shimmering skin of the water's surface is blue sky and air; below it a liquid world stretches away beneath you, seemingly without dimension.

Now imagine you have begun to sink. As the waves close over the top of the sphere, light and colour suffuse the water. Yet as you descend that changes. Even in the clearest water the amount of solar energy declines by almost half in the first metre or so; at 10 metres less than a sixth remains, at 100 metres only slightly more than 1 per cent. At first you might not notice this reduction, or only register it as a gradual dimming: the human ability to see across conditions ranging from near darkness to brightest day is only possible because our eyes compress gradations of light. Even if you were only half-aware of the drop in the light's intensity, you would certainly notice colours beginning to fade, as longer wavelengths are absorbed by the water. Reds are the first to disappear, vanishing by about 10 metres below the surface, followed by oranges, and then yellows, until finally, at about 200 metres, where the amount of light drops to 1 per cent of what it is at the surface, only greens and blues remain.

If you are close to land it is possible you will have hit the bottom by now: on the continental shelves that fringe the major landmasses the sea floor is seldom more than 200 metres below the surface. Yet given these relatively shallow waters make up less than 10 per cent of the ocean's total area, it is more likely you will continue to sink.

Although only the tiniest fraction of the Sun's light penetrates below 200 metres it is not dark here, or not quite. On a bright day in clear water it might be closer to a deep dusk. But this boundary, where the layer of the ocean known as the epipelagic, photic or sunlight zone gives way to the mesopelagic or twilight zone is also the boundary beyond which photosynthesis is no longer possible. Above it, in the sunlight zone, seaweed and phytoplankton thrive, converting the Sun's rays into energy that is then consumed by other animals. Below it, life relies on energy from above. For many of the animals in the mesopelagic the solution is to join the diel migration and head upwards to feed in surface waters under the cover of darkness. Others pursue more opportunistic strategies, either feeding on the gentle rain of dead plankton and other detritus known as marine snow that drifts slowly down from above, or preying upon their neighbours.

The loss of the lifegiving potential of the Sun is not the only boundary you cross as you leave behind the light. As you have been descending the water has been growing slowly cooler. At around 200 metres the temperature begins to fall rapidly as you enter the transition zone between the Sun-warmed waters of the epipelagic and the colder, denser water below. Salinity also changes, especially in warmer regions, where evaporation rates are high, as the salty surface waters give way to less saline water.

Look up as you descend, and you will still see the faint glow of the daylight far above. But as you drop deeper into the gloaming of the mesopelagic the light suffusing the water around you grows ever weaker until finally, perhaps a kilometre down, it disappears entirely, and you enter the bathyal or midnight zone. Few mammals descend this far: you might glimpse the shadowy bulk of an elephant seal, whose specially adapted eyes allow it to see in near-darkness, or hear the shuddering creak of a sperm whale closing in on a squid, but otherwise the world of the warm-blooded now lies far overhead.

Yet while the Sun's rays do not penetrate these waters, that does not mean they are lightless. The organisms that dwell here produce their own illumination, generating dancing flashes and clouds of bioluminescence that spark and glow in the darkness. These displays are sometimes associated with attracting prey: tiny anglerfish famously dangle a glowing lure in front of themselves in order to tempt unwary prey into range of the curving teeth that crowd their nightmarish mouths. More often they are defensive. Many species of copepods, tiny crustaceans a millimetre or so in length, startle predators by flashing brightly when threatened, or distract them by producing clouds of light. Other species use bioluminescence as camouflage, producing soft light on their undersides that allows them to disappear into the glow from above when seen from underneath. Still others use it to communicate: Humboldt squid, highly intelligent, fast-moving predators more than a metre in length that hunt in packs of 1000 or more, use the chromatophores in their skin to produce complex displays of red and white light that travel up and down their bodies. Researchers have identified at least twenty-eight distinct signals, and believe the displays are a form of communication, allowing the squid to hunt cooperatively and to interact with potential mates. These signals can be broken down and recombined in multiple patterns to form different messages, although the scientists who discovered the behaviour emphasise it is too early to say whether the squid are using language as we might understand it.

In recent years it has become apparent bioluminescence is ubiquitous in the deep ocean, occurring in organisms ranging from bacteria up to the mysterious and elusive colossal squid. One researcher I spoke to described bioluminescence flowing over the top of their submersible as they ascended, like a waterfall of light. Others believe it may explain the mysterious flashes and flares of light sailors often report glimpsing far beneath the water, and that Micronesian and Polynesian navigators once used to find their way across the expanse of the Pacific. Known as *te lapa* by the people of the Santa Cruz Islands and *ulo a'e tahi* – 'to burst forth with light' – in Tonga, it is possible this underwater lightning is caused by bioluminescent plankton reacting to the waves that move along the boundaries between the ocean's layers, or by the movement of large creatures such as whales through clouds of plankton far below the surface.

As you descend the temperature continues to decline until, somewhere between 2000 and 4000 metres below the surface, it tips below 4 degrees, heralding the boundary of the abyssal, 4000 metres down.

Cold and lightless, the abyssal takes its name from the Greek word *abyssos*, meaning unfathomable or bottomless. It is not bottomless, of course – indeed it is likely that at some point around here it is possible you might strike the ocean floor. That could mean fetching up on one of the mid-ocean ridges that snake their way through the ocean basins like giant sutures. These submarine mountain ranges, which can rise up to 3 kilometres from the surrounding sea floor, form where the tectonic plates are gradually pulling away from each other, causing molten rock to bubble forth from the Earth's interior. Although often referred to individually they are really part of a single system, an enormous mountain range more than 65,000 kilometres long that wends its way around the planet's surface. Or perhaps you might fetch up on the side of one of the seamounts that rise from the ocean floor here and there.

But it is more likely you will come to rest somewhere upon the abyssal plain, the vast, sparsely populated expanse of ocean floor that stretches from the base of the mid-ocean ridges to the continental shelves and accounts for 70 per cent of the ocean floor and almost half the surface of the entire planet. With only the slow rain of nutrient from far above to sustain it, life moves slowly here. Look closely, however, and you will find fish, octopus, brittle stars and starfish picking their way across the sea floor, while complex communities of worms, crustaceans and other invertebrates cling to the few pieces of solid substrate or lurk beneath the largely featureless sediment.

Yet while you have already travelled many kilometres downwards, there is still further to go. Criss-crossing the ocean floor is a system of trenches. These trenches trace out the fault lines where the tectonic plates collide: as one plate dives under the other they form subduction zones, essentially giant canyons where the abyssal floor falls away thousands of metres into the darkness.

Known as the hadal zone – the name refers to Hades, the Greek god of the underworld, and translates roughly as 'the unseen', or 'abode of the dead' – these trenches are among the most extreme environments on the planet. Immense geological forces ripple through these fissures in the Earth's crust, causing earthquakes and sediment slides, and the pressure can be more than 1000 times that at the surface – the equivalent of a small car pressing on your fingertip – and temperatures a mere degree above zero. As a result the diversity of life declines as you descend: hard-shelled creatures disappear by about 5000 metres, the chemical processes needed to make their shells no longer possible; and although you might glimpse ghostly snailfish and cusk eels at 8300 metres or slightly more, even they cannot survive much deeper than that – the cells of hadal fish are filled with a compound known as trimethyl-amine oxide, or TMAO (also, incidentally, the chemical responsible for the distinctive and extremely unpleasant smell produced by rotting fish), which helps keep the proteins in their bodies stable, yet below about 8500 metres the concentrations of TMAO required are so high their cells begin to lose integrity. Despite that, there is life even here, and as you reach the bottom of the deepest trench, almost 11,000 metres below the surface, you will see pale amphipods and other creatures moving here and there.

THE EXTREMITY OF CONDITIONS in the ocean's depths mean that until recently only a handful of human beings had ventured down into the hadal zone, and even then only briefly. In 1960 Swiss oceanographer Jacques Piccard and American naval officer Don Walsh became the first to reach the ocean's deepest point, descending almost 11 kilometres to the bottom of the Challenger Deep in the Mariana Trench in their bathyscaphe *Trieste*. It would be more than fifty years before their feat was replicated, this time by the film director James Cameron, who spent two and a half hours at the bottom of the trench in his submersible *Deepsea Challenger* before cutting his mission short after a malfunction affected the craft's manoeuvrability. Other notable missions have included a French dive to 9560 metres in the Kuril-Kamchatka Trench in 1962, and a series of missions by the Chinese submersible *Jiaolong* in the 2010s, including one dive to more than 7000 metres in 2012.

That changed in 2014, when American entrepreneur Victor Vescovo conceived of what would become the Five Deeps Expedition, a scheme to descend to the deepest point of each of the five oceans. Vescovo charged US company Triton Submarines with developing a submersible capable of repeated journeys into the hadal zone. The result, the DSV *Limiting Factor*, eschews the sphere and pontoon, or vaguely helicopter-like form factor, usually associated with small submersibles, instead opting for a white, lozenge-like design that looks rather like an upright AirPod case. This unusual shape reflects the fact its main trajectory is not horizontal but vertical, allowing it to plunge nearly four hours downward into the darkness, and to sustain its two-person crew for up to sixteen hours at a time. Alongside the *Limiting Factor*, Vescovo also recommissioned a former US Navy surveillance ship to act as the project's support vessel, renaming it the DSSV *Pressure Drop* (Vescovo is a science fiction fan, and the name is a nod to the gnomic names of the starships in the late Iain M. Banks' Culture novels). After a race to finish outfitting the vessels, the expedition began testing the submersible, and finally, four days before Christmas in 2018, Vescovo undertook the first of the signature descents, diving 8376 metres to the deepest point in the Atlantic, the bottom of the Puerto Rico Trench. It was a feat he would repeat multiple times over the next nine months, taking the *Limiting Factor* to the deepest points in the Southern, Indian, North Pacific, South Pacific and Arctic oceans.

Alan Jamieson was chief scientist on the Five Deeps mission. In that capacity he undertook multiple dives in the *Limiting Factor*, including one in the Mariana Trench to 10,720 metres. Now a professor in the University of Western Australia's Oceans Institute and director of the Minderoo-UWA Deep-Sea Research Centre, Jamieson is a ruggedly handsome Scot who looks a little like an adventurer from central casting. On the morning I speak to him for the first time he has just returned from a five-week trip to the northern Pacific on the *Pressure Drop*, during which he carried out a series of dives, including one to the bottom of the Ryukyu Trench, 7340 metres below the surface. Although he has only been back for a little over twelve hours, he is bursting with the energy I will come to realise is his default mode.

Unlike many deep-sea specialists, Jamieson did not begin his career as a marine biologist or oceanographer. He studied industrial design in Scotland and fell into a job working as a mechanical engineer. 'Because it was in Aberdeen it's all oil and gas, so there was lots of cool underwater tech.' After completing what he calls an 'accidental' PhD on the use of autonomous instrument platforms to support oil and gas operations, a chance conversation sent his life on a new tangent. 'We were drinking in the pub when somebody asked which of us had taken the deepest photograph. And I found myself wondering, why does nobody send equipment to 11,000 metres? Why does everybody stop at 5000?'

Back in the lab, Jamieson began to explore the possibilities of cameras that could go twice as deep as the ones he was already working on, as well as machines that could carry them to the ocean's depths and return them to the surface. Less than five years later he was aboard a ship in the Pacific, preparing to test the first hadal landers – autonomous devices capable of recording video and collecting samples before returning to the surface – in the Kermadec and Tonga trenches.

These first missions proved the technologies Jamieson had developed were capable of operating at extreme depth. But they also resulted in immediate and significant scientific outcomes, delivering the first video of snailfish – pale, tadpole-like fish that live in the deep ocean; discovering a species of prawn living at far greater depths than thought possible; and capturing specimens of amphipods – crustaceans that look a little like a cross between a flea and a shrimp – from 10,000 metres down. The next few years brought more discoveries. Perhaps the most significant of these was footage of a group of snailfish 7703 metres down, at that point the deepest any fish had ever been observed, and prawns living 2 kilometres deeper than previously known. The discovery of fish and other organisms at such depths altered our understanding of the ecology of the trenches, suggesting they are not places where a few fish eke out a living, but rather home to active populations of multiple species.

The deep-sea science community is surprisingly small, yet still, Jamieson's invitation to join Five Deeps came as a surprise. 'One day in 2017 I got a call from Triton Submarines saying they needed a scientist, and since I'd written what was literally the only book on the hadal zone, they wanted me. I figured I'd be crazy not to tag along.'

Over the next two and a bit years, Jamieson took full advantage of the possibilities for scientific research Five Deeps provided, deploying landers into the deep ocean hundreds of times to recover samples and record video as well as conducting multiple dives in the *Limiting Factor*. But the experience was not without its heart-stopping moments. Some of these were just the run-of-the-mill challenges associated with working in deep water far from land: in early 2019, in the freezing waters of the South Atlantic, Jamieson lost two hadal landers in a single night. Others were related to the submersible. On Vescovo's first dive into the Tonga Trench, one of the batteries on the *Limiting Factor* malfunctioned 10,800 metres below the surface, melting wiring and impairing the vessel's electrical system; on another trip down into the Puerto Rico Trench Jamieson and the pilot had to return to the surface when a huge amount of water flooded in 2000 metres down, shorting out the controls and sending the vehicle into a spin. 'Normally a submersible wouldn't even get delivered until it had done at least fifty test dives,' says Jamieson. 'By the time we finished Five Deeps the *Limiting Factor* had only done forty-two dives in total.'

Jamieson says these sorts of incidents are difficult to avoid when working with submersibles, and admits they can also have a psychological cost. On a recent dive the *Limiting Factor* made an alarming clunk 9000 metres below the surface. Although Jamieson and the pilot knew at once something wasn't right – 'the sub makes all kinds of clicks and groans on the way down, but this was different' – they didn't panic. Instead the two of them went into a 'heightened sensory mode' in which they focused on analysing the problem and guiding the craft back to the surface. It was only afterwards that the full significance of the situation really registered. 'Once you're back on board everybody is crowding around you, so it's only later, when they all leave you alone, that you have time to absorb what just happened, and think, "Oh hell, that was bad".'

These experiences have not diminished Jamieson's sense of having encountered something profound during his time in the deep. As he seeks to describe the experience of being kilometres below the surface, he falls quiet and looks away for a moment or two. 'It's very quiet, very peaceful. It's humbling.'

This belief that there is something to be learned from visiting the deep that cannot be gained with drones and robots is central to Jamieson's work. 'As Patrick from Triton Submarines used to say, would you want to be there to witness the birth of your child, or rather watch it on video?' He laughs, then pauses before continuing. 'But it's not just that. There are things about being in that environment that make sense to have seen rather than just having analysed on video. And it's hard to put them into words.'

For Jamieson these go beyond simply being the first, or going somewhere nobody has been before. 'About a year after I started the sub project I basically had to withdraw from all other science because the scientists were doing my head in. There's such a sterile conformity to what they think is acceptable and what's not. But I was also working with Victor, who was only really interested in the bragging rights that come with diving to the deepest points, and that was very hard to absorb at the time. Because it's such a waste. So we found a middle ground. He did his dive, and if there was time we did a scientific dive as well.'

There is no question Jamieson's work in the *Limiting Factor* has led to discoveries that are unlikely to have been possible without a submersible. On a recent dive into the Boso Triple Junction, where the Izu-Bonin Trench intersects with the Japan and Sagami trenches off the coast of Japan, he surveyed a meadow of crinoids – delicate, flowerlike organisms related to sea urchins and starfish that anchor themselves to the bottom and use their petals to feed – more than 9000 metres below the surface. 'It was just absolutely stunning. You get that far underwater and there's this beautiful field of what look like yellow flowers.' When I ask him how far the meadow stretched, he laughs delightedly. 'I have no idea! Probably hundreds of miles.'

Jamieson suspects the presence of so many crinoids may be because the Boso Triple Junction is less seismically active than many trenches. 'It's probably because there aren't lots of earthquakes, so the slope isn't collapsing all the time, which doesn't bode well if you're a filter feeder that lives stuck to a rock.' But it also points to what he believes is one of the most remarkable things about life in the trenches. 'What we keep finding in the trenches is this amazing relationship in which unbelievably high-energy tectonic systems are driving some of the most delicate animals on the planet. There are things like glass sponges, which are just incredibly delicate, clinging onto a rock face that's being torn apart by seismic energies you can't even fathom.' He shakes his head in disbelief. 'The juxtaposition is incredible.'

For Jamieson this suggests we have misunderstood what makes deep-sea environments tick. 'Being able to cope with the pressure is probably the easiest ticket to the party. Either you can withstand it, or you can't. Same with adaptation to very low temperature. But the big challenges are really about being able to make do with extremely low and sporadic food supply, and surviving with a very low concentration of individuals. Because if you live on the abyssal plain in the Pacific, your population occupies half the planet, but the next female might be a hundred miles away. Sometimes we survey the sea floor for days and days on end, and we find one individual of one species, and all I can think is, who the hell are your parents? Because it doesn't make any sense! There will be one beautiful anemone sat on a rock, all on its own. And it's not as if it can move. But once you start to think about it, you have to wonder how old this thing is. Because assuming it only spawns once a year, it might take a hundred years for one of them to stick to a rock. It's mind-blowing. But it also shows that boiling the whole thing down to depth is very oversimplified.'

JAMIESON'S OBSERVATIONS about depth and pressure suggest a different way of thinking about the deep ocean and its relationship with the rest of the planet. The deep has long been treated as somehow separate from the surface world, a shadowy non-place populated by alien creatures and the effusions of nightmares. While this is partly a response to its inaccessibility and the difficulty of studying it, it also reflects a deeper psychic tendency. As the writer Robert Macfarlane has observed, humans are creatures of the air and light, and we have often regarded the spaces beneath our feet with abhorrence, associating them with death, entombment, and the unseen and unnameable. And while what Macfarlane calls the under-land might be a place of ritual power as well as a place of burial, the ocean's depths are more frequently equated with loss and forgetting, albeit often of an uneasy kind. Just as many contemporary cultures associate dreams with liquidity, and the repression of memory with submergence, the ancient Greeks consigned corrupted substances and the bodies of enemies and criminals whose crimes were unfor-givable to the ocean's depths as a way of erasing them from the world. And some West African cultures believed the resisting tension of the ocean's surface to be a veil through which the dying would pass on their way to the next world (the loss of those abducted by slavers thus became a journey into the world of the dead, no doubt amplifying the unbearable grief of seeing so many ripped away from their home and families, never to be seen again).

Although those versed in traditional wayfinding techniques often understood the space of the ocean in more complex ways, the idea of the deep as an unknowable non-place was also embedded in navigational practices. For European sailors plying the waters of the Atlantic, Mediterranean and Indian oceans, all that really mattered was knowing where potential obstacles and risks such as reefs and sandbars lay – a way of thinking that transformed the ocean's depths into a blank irrelevance.

It was not until the early nineteenth century that a more detailed scientific understanding of the deep began to take shape. In part this was a result of the growing reach of the colonial powers: as the commercial and territorial aspirations of Europeans and Americans expanded to encompass the globe, the need for more accurate and more detailed knowledge of the ocean grew as well. But it also grew out of the experiences of whalers, whose voyages were now taking them far out into the open waters of the Atlantic and Pacific oceans, and leading to an appreciation of the great depths to which whales would often dive.

This increasing interest in the deep ocean took on a new urgency in the 1850s, when British and American entrepreneurs began to lay the first submarine telegraph cables across the Atlantic. The technical challenges of these ventures demanded a more detailed understanding of the ocean floor, and led to the first bathymetric maps. But it was not until the *Challenger* expedition circumnavigated the globe on its pioneering scientific survey of the world's oceans in the 1870s that the true extent of the deep ocean finally started to emerge. In the northwest Pacific, where the Mariana Trench plunges downwards into the planet's crust, HMS *Challenger* recorded depths in excess of 8000 metres. Perhaps even more startling to the scientists of the day, though, was *Challenger*'s discovery of tiny shells – and therefore living things – more than 7 kilometres down.

Across the century and a half since the *Challenger* expedition our understanding of life in the depths has offered many surprises. One of the most important of these involves the existence of thriving communities of living things clustered around hydrothermal vents in the ocean floor. These vents form where cracks in the Earth's crust allow seawater to come into contact with liquid magma. At the surface, water exposed to magma would simply boil away, but in the deep the pressure prevents this and the water is instead expelled back into the ocean in a superheated geyser. Dubbed smokers, these jets can exceed 400 degrees, and bear a stream of minerals upwards from the Earth's mantle; as the water cools these minerals solidify, forming structures that can be dozens of metres high and can grow as much as 30 centimetres a day.

The first hydrothermal vent was found in 1977 by scientists surveying the seabed 2500 metres below the ocean on the Galapagos

Rift, between Ecuador and the Galapagos Islands, who detected a temperature spike near the ocean floor, along with a shimmer of superheated water. When the scientists reviewed the photos their submersible had taken, they were amazed to find a thriving community of living creatures. In an article published soon afterwards the scientist Robert Ballard marvelled that the photograph taken just seconds before the temperature anomaly showed only barren, fresh-looking lava terrain in 'an endless variety of sculptured pillow forms'. But for 'thirteen frames (the length of the anomaly), the lava flow was covered with hundreds of white clams and brown mussel shells. This dense accumulation, never seen before in the deep sea, quickly appeared through a cloud of misty blue water and then disappeared from view. For the remaining 1,500 pictures, the bottom was once again barren of life.'

Since that first discovery more than 600 fields of vents have been identified, all teeming with living organisms. Specially adapted colonies of mussels and other shellfish cling to their seething columns alongside fields of feathery worms and starfish; crabs and shrimp dart here and there, feeding in the cloudy water. Such richness of life should be impossible in the darkness of the ocean's depths – without sunlight there is no photosynthesis. But the creatures that thrive around the vents do not draw upon the Sun's energy. Instead, they rely on chemosynthetic microbes capable of transforming the chemicals produced by the vents into energy.

The discovery of animals around hydrothermal vents has led to a dramatic broadening of our understanding of the sorts of environments in which life can survive. This has significant implications for the search for extraterrestrial life – if life thrives in such environments on Earth it is plausible it might flourish in similar conditions in the oceans of ice moons such as Enceladus and Europa. It has also shifted assumptions about where life on Earth began: perhaps it was not in a shallow pool, but somewhere in the depths of the primordial sea.

In other words the deep ocean might not be a place of death and forgetting, but rather the birthplace of life on our planet.

Hydrothermal vents are not the only parts of the deep ocean that are revealing unexpected biological diversity. Tim O'Hara is a senior curator at the Melbourne Museum. In 2017 and 2021–22 he headed up missions to explore the sea floor off the continental margin along the east coast of Australia and in the depths of the Indian Ocean to Australia's west. In a reminder of how little is known of deep-sea environments, these missions were the first attempts to map these regions in any detail, and as such delivered a number of surprises, not least of which was the ruggedness of the continental slope. 'Australia is an old continent, so I was expecting the continental slope to be flat and more or less covered in mud,' says O'Hara. 'But in fact it was full of canyons and knolls and hills, and particularly around Sydney was just bare rock, because the current just comes barrelling up from Antarctica and strips all the sediment away.' Similarly, the 2021 expedition revealed previously unknown seamounts and 100-million-year-old volcanic calderas more than 2.5 kilometres below the surface, one of which the team christened 'The Eye of Sauron' on account of its resemblance to the fortress of Sauron in Tolkien's *The Lord of the Rings*.

For O'Hara, though, the real revelation of both trips was the incredibly rich biological world their sampling revealed. Even several kilometres below the surface the flanks of the seamounts are rich with life, including corals, crustaceans and a multitude of bizarre-looking fish. Although it will take decades for scientists to finish cataloguing the species the expeditions found, it is likely as many as 30 per cent will turn out to be new to science. O'Hara sees this as a reminder that even the most sophisticated deep-sea surveys are little more than isolated snapshots of a much larger world. 'The deep sea is vast, and when we do these abyssal studies, all we're doing is shining a light on one tiny little spot. And although there are other people shining lights on other spots, what we end up with is a constellation of little points, not a true picture of how biodiversity is spread through the ocean.'

Given the limited nature of this data, it might seem impossible to develop a sophisticated understanding of the distribution of biodiversity in the deep. But incredibly, that is precisely what has begun to emerge out of O'Hara's research.

The story begins with brittle stars, relatives of starfish that use their long, spiny arms to move across the sea floor. Mostly found in deep water, and spread through every ocean from pole to pole, brittle stars are also extremely diverse, with over 2200 species classified to date, and many more likely yet to be discovered. As part of a project to understand the distribution of these various species and the relationships between them, O'Hara and his team assembled genetic material from museums all over the world. This allowed them to map where different species were found. But as their map grew more detailed, O'Hara realised that it didn't just offer a way of visualising the current distribution of brittle stars. Instead, by tracing the development of the various species back through evolutionary time, it was possible to chart rates of diversification in different parts of the world.

Given the highest levels of biodiversity are found in tropical waters, it would seem logical to assume they are regions in which diversification is most rapid. Yet O'Hara found the opposite is true, and speciation rates in the tropical deep sea are surprisingly low. The rich biodiversity found near the equator has accumulated gradually, over tens of millions of years, largely due to relatively low rates of extinction. Polar regions, by contrast, are far less diverse, yet changing rapidly. 'About 15 million years ago there was a precipitous drop in global temperatures and lots of things went extinct, especially in polar regions, where life suddenly got a lot harder. What that meant was there was suddenly a wealth of niches to speciate into, so you saw lineages from the deep sea, where it's already cold, move upwards and radiate out into lots of different habitats.'

O'Hara is now working on a far larger project that aims to use genetic analysis to map the history of biodiversity in the oceans over the past 100 million years. 'My dream is to be able to say, "We can see that 20 million years ago all the animals in the Atlantic flooded downwards, and then the circumpolar current swept them around so they populated Tasmania," and so on. Because if we can do that, we can make an animated map that shows the swirling movement of biodiversity across tens of millions of years.'

WOULD SUCH A MAP CHANGE the way we imagine the deep? It seems likely it would, if only because it would make it clear that the ocean's depths are not an alien realm, but intimately entangled with every other part of the planet. Just as the ocean's currents help regulate the Earth's climate by transporting icy water through the deep ocean from the poles towards the tropics, and the constant movement of biological matter from the surface to the sea floor helps drive the carbon cycle by locking away prodigious quantities of carbon dioxide in ocean sediments, so too the changing distribution of species across millions of years is at least partly about the movement of organisms from the deep upwards, towards the surface.

This interconnection is even apparent at geological scales: where the tectonic plates pull apart or dive under one another the interaction between the mineral-rich rock and magma produced transports minerals from the Earth's mantle upwards, helping maintain the chemistry of the ocean. Similarly estimates suggest that the amount of water that seeps downwards into the mantle, only to be expelled through hydrothermal vents, is so great that over cycles of approximately 10 to 20 million years the entire volume of the ocean passes through them, stripping out magnesium, sulphur and other elements that are deposited in the oceans by rivers, and replenishing ocean salts and other minerals.

More importantly, though, it might provide an antidote to the tendency to treat the ocean – and particularly the deep ocean – as a convenient place to dump waste too dangerous or expensive to store on land. In the years after both World War I and World War II, the British, American, Soviet, Australian and Canadian governments consigned hundreds of thousands of tonnes of obsolete chemical weapons to the depths in waters all over the world, either piecemeal in drums or by scuttling entire ships laden with mustard gas and nerve agents such as sarin. Although public outcry meant the practice was brought to an end in 1972, hundreds of fishers in Europe, the United States and elsewhere have been hospitalised after hauling solidified lumps of mustard gas or shells containing the substance to the surface in their nets.

The ocean's depths have also been used as the final resting place for large amounts of nuclear material. A 2019 study found at least 18,000 radioactive objects scattered across the bottom of the Arctic Ocean, many of them dumped there by the Soviet Union. These objects include vessels such as the *K-27*, a 118-metre submarine powered by an experimental liquid-metal-cooled reactor, which was scuttled in 1982 with its reactor still on board (when the explosive charges that were supposed to sink the *K-27* failed to fully detonate, it had to be rammed with a tug); the wreck of the K-141 *Kursk*, which sank in the Barents Sea in 2000 during a naval exercise, killing all 118 on board and bearing its reactor and fuel to the bottom; and the *K-159* attack submarine, which sank near Murmansk while being towed from Gremikha to Polyarny in 2003 with 800 kilograms of spent uranium fuel on board. The head of Norway's Nuclear Safety Authority says it is only a matter of time before these objects begin to release their toxic legacy into the water; others have called the situation a 'Chernobyl in slow motion on the sea floor'.

While the Soviet Union dumped more nuclear waste on the sea floor than any other country, it was certainly not alone. Between 1948 and 1982 the British Government consigned almost 70,000 tonnes of nuclear waste to the ocean's depths, and the United States, Switzerland, Japan and the Netherlands are just a few of the nations that have used the ocean to dispose of radioactive material, albeit in much smaller quantities. And while international treaties now prohibit the dumping of radioactive material at sea, the British Government is currently exploring plans to dispose of up to 750,000 cubic metres of nuclear waste, including more than 100 tonnes of plutonium, *beneath* the sea floor off Cumbria. British officials argue this sort of geological disposal offers a way of keeping waste stable and secure over hundreds of thousands of years, although incidents such as the 2014 leak of radioactive material at a waste disposal facility half a kilometre beneath salt beds in New Mexico suggests that like many of the assurances offered by the nuclear industry, this claim should be approached with great caution.

The dumping of nuclear waste in the ocean is only one part of a far larger story of carelessness and greed. Human waste in the form of plastics and other objects is everywhere in the deep ocean, a fact that is made brutally apparent by the Japan Agency for Marine-Earth Science and Technology (JAMSTEC) Deep-sea Debris Database, which documents the presence of tyres, fishing nets, sports bags, mannequins, beach balls and baby's bottles spread across the sea floor at depths of many thousands of metres. In some regions the number of such objects exceeds 300 per square kilometre.

This tide of garbage has even reached the deepest and most remote parts of the ocean: when Victor Vescovo arrived at the bottom of the Mariana Trench in 2019, he didn't just encounter previously unknown species of amphipods, he also found a plastic bag and candy wrappers, while another expedition spotted a balloon decorated with the faces of Elsa and Olaf from the film *Frozen*. Possibly more disturbing, though, is the growing accumulation of microplastics in the ocean depths. Ranging from a few millimetres to mere nanometres in size, microplastics are created in a variety of ways. Some are the result of larger plastic objects breaking down in the water. Others, such as the tiny plastic beads used in face scrubs and other products, are deliberately engineered. Increasing amounts are also a result of the use of artificial fibres such as polar fleece, many of which shed huge quantities of tiny filaments every time they are washed.

In the upper layers of the ocean microplastics have invaded the, food chain, collecting in higher and higher concentrations as one moves upwards through the layers of predation. In areas such as the eastern Pacific, there is now more zooplankton-sized plastic than plankton, meaning animals such as whales and birds are consuming microplastics in large quantities, leading to malnutrition and damage to many organs as the microplastics collect in their tissues.

The volume of plastic in the surface layers of the ocean pales in comparison to the amount in deeper waters. Studies suggest that as much as 99.8 per cent of the more than 11 million tonnes of plastic that enters the ocean every year disappears into deeper waters. Larger plastic objects simply sink, but the millions of tonnes of microplastic that makes its way into the deep ocean every year takes more circuitous routes. Some of it drifts downwards with the marine snow, expelled in the faeces of fish and other animals, or congealed in fragments of decaying zooplankton and algae: one model suggests this process may transport more than 400,000 tonnes of plastic into the deep ocean every year. The diel migration bears microplastics downwards as well, as animals from deeper waters consume plastic fragments or animals that have themselves consumed them near the surface, and ferry them back down as they return to the mesopelagic.

Microplastics are also transported into deeper waters by the network of thermohaline currents that circulates water through the ocean: borne downwards by the cooler, denser water that flows across the ocean floor, trillions of tiny plastic fragments settle into the sediment drifts that accumulate in the canyons and trenches far below the surface. Indeed microplastics are now ubiquitous in ocean sediments, which can contain as many as 1.9 million pieces of microplastic per square metre, many of them made up of minute micro-filaments from artificial fibres. These filaments are particularly dangerous because they are so easily consumed by filter feeders and other animals. This problem that is only amplified by the fact that the same currents that deliver plastics to deeper waters also transport oxygen and organic matter, meaning the places where the drifts accumulate are often home to communities of deep-sea coral and other species that depend upon this influx of nutrients, and are now receiving huge amounts of inedible plastics instead. One study recently found plastic microfilaments coating the surface of deep-sea corals retrieved from seamounts and undersea canyons at a dozen locations in the North Atlantic, Mediterranean and Indian oceans.

Nor is plastic the only thing that drifts downwards. In 2019 Chinese scientists discovered radioactive carbon-14 from the detonation of nuclear bombs in the 1940s and 1950s in the bodies of amphipods living at the bottom of the Mariana Trench, borne into the deep not by ocean circulation, but in the rain of organic matter from above. More recent studies have found radioactive caesium from the Fukushima nuclear disaster in sediment more than 7000 metres down in the Japan Trench.

Many chemicals accumulate in much the same way. Some of the most worrying of these are what are known as persistent organic pollutants (POPs), such as polychlorinated biphenyls (PCBs). PCBs were originally used for cooling and insulation in the 1920s, but by the 1940s they were being incorporated into paints, adhesives, the PVC coatings on electrical wires and a plethora of other products.

The widespread use of PCBs meant large quantities were released into the environment, but the effect of that did not become apparent until the 1950s, when Danish scientist Sören Jensen found traces of them in pike caught in Sweden. Over the next two years Jensen detected trace of PCBs everywhere: in fish, in birds, even in the bodies of his wife and daughter.

In the decades since Jensen's discovery PCBs have been banned or regulated in many countries. But that does not mean they have disappeared. Instead, as Alan Jamieson's discovery of high concentrations of them in the bodies of amphipods recovered from the bottom of the Mariana and Kermadec trenches demonstrates, they have simply migrated into deeper waters.

Exactly what this will mean is not yet fully understood. PCBs are highly toxic in even minute doses, causing cancer, liver damage, and deformities in many species as well as disturbing hormonal balances in fish, birds and mammals, and have been linked to neurological disorders in birds. Because PCBs collect in fatty tissues they also accumulate as they move up the food chain, thus concentrating in the bodies of long-lived, high-level predators such as sharks, seals and whales, where they have been implicated in mass die-offs of dolphins, and are known to result in increased mortality in whales and dolphins, who transfer high concentrations of them to their young in their milk. Worse yet, PCBs break down extremely slowly when kept out of the sunlight, meaning they can linger in the deep ocean and the bodies of animals for decades or even longer.

Like the slow decay of the hundreds of thousands of tonnes of nuclear waste spread across the sea floor, the toxic legacies of human industry written into the bodies of ocean creatures are a reminder that the deep is not a place of forgetting, but an ark of memory. And while its repositories contain a record of the spasm of environmental destruction over the past century or so, they also stretch back far beyond human timeframes. A surprising amount of our understanding of the Earth's past climate is drawn from studies of the shells of ancient foraminifera, tiny, single-celled organisms often less than a millimetre in diameter that thrive in the mud on the sea floor or drift with other plankton near the surface. As they die, foraminifera are layered into the sediment that slowly accumulates on the sea floor, so that by drilling down scientists are able to use the different isotopes contained in their shells to reconstruct not just ocean conditions, but changes in atmospheric carbon dioxide, ocean circulation, and the extent and rate of change of the ice caps and plant life over the past 100 million years or more.

More remarkably again, the deep ocean may even serve as a biological time capsule, preserving living things across seemingly impossible spans of time. In 2010, Japanese scientists sank a drill into the sea floor 6 kilometres below the surface near the centre of the South Pacific Gyre, the rotating current system that circulates water and other materials counterclockwise through the South Pacific. Better known these days for its role in forming the South Pacific Garbage Patch, the gyre is also notable because its centre overlaps with the Oceanic Pole of Inaccessibility. Also dubbed Point Nemo, the Oceanic Pole of Inaccessibility is almost 2700 kilometres from land in any direction, making it the most remote part of the ocean on the planet.

The remoteness of Point Nemo inspired the writer H.P. Lovecraft to make the ocean below the home of the cosmic monstrosity Cthulhu; more recently the area has served as a space cemetery, the final resting place of hundreds of satellites and space vehicles, sent there to drown as far from land as possible. Its remoteness and the currents that swirl about it also make it the least biologically productive part of the ocean; isolated in the eye of the gyre, the nutrients created by what little life there is above drift down so slowly that it takes a million years for a metre of sediment to accumulate on the sea floor.

By drilling down into this sediment, the scientists were able to peer back in time, effectively recovering a record of ocean conditions stretching back a million years for each metre. Yet the cores did not just contain the remnants of marine snow and inorganic matter; instead, a hundred metres down, in sediments laid down 100 million years ago, they discovered microbes that had become trapped as it was deposited. At first these were assumed to be fossils, part of the detritus trapped in the sediment. But when given food and oxygen the microbes sprang back to life and began to feed and reproduce.

Questions remain about the true age of these microbes – it is possible they may have been reproducing incredibly slowly deep in the sediment. But if they are what they appear to be – which is microbes that had been dormant since they were trapped in the sediment – then they are not just living fossils, but something even more extraordinary: an organism that was born when dinosaurs and ammonites still swum through the waters above, a living thread connecting the present to the distant past.

THESE EXAMPLES ARE A REMINDER of the dangers of treating the deep as somehow separate from the rest of the planet. The animals that inhabit the depths are not fundamentally different to those at the surface, nor are they immune to the things that harm animals at the surface. But more significantly, the tendency to treat the deep as an alien realm obscures the impact of human activity upon it.

This is visible in the intensifying efforts to extract oil and gas from beneath the ocean floor. Despite unambiguous advice from the United Nations, the IPCC and scientific organisations around the world that to have any chance of averting catastrophe there can be no new fossil fuel projects, oil and gas companies are continuing to sink hundreds of millions of dollars a day into new and expanded production. In 2021 the United Nations Environment Programme (UNEP) found that the fossil fuel projects then planned would result in the production of more than twice the amount of oil, coal and gas consistent with limiting global heating to 1.5 degrees, and almost half as much again than is consistent with 2 degrees. More recent analysis shows this projected new production continues to expand even as the effects of the climate crisis begin to accelerate.

A significant part of this expansion is tied up in what have been dubbed 'carbon bombs': gigantic projects with decades-long lifespans that will each produce the equivalent of more than a billion tonnes of carbon dioxide. Almost two thirds of these projects are from what are described as 'unconventional' sources, and in particular deep-water drilling. One of the largest of these is Australian company Woodside Energy's Burrup Hub, located near Karratha, halfway up the Western Australian coast. The facility will allow Woodside to tap two enormous gas fields several hundred kilometres offshore and transport the gas back along submarine trunklines to a huge processing facility on the Burrup Peninsula. That the project is even being considered speaks volumes about the continuing influence of the fossil fuel industry on the Australian Government, as well as the ongoing lack of regard for the environmental and cultural consequences of mining and development. Not only does the project pose risks to ancient rock art on the peninsula and create the potential for gas leaks that would devastate marine ecosystems as far away as East Timor, destroying fragile coral reefs and whale, dugong and turtle populations, the gas produced by the project will pump the equivalent of more than four billion tonnes of carbon dioxide into the atmosphere. It also offers a reminder of the way the remoteness of such projects from major population centres and the unimaginability of the deep obscure their impacts.

The push to drill in ever deeper waters is not the only way in which industry is using the remoteness of the deep to disguise their effect upon it. One Friday afternoon I meet Alan Jamieson and several of his colleagues for what they refer to as 'Beer Club': an informal weekly gathering in the car park beside the Marine Science building at the University of Western Australia. After a few drinks we head inside, and after showing me the jars full of pale amphipods that stand in the display cases in the corridor, Jamieson unlocks a colleague's office and removes a small white box from a desk drawer.

Inside are half-a-dozen lumps of blackish rock, together with a number of smaller chips and granules. Roughly spherical in shape and ranging from the size of a marble to a golf ball, they look a little like a mismatched collection of barbecue briquettes. Picking one up I turn it in my hand. It is lighter than I expected, its weight closer to that of a clump of dirt than the small stone it resembles, and where the black exterior has been chipped away the interior is brown and slightly crumbly.

These are deep-sea nodules – lumps of iron and manganese oxide that form on the seabed and are rich in metals such as copper, nickel, cobalt and titanium, as well as lithium and rare earths such as yttrium. At the centre of each is a nucleus of some sort – perhaps a tooth or shard of bone or shell that settled to the sea floor millions of years ago – around which the metal compounds dissolved in seawater have slowly accreted, a little like an oyster enveloping a grain of sand to create a pearl. This process is almost unimaginably slow, even in geological terms: many nodules will grow only a few millimetres every million years, meaning the nodule I am holding may have taken tens of millions of years to reach its current size. Yet while they grow slowly, they are not rare: instead trillions of such nodules lie scattered across the vast expanse of the abyssal plain, in some places so thickly they almost entirely cover the seabed.

The mineral wealth contained in these nodules and similar deposits that accrete on the walls of seamounts and the mineral-rich columns that form around hydrothermal vents is staggering: the region in the North Pacific known as the Clarion-Clipperton Zone alone is estimated to contain almost twice as much nickel and more than three times as much cobalt as there is on land.

The notion of somehow exploiting this untapped resource is not new, although early interest in the idea is also tangled up in one of the odder episodes of the Cold War. In 1968 the Soviet Union lost the diesel-powered ballistic-missile *K-129* submarine in the Pacific. After the Soviets abandoned the search for the missing vessel the United States managed to locate it, 5 kilometres beneath the surface, 2400 kilometres northwest of Hawaii. For the CIA and the Department of Defense the submarine offered an unprecedented opportunity: not only would it give the United States access to the Soviets' nuclear missile technology, it would allow them to study their submarine designs, code books and other classified information.

The problem was that if the Soviets discovered the United States had plans to recover *K-129* it ran the risk of destabilising the fragile détente between the two superpowers. In search of a cover story the CIA turned to reclusive billionaire Howard Hughes. With his assistance they outfitted a specially designed ship, placed stories in the press about Hughes' intention to extract manganese nodules from the sea floor, and even carried out a fake launch ceremony.

Exactly how successful the mission was remains contested: although the CIA recovered *K-129* from the sea floor, the submarine broke up while being winched to the surface, which is believed to have resulted in the loss of much of the most important material. What was indubitably a success, however, was the cover story, which didn't just manage to keep the mission a secret – in the end the operation only came to light after a break-in at Hughes's corporate headquarters in 1975 – it also helped stimulate real research into the idea of deep-sea mining.

Recent years have seen a concerted push to make deep-sea mining a reality. Nickel, cobalt, copper and manganese all play critical parts in the engines and batteries used in electric vehicles, wind turbines and mobile phones. As demand for these technologies has soared, securing reliable access of these metals has become an existential concern for many of the world's largest and most powerful corporations. And that need is only going to intensify. World Bank research suggests that to have any chance of keeping global heating under 2 degrees, the production of nickel will need to double and production of cobalt will need to increase fivefold by 2050, while the International Energy Agency estimates mineral production will need to rise sixfold if the world is to reach net zero by 2050.

Deposits of many of these minerals on land are highly geographically concentrated, and the race to extract them has resulted in immense human and environmental damage. In Indonesia, which has around a fifth of the world's nickel, huge Chinese-owned mines that supply nickel to Tesla have clear-felled great swathes of rainforest in Sulawesi and nearby islands and driven Indigenous people and local communities off their land. Meanwhile dust and other contaminants from the processing plants and the coal-fired power plants that fuel them – although the nickel is destined for electric vehicles, its production is powered by fossil fuels – have poisoned rivers and reefs and led to soaring rates of eye and respiratory disease in nearby communities.

Cobalt mining is even more destructive. More than three quarters of the world's terrestrial reserves of cobalt are held by just four countries – Australia, Indonesia, Cuba and the Democratic Republic of the Congo (DRC) – and two thirds of that lies within the borders of the DRC. In Cuba, run-off from the massive Moa mine has resulted in a 'lunarlike landscape', and contaminated rivers and coastlines for kilometres. Meanwhile in the DRC, cobalt mining has resulted in an ongoing human and ecological tragedy. In a horrifying repetition of the exploitation of the Congo by the Belgians, Congolese men, women and children are being forced to labour in brutal and dangerous conditions while foreign-owned companies siphon off the country's wealth and lay waste to its environment. Health problems due to exposure to cobalt are widespread, as are threats and violence against those who seek to publicise conditions in the country. Writer Siddarth Kara, who spent a number of years investigating the human cost of cobalt mining in the DRC, says, 'Across twenty-one years of research into slavery and child labour, I have never seen more extreme predation for profit than I witnessed at the bottom of global cobalt supply chains.'

Advocates of deep-sea mining tout it as a less environmentally destructive and socially costly way to meet the soaring demand for these minerals. The website of one company describes nodules as 'a battery in a rock', and argues they offer a 'way to compress . . . environmental and social impacts of producing critical metals', thereby offering 'the lightest planetary touch', 'no disruption to Indigenous communities, no deforestation and no child labour'.

Yet these claims ignore serious environmental concerns. Most current proposals for deep-sea mining rely on what are effectively giant vacuum cleaners that blast the sea floor with jets of water to lift the top few centimetres of sediment and then filter out the nodules so they can be transported up a pipe to the surface. This is likely to have catastrophic effects on the communities of animals that live on the sea floor. In the sprawling emptiness of the abyssal plain, nodules are frequently the only solid objects, meaning they provide critical anchor points for species such as anemones, sponges and crinoids that cannot survive in the sediment that covers the plains. Without them, these creatures will simply die.

The disturbance of the sea floor will also harm species that live beneath it, and produce clouds of sediment that will spread for hundreds of kilometres. Meanwhile the din of the mining machines could travel up to 500 kilometres, disrupting feeding and social behaviours in species that rely on sound to navigate the lightless world of the abyss.

Properly assessing the scale of these impacts is difficult because our understanding of deep-sea ecosystems is sketchy at best. In studies of biological communities in the Clarion-Clipperton Zone, between 70 and 90 per cent of the species recovered were new to science; in less well understood areas the figure is likely to be far higher. Indeed many experts argue that we know so little about these ecosystems, or the likely impacts upon them, that it is impossible to make a coherent assessment of the potential costs and benefits of mining, while others argue developing the knowledge base necessary will take decades.

Despite this, deep-sea mining may well become a reality sooner rather than later. Although some areas of the abyssal plain lie within the exclusive economic zones of individual nations, most of its expanse is in international waters and is regulated by the International Seabed Authority (ISA). Set up under the United Nations Convention on the Law of the Sea and headquartered in Jamaica, the ISA has a mandate to act for the benefit of humankind. To that end it has spent much of its thirty-year history attempting to develop a framework to regulate mining on the ocean floor and quietly issuing exploration permits. Yet in recent years the independence of the ISA has come under increasing scrutiny. In an investigation by the *New York Times* that revealed uncomfortably close links between the ISA and mining companies, the ISA's secretary-general mocked the concerns of environmentalists, whom he described as propagandists ('Everybody in Brooklyn can . . . say, "I don't want to harm the ocean." But they sure want their Teslas.'). Investigations by Greenpeace and others have also raised serious concerns about the number of ISA staff with close links to the mining industry, and in particular about the makeup of the body's powerful Legal and Technical Commission, which is dominated not by people with scientific or environmental credentials, but by representatives of the countries that hold the exploration contracts. Similarly, nearly a third of the contracts issued to date have been to private companies, and in the case of the contracts relating to the unusually rich Clarion-Clipperton Zone, half of the contracts are held by just four companies, most of whose corporate structures guarantee the bulk of any profits will be siphoned off into the Global North. But perhaps the clearest sign of the ISA's capture by the mining industry came in 2019, when the places of Belgium and Nauru at the ISA Council meeting were filled by executives from seabed-mining corporations instead of government representatives.

The tensions created by this corporate dominance came to a head in 2021, when Nauru, which has close ties to the Metals Company, announced it was invoking a rule that would allow mining to commence if the ISA did not finalise its regulatory framework within two years. Despite speculation Nauru would announce it was pushing ahead with granting licences once the deadline expired, that did not happen. Instead the 2023 meeting agreed it would attempt to have regulations in place by 2025.

What this will mean in practical terms is unclear: this new deadline is not binding, and nations such as Norway, the United Kingdom, Russia, South Korea and in particular China, which is now one of the biggest financial contributors to the ISA, are pushing hard to open up mining. Meanwhile a number of countries are pushing ahead with plans to allow mining within their territorial waters: in September 2023 the *Sydney Morning Herald* revealed the existence of a project to mine black smokers in the Bismarck Sea off Papua New Guinea, backed by the company of Russian oligarch Alisher Usmanov and Omani oil billionaire Mohammed Al Barwani, and similar projects are being pursued in Tonga, Namibia and elsewhere.

Simultaneously, however, opposition to deep-sea mining is growing. Environmental groups such as Greenpeace have initiated highly visible campaigns against it that have resulted in corporations such as Google, Samsung, BMW and Volvo signing pledges declaring they will not use metals sourced through deep-sea mining. These campaigns also played a part in convincing a range of nations to call for a moratorium: at the 2023 meeting of the ISA, twenty-one nations, including France and Costa Rica, called for a precautionary pause on mining.

IF THE RUSH TO MINE THE DEEP sea ahead of a proper assessment of its environmental risks seems to bear an uncomfortable resemblance to previous resource scrambles, that is because it does. Despite the assiduous attempts of many in the industry to promote deep-sea mining as a solution to the climate crisis, it is driven by the same extractive logic as mining on land, is facilitated by the same webs of corporate influence, and looks likely to result in the same imbalances of power and unequal ecological exchanges.

What then is the solution? Perhaps a good place to begin would be to reject the false binary of the argument that the only way to supply the metals needed to power the transition to a low-emission economy is by mining the sea floor. Demand for metals such as cobalt is soaring, but many companies are already working to develop new battery technologies that do not require cobalt. Moving towards a more circular economy by improving recycling of batteries and extending the life of electric vehicles and other devices would also lower demand: Apple has already announced plans to employ only recycled cobalt in all Apple-designed batteries. Changes to the transport mix in many countries will also reduce the need for cobalt and other metals: according to the most recent IPCC Report, an accelerated rollout of public transport, bikes and e-bikes will do more to bring down emissions than the rollout of electric cars.

Equally importantly, we should ask why the only alternative to destructive practices on land is destructive processes beneath the ocean. And why the immiseration of people in the Congo and elsewhere is regarded as an inescapable fact of life. Companies such as Samsung and Tesla that manufacture the devices that use cobalt and other minerals to power their batteries rely upon third-party auditing systems to promote their corporate commitment to international human rights. But these auditing systems are a farce. Not only are the supply chains too murky to allow any such certainty, the organisations that purport to provide third-party assessments of cobalt supply chains and to monitor mining sites for child labour have almost no presence on the ground. Across multiple visits to the DRC Siddharth Kara never once saw or heard of any activities related to these organisations. Nor did he see anything that remotely resembled 'corporate commitments to human rights standards, third-party audits, or zero-tolerance policies on forced and child labour'.

These companies cannot not know this. Nonetheless they cleave to the fig leaf of plausible deniability to absolve themselves of responsibility for the lives ruined in their pursuit of profit. And so, by extension, do those of us who buy and use their devices. The reality is that every battery that is made using cobalt from the DRC potentially contains metals produced using child labour. It is likely that includes the phone in your pocket, the laptop you work on, the car you drive. Perhaps one might ask why the companies that are so exercised about the environment at the bottom of the ocean are less exercised by the lives of children in the Congo – and why you and I are as well?

These observations are not intended as an argument in favour of deep-sea mining, or against the fastest possible transition to a renewable economy. Instead they are intended as a reminder of the degree to which the systems of extraction and exploitation that have brutalised the Global South for nearly 600 years continue to shape our world. Allowing these same extractive processes to be let loose on the ocean floor will not end cobalt mining in the DRC or nickel mining in Indonesia, it will only cause a new disaster far beneath the waves.

Instead what is necessary is an energy transition founded on principles of justice and constrained by ecological limits. Rather than destroying fragile and poorly understood deep-sea ecosystems in the pursuit of private profit, governments and corporations should be developing policies and products that reduce impacts on the environment by enabling a more circular economic model based on reducing demand for these materials by improving energy efficiency and promoting recycling and the development of alternative technologies. Similarly it is long past time that governments and consumers in the developed world insisted upon effective controls upon the human and environmental costs of the production of the minerals that power our lives. The violence and despoilment that powers the devices we carry in our pockets only continues because we choose to turn a blind eye.

IT IS BARELY 250 YEARS since Western science first began to comprehend the true age of the Earth. This growing awareness of time's immensity transformed European assumptions about humanity's centrality to the story of the planet so profoundly that the historian Tom Griffiths has compared the significance of the discovery of deep time to the Copernican and Darwinian revolutions. A fuller appreciation of the scale and significance of the deep ocean requires a similar shift in our understanding of life on our planet. For while as creatures of the light and air it is easy to assume our planet is defined by terrestrial environments, in fact the opposite is true. The deep ocean is the largest environment on Earth, making up some 95 per cent of the ocean biosphere and, depending on how you measure it, close to 90 per cent of the liveable space of the planet as a whole.

Along with the growing evidence of the scale of human impacts on the deep ocean, this makes it clear it is no longer possible to treat the deep as somehow separate from human activity. Nor is it possible to regard it as simply a new frontier to be exploited. Instead just as the discovery of deep time altered the way Western culture understood humanity's place in the larger story of Earth's history, recognising that the deep is intimately entwined with the rest of the planet demands a shift in our understanding of the true scale and complexity of our planet's biosphere, and by extension, the fact that the future of not just human life, but all life on Earth, is inextricably connected to the deep.

8

CARGO

*'Put your freight in a large ship; for the greater the lading,
the greater will be your profit.'*
Hesiod

ALTHOUGH I HAVE SPENT WEEKS reading about it, I am still not prepared for the sheer scale of the port. Driving out along the motorway from Rotterdam, hemmed in on both sides by lines of trucks, I pass through kilometre after kilometre of refineries and warehouses, their anonymous shapes spreading out across the flat landscape as far as the eye can see. In the distance plumes of steam and smoke rise in pale columns from factories and power stations; occasionally I glimpse river ships laden with containers hurrying along the Hartelkanaal towards the city, their gunnels alarmingly low in the water. Closer by wind turbines tower above the road, their pale blades turning slowly and silently overhead.

After half an hour the road swings around, factories giving way to serried rows of huge power lines. I pass a rusting power plant, coal heaped in great piles around it; beyond it, container cranes rise. Finally I come to a halt near the centre of the Maasvlakte, a bulge of land by the mouth of the river Maas that protrudes some 8 kilometres out into the North Sea. As recently as the 1960s, this place did not exist – what is now land was open water and a notoriously hazardous sandbank. But as seaborne trade increased across the 1950s, the wharves closer to the centre of Rotterdam began to become crowded, and so the idea of creating a new port closer to the coast was born: by the early 1970s the first part of the Maasvlakte – which covers some 40 square kilometres – was operating. In 2013 construction began on Maasvlakte 2, an extension that added another 20 square kilometres, as well as facilities for newer and bigger ships. Maasvlakte 2 is protected from the North Sea by an 8-kilometre-long dyke that loops around the wide canals of the harbour itself, its low ridge surmounted by a long line of huge wind turbines.

There are plans afoot to expand the port further in the near future, but even without the addition of those new facilities Rotterdam is the busiest and most important port in Europe. Each year almost 30,000 ocean-going ships and nearly 100,000 inland vessels move close to half a billion tonnes of cargo through its wharves and terminals. Food, components and consumer goods arrive from Asia, Africa and the Americas; once unloaded they are transported inland along the arterial networks of road and rail that connect the port with the rest of Europe, or travel up the rivers to Germany and beyond; meanwhile chemicals and other goods manufactured in Europe are shipped outwards. Much of the coal and oil that drives Europe's power plants and vehicles arrives here as well – some 30 million tonnes of coal, and close to 100 million tonnes of oil, as well as raw materials such as iron ore.

Merely reciting these statistics does not capture the scale of the place, however. On the channel known as the Prinses Amaliahaven I pass ships such as the MSC *Kalina*, 366 metres in length and capable of carrying some 10,000 containers atop her shallow black hull, and the *CMA CGM Concorde*, a green and blue monster some 400 metres long that can carry 23,000 containers. Almost as long as the Empire State Building is tall, and too big to fit through the Panama Canal, ships the size of the *Concorde* mostly ply the route from Europe to China and back; as our boat motors alongside it, the *Concorde* seems to go on and on. Meanwhile, on the wharf, machines the size of office buildings move back and forth, lifting containers and shunting goods.

There is something curiously awe-inspiring about the inhuman scale of the ships and the machines and their constant motion. Watching the containers rise and fall it is easy to imagine that if the world were to end tomorrow, this would continue without us. And that feeling that this operation exceeds human scales of reference is not wholly mistaken. For what is visible here is nothing less than the underlying machinery of the global economy, the constant movement of materials and resources that not only supply the food, energy and goods upon which most of us rely, but also underpin the processes of consumption and exploitation that have resulted in the crisis that is overtaking our planet.

IN THE NORMAL COURSE of events most people in the developed world don't tend to think about where the extraordinary overabundance of stuff that fills our lives comes from. But in most cases it involves a ship. Look around yourself. Are you sitting on a chair? Chances are it came by sea, flat-packed from an Ikea factory or a production line somewhere in China. The components it was assembled out of will have spent time on ships as well: the metal components will have begun life as ores that were shipped from a mine to a refinery and then shipped on again for manufacture, before being transported once again to the factory where the chair was created. The crude oil that was used to make the polyurethane in the seat was probably shipped from the Middle East to wherever it was processed, then moved on again to be transformed into foam.

If they are woven out of wool or cotton, it is possible they have been shipped several times, first from the place the raw materials were produced to somewhere they could be transformed into fabric, then again to wherever the clothes were made. Most synthetic fibres begin their life as oil, which will probably have been shipped from the Middle East or Russia to China to be transformed into fibre, and then on to a factory to be woven into cloth. The zips and buttons and solid components of your shoes will have been shipped as oil or metal ore to a factory, then shipped on after manufacture to China, Europe or Bangladesh in the case of your clothes, or China, India or Vietnam in the case of your shoes. Once complete they will have been placed in another container and shipped to wherever you are. Each stage in this process will have taken a few weeks at most.

Are you reading this on an e-reader? It was probably made in China, quite possibly incorporating chips from Korea and plastics made from oil shipped from the Middle East. Listening to it as an audiobook? Smartphones incorporate sixty or more metals and hundreds of components, all of which will have had to be mined and refined and engineered in different places before the phone itself was assembled somewhere in East Asia. Its battery will probably have been built in Taiwan using lithium from Australia or Chile and cobalt from the Congo or Indonesia, and your headphones are likely to have been assembled in China, Vietnam or India out of components made in a dozen other places.

These fantastically complex supply lines and business models are only possible because of shipping. More than 80 per cent of everything we use or consume involves a ship. The petrol that powers our cars travels by ship, and so do the chemicals that underpin industries from agriculture to cosmetics. In 2022 international shipping transported almost 11 billion tonnes of goods – and this amount is growing, fast. Over the past fifty years international maritime trade has more than quadrupled, rising from a mere 2.6 billion tonnes in 1970 to nearly 11 billion tonnes in 2021, a figure that would have been closer to 12 billion without the interruptions of the pandemic and the war in Ukraine. And there is no sign of this boom stopping any time soon. In 2019 the Organisation for Economic Co-operation and Development (OECD) predicted demand for global shipping will triple by 2050.

There is nothing new in the role of maritime trade in shaping societies. Just as the violence of colonial exploitation depended upon the movement of people and resources across the ocean, so too the staggering growth in the world economy over the past seventy years is intimately bound up in the expansion of the movement of goods and energy by ship. This expansion has had several distinct phases. The first began in the years immediately after World War II, when oil began to flow out of the Middle East. Together with an increase in the movement of raw materials such as iron ore, this explosion in cheap energy drove the rapid growth of the economies of Europe, Japan and America as demand for cars and other luxuries led to an expansion of industrial capacity.

Yet while trade in oil and raw materials continued to grow, trade in manufactured goods remained relatively static, held back by the high cost and complexity of moving goods between countries. The transformation of this process was facilitated by a revolution in logistics underpinned by the shipping container.

The history of the shipping container is complex, but it depends in no small part upon one man, the American entrepreneur Malcom McLean. Born in 1913 in North Carolina, McLean began his career running a trucking business and quickly became known in that industry for his ruthless focus on cost and his appetite for risk. In the early 1950s McLean came up with the idea of avoiding congested highways and tolls by having his trucks drive straight onto ships and leave their trailers there. The ships could then move the trailers to the next port, where new trucks would be waiting to pick up the trailers.

It quickly became apparent that McLean's initial concept was too cumbersome, but instead of abandoning it, he went back to the drawing board. Giving away the notion of transporting whole trailers he instead began to explore using specially designed containers, and designing cranes capable of loading and unloading them as quickly as possible. As his plans advanced he bought two decommissioned World War II tankers that he had refitted to hold his containers. Finally on 26 April 1956 the first of these tankers, the *Ideal-X*, departed New Jersey with fifty-eight 35-foot containers on board; five days later it docked in Houston, and was unloaded.

The speed with which the containerised cargo was able to be moved would have been remarkable in its own right: whereas loading a ship the size of the *Ideal-X* with loose cargo could take several days, McLean's cranes took less than eight hours to transfer the fifty-eight containers aboard. But the savings were even more mind-boggling: while loading loose cargo cost around US$5.83 a tonne at the time, McLean's system could load a tonne for a mere 15.8 cents.

It would take another decade for the various components of containerised shipping to fall into place. But by 1966 container ships were plying international routes, and by 1968 the shipping industry had agreed to a set of standard sizes and specifications, making containers truly interoperable and unleashing mass investment in the technology.

The effects of this process were felt almost immediately. As freight costs plummeted, trade boomed. Over the decade or so after the introduction of the container, international trade grew more than twice as fast as manufacturing production and two and a half times as fast as global economic output. Some of the most rapid growth occurred in East Asia, where container ports were quickly constructed: in Japan, exports of televisions and tape recorders almost doubled between 1968 and 1971; ocean-borne exports from Korea to the United States trebled between 1969 and 1973, largely due to the ability of the Korean rag trade to undercut the price of American-made clothing; and exports from Taiwan more than trebled between 1970 and 1973. Even Australia, which had spent more than a century riding on the sheep's back, experienced a surge, as exports of manufactured goods rose 16 per cent annually through the second half of the 1960s.

But containerisation didn't just reduce costs. It also allowed manufacturers to move goods swiftly, seamlessly and reliably across the ocean, creating new efficiencies and enabling new ways of managing inventory and supply chains. No longer bound by national borders and assisted by the ongoing dismantling of tariffs and other trade barriers, corporations were able to seek out savings by moving manufacturing production offshore. This increased mobility and fragmentation of production saw declines in the power of organised labour in developed countries, permanently altering the lives of many traditional blue-collar workers and their communities, and helping drive growing rates of inequality.

This transformation of the global economy has played a central role in the rapid rise in GDP over recent decades. But it has also helped drive the massive increase in the material footprint of human society, and in particular the surge in carbon emissions. Across the 1990s, global emissions increased at around 1 per cent a year. But as barriers to trade were dismantled, and manufacturing relocated from the developed world to developing countries, emissions shot up, rising 3.4 per cent a year between 2000 and 2008.

Biodiversity loss from deforestation has also accelerated rapidly. In Brazil, great swathes of the Amazon have been cleared to make way for beef and soybean production, while in Malaysia, Indonesia and elsewhere rainforest is being felled to make way for palm oil. And, in a particularly bleak irony, forests are also being cleared to make way for timber plantations capable of feeding the insatiable demand for cardboard to package goods sold by Amazon and other online retailers, many of whose business models rely upon the same logistical systems that underpin global shipping. In the words of Naomi Klein, 'the twin signatures of this era have been the mass export of products across vast distances (relentlessly burning carbon all the way), and the import of a uniquely wasteful model of production, consumption and agriculture to every corner of the world (also based on the profligate burning of fossil fuels). Put differently, the liberation of world markets, a process powered by the liberation of unprecedented amounts of fossil fuels from the earth, has dramatically sped up the same process that is liberating Arctic ice from existence.' Or in the words of the filmmakers Allan Sekula and Noël Burch, the 'mobile and anonymous' cargo containers are 'coffins of remote labour-power, carrying goods manufactured by invisible workers on the other side of the globe'.

The increasing mobility of capital enabled by shipping has also turbocharged the unequal exchange of wealth between the Global North and the Global South. As processes such as cobalt mining in the DRC and land-clearing in the Amazon vividly illustrate, the growth of the world economy over recent decades has depended upon the developed world's appropriation of cheap labour, land and raw materials from the developing world. Some estimates of the monetary value of this hidden transfer in value from the periphery of the world economy to the centres of consumption in the Global North suggest that between 1990 and 2015 US$242 trillion flowed from the developing world to richer nations. This amount would have been enough to end extreme poverty seventy times over. The scale of this transfer is also visible in the explosion in the consumption gap between the developed and the developing world: in per capita terms fully 81 per cent of the growth in material use across the same period results from increased consumption in developed countries.

THE MOVEMENT OF GOODS that drives these processes of exchange is underpinned by a web of fantastically complex but largely invisible supply chains and interlocking systems of financialisation, production and transportation. The geographer Deborah Cowen has described global logistics as 'an extraordinary apparatus of management that is neither just public nor private and neither military nor civilian but something else'. It is a formulation that captures something of the tentacular nature of the shipping industry, the way it sits at the centre of contemporary life, and the forms of violence it enables and enacts.

In 2008 the Swedish artists Erika Magnusson and Daniel Andersson decided to try to capture the 'almost inconceivable global logistics' behind the 'anonymous clutter' that surrounds us by tracking the path of a single item back through the supply chain.

The resulting film, *Logistics*, follows a pedometer from the store in Stockholm where it was purchased back to the factory in China where it was created in real time. For a time all 857 hours – or five weeks – of it were available on Vimeo, although eventually it was taken down because of the prohibitive cost of hosting something of such length, but there is a 72-minute highlights reel on YouTube that captures something of its mesmerising power. Soundlessly the camera travels down a darkened road, the lines and streetlights ticking past hypnotically, then down a trainline through pine forests to a container port and onto a container ship. Days pass, one after another, blue skies giving way to the haze of the Suez Canal and the Indian Ocean, the dull gleam of the sun on the ocean to magenta sunsets and huge tropical cloud formations, until finally we are back on a road in China, travelling towards the factory where the pedometer was made.

At one level *Logistics* is complicit in the erasure it seeks to make visible, for while it follows the path of the pedometer, it ignores the multiple flows of components and materials that went into the device's construction (Magnusson and Andersson have said their dream is to follow the parts of the pedometer back to the mines and oil wells from which their raw materials originally came). Yet in its inhuman scale the film reveals something that is ordinarily hidden.

As Marx observed, capitalism aspires to annihilate time and space, to render the circulation and exchange of goods as frictionless as possible. Part magic trick, part something darker, this process demands all impediments to the flow of capital be eliminated. Some of these impediments are physical: over the past fifty years the ports that were once the lifeblood of cities such as Sydney and New York have shifted away from urban centres. Largely automated, and hidden behind fences and security cameras, they now exist in a liminal, transnational space. This process has also reshaped landscapes, leading to the creation of structures such as the Maasvlakte, or transforming wetlands and mangroves into harbours and ports designed to facilitate the extraction of resources such as oil, fossil fuels and timber, or the seamless movement of goods between factories.

Other impediments to the flow of capital are human. Life at sea has always been violent and perilous. And while conditions have improved somewhat over recent decades, shipboard life remains hard, isolating and dangerous. This was made starkly apparent during the first months of 2020. As countries around the world began to close their borders, almost 400,000 seafarers found themselves effectively trapped on their vessels, unable to disembark or return home due to the absence of international flights and quarantine restrictions. Crews were required to keep working without hope of relief, and often without pay. Exhausted, isolated, and with limited ability to contact friends and family at home, many struggled with depression and anxiety. Meanwhile the profits of shipping companies soared: between the beginning of the pandemic and the beginning of 2023, the shipping industry raked in as much as it had made over the previous sixty years.

Simultaneously the creation of new resource frontiers and manu-facturing hubs has reshaped communities on land. Nowhere is this more visible than in the Middle East. In her study of the flow of oil out of the Arabian Peninsula, political scientist Laleh Khalili describes the waves of migration that have transformed its societies to feed the demand for labour in construction and the ports, resulting in migrants making up more than half the population in Bahrain, two-thirds of the population in Kuwait, and close to 90 per cent in the United Arab Emirates and Qatar. The bulk of these foreigners are South Asian – Indians, Nepalis, Pakistanis, Sri Lankans and Bangladeshis, but there are also significant numbers of Filipino, Egyptian and African expa-triates in these countries. For some this has opened new opportunities, but for many it means living precarious lives in dangerous conditions.

Other impediments are legal, involving the creation of new legal structures to facilitate the free movement of goods and capital across borders, to carve out exceptions to labour laws, or to create new

laws to protect supply lines from disruption. When climate protesters disrupted trains bound for Sydney's Port Botany in 2022 as a way
of highlighting the railway's role in the 'exploitation of global supply
chains . . . and the destruction of the planet', the New South Wales
Government swiftly enacted new laws under which such protests
could be punished with two years in prison, and two German nation-
als involved in the action were deported.

The removal of these impediments is not costless. The construction of ports and terminals destroys ecosystems, automation denies
communities jobs and pushes down incomes, deregulation of labour
and surveillance creates new forms of exploitation. The movement
of goods also creates opportunities for other kinds of violence:
a couple of days before my visit to Rotterdam I spend an afternoon
with a young Australian lawyer who studied in the Netherlands and
now works at the International Criminal Court. As a student she
volunteered with an organisation that supports female victims of
people-trafficking, most of whom arrive in the Netherlands through
the port, locked inside containers. She tells me many of them are
from poor countries in Central America and the Caribbean such
as Nicaragua and Haiti, who go to places like Chile and Argentina
for work and are told they can find better work in the Netherlands.
Once they arrive the traffickers take their passports and force them to
work, usually in the sex industry. But people trafficking is not the only
illegal activity that depends upon the movement of goods through the
port. She adds that drugs and other illegal goods flow into Europe
through Rotterdam as well, leading to violence and corruption.
'Two kilometres from where I used to live in Rotterdam, they found
a soundproofed underground bunker with a dentist chair filled with
torture equipment that was being used by an organised crime ring.'

The invisibility of shipping also disguises other dangers. Over the
past few decades the number of large ships that sink or go missing
has fallen by three quarters, largely due to improvements in tracking

and monitoring. Nonetheless, around fifty vessels are still lost every year – an average of one a week. About half these ships sink; others run aground or are destroyed by fire.

Most of the time these events barely register with the general public; those that do often involve oil spills or similar disasters. Again, these happen more often than one might suppose: in 2022 alone there were seven major oil spills, including an incident in which the tsunami caused by the eruption of the underwater volcano Hunga Tonga–Hunga Ha'apai struck a tanker that was unloading at a refinery in Peru. This resulted in the release of more than 10,000 barrels of crude oil and the devastation of ecosystems along almost 80 kilometres of the coast. More destructive again was the explosion of the refinery ship FPSO *Trinity Spirit* off Nigeria, which killed at least five and released over 60,000 barrels of oil.

Nor is it just oil and other chemicals that spill into the ocean. In 2021 more than 3000 containers were reported lost at sea. It is likely this figure understates the true number lost, possibly significantly – shipping companies that lose containers also lose customers – but what is not in question is that the number of containers being lost each year is increasing, possibly because of the increasing size of container vessels and the rise in extreme weather and wave activity due to rising ocean temperatures. While most containers sink, enough remain floating long enough to pose a genuine risk to smaller vessels. Others break open, spilling their contents into the water. Perhaps the most famous incident of this sort involved the loss of almost 29,000 rubber ducks in the northern Pacific in 1992; over the following years the ducks washed ashore in Alaska, Hawaii and Australia, while others drifted through the Arctic and ended up on beaches in Ireland and on the Atlantic coast of the United States.

Piracy also remains an issue. Between 2007 and 2011 around 3500 seafarers were taken hostage by Somali pirates. Many were held for months or years, hundreds were tortured or abused, and at least sixty-two died. And while a concerted international effort has reduced the number of attacks off Somalia, piracy remains a threat in the Gulf of Guinea, where 130 crew were kidnapped in twenty-two separate incidents in 2020. Once again, international efforts have reduced attacks on cargo ships, although the pirates have transferred their attention to local fishing boats. Piracy is also on the rise in the straits of Singapore, and continues to be an issue off Peru. And in a twenty-first century twist on piracy, cyberattacks on shipping companies are increasing in frequency; while most are still directed at facilities on shore, the growing level of integration across fleets means the risk to ships is intensifying.

It is possible one of the reasons these accidents and attacks usually pass almost unnoticed is that the crews who ensure the immense machine of global shipping continues to function are mostly not European or American; instead the majority come from Asia, in particular the Philippines, China and Indonesia, or Russia. Alternatively it may just be a function of the larger invisibility of the industry as a whole. Either way, between these disasters and other incidents at sea and in ports, several hundred seafarers are killed or go missing every year.

This invisibility also helps disguise the environmental cost of shipping. Most large ships are powered by heavy fuel oil (HFO). Also known as bunker fuel or bunker oil, HFO is a thick, black, tarlike substance that is essentially the sludge left over from the distillation process that extracts gasoline, diesel and other products from crude oil. HFO is cheap, and far denser with energy than many other fuels. But it is also incredibly dirty. Burning it releases large quantities of sulphur dioxide, nitric oxide and nitrogen dioxide, contributing to acid rain and respiratory illness, as well as toxic particulates that are implicated in at least 60,000 premature deaths a year, most of them concentrated in areas around ports and near major shipping lines.

Despite these destructive effects, this poisonous fug has helped shield the planet from the heating effect of fossil fuels: after new rules limiting the amount of sulphur in shipping fuel were introduced in 2020 it reduced global emissions of sulphur dioxide by about 10 per cent. Because sulphur dioxide reflects the Sun's energy, the removal of this insulating layer of sulphur dioxide is likely to have been one of the factors driving the startling spike in sea surface temperatures during 2023. Elsewhere shipping exhaust is speeding up the melting of the ice in the Arctic: as the fine coating of soot settles on the ice it traps the Sun's rays, raising temperatures and causing the ice to melt faster (ironically, most of the black carbon is being produced by ships moving fossil fuels).

YET THESE IMPACTS pale in comparison to shipping's contribution to global heating, and not just because almost 40 per cent of the goods transported back and forth across the ocean are fossil fuels. Shipping currently produces around 700 billion tonnes of carbon dioxide a year, or slightly less than 3 per cent of global greenhouse emissions. To put that figure in perspective, if the shipping industry was a country, it would be the seventh largest emitter – just behind Japan and Iran, and well ahead of Germany, Saudi Arabia and Indonesia. And while the impact of the pandemic and the war in Ukraine have temporarily interrupted the industry's expansion, the emissions produced by shipping are growing fast, rising almost 40 per cent between 2000 and 2019.

Despite this, the shipping industry has until recently largely resisted reforms that would reduce its impact on the environment. Faced with the complexities of ship ownership, and the fact most shipping takes place in international waters, the 1997 Kyoto Protocol handed responsibility for reducing the industry's emissions to the body that regulates global shipping, the International Maritime Organization (IMO). In a perfect world this would have led to change. Instead the IMO refused to act, actively opposing not just energy efficiency standards and emissions targets that might have reined in the industry's soaring carbon emissions, but even rules that would allow its emissions to be effectively measured.

The reasons for this are not difficult to discern. Although nominally a regulatory body, in reality IMO policy is shaped by the industry it is supposed to be regulating. Representatives of shipbuilders, mining and fossil fuel companies, shipping companies and lobbyists are permitted to attend meetings and take part in deliberations, either in the capacity of advisers or, in some cases, actually taking over seats from individual countries. Trade associations such as the International Chamber of Shipping (ICS) and the World Shipping Council, many of which have a history of working to block emissions targets and other environmental regulation, also attend meetings in order to lobby for their members. This corporate stranglehold on IMO policy is backed up by a total lack of transparency: reporting of meetings is tightly controlled, and journalists who breach the rules can find themselves suspended or barred from meetings. As the editor of *Climate Home News*, Megan Darby, tweeted after she was banned from IMO proceedings for two years for quoting comments made by a representative of the Cook Islands with a long history of opposition to environmental regulation, 'It's mind-boggling how corrupt and unaccountable the IMO is, hidden under a veneer of sheer tedium.'

The first cracks in this clubby refusal to address the question of emissions came in 2015, when Tony de Brum, the foreign minister of the Marshall Islands, decided the time had come for the industry to face up to its obligations. De Brum, who died in 2017, was a remarkable figure, who exerted an outsize influence on the planet's future. Born in Tuvalu in 1945, he became a lifelong advocate for the elimination of the 'senseless threat' of nuclear weapons after experiencing the effects of American nuclear testing firsthand as a boy: 'I was fishing with my grandfather. He was throwing the net and suddenly the silent bright flash, and then a force, the shockwave. Everything turned red: the ocean, the fish, the sky, and my grandfather's net. And we were 200 miles away from ground zero.' De Brum later played a central part in securing the Marshall Islands' independence from the United States, before becoming a key figure in the climate movement, a process that culminated in his leading role in forcing developed nations to agree to limit temperature rises to 1.5 degrees Celsius in the Paris Agreement.

Climate change is an existential issue for the Marshallese, whose low-lying islands are already suffering the effects of rising sea levels. But while the tiny nation has little in the way of industry, in 1990 it signed an agreement with American-based company International Registries, Inc. (IRI) that allowed IRI to register ships under the Marshallese flag. This venture has been a steady source of income for the Marshall Islands, and a bonanza for IRI, which has become the third largest ship registry in the world.

De Brum and his delegation assumed this would give them some standing at the IMO. But when they arrived at the IMO's Marine Environment Protection Committee (MEPC) in London they found their seat had already been filled by representatives of IRI. When they insisted on their right to be at the meeting, IRI and representatives of the IMO responded with disbelief and contempt. One Marshallese diplomat was even refused entry until he went back to his hotel and found a tie.

Despite the IRI's resistance, de Brum succeeded in delivering his message, telling the assembled dignitaries that 'time is short, and it is not our friend'. While this message found a receptive audience among some – in particular European nations – many others flatly rejected the idea the IMO should be taking action to reduce emissions. Indeed it was not until 2018 that the IMO finally acquiesced, first by agreeing to reduce emissions to 50 per cent of 2008 levels by 2050 and, in 2020 and 2021, to measures that would reduce the carbon intensity of shipping by 40 per cent by 2050.

These targets were revised upwards again in 2023, this time to reductions of 20 per cent (and if possible 30 per cent) by 2030, 70 per cent (and if possible 80 per cent) by 2040, and net zero by or around 2050. While these targets continue to fall short of what is required to hold global heating to 1.5 degrees, they do demand significant reductions in the emissions produced by individual ships, especially if the industry continues to expand. In combination with other initiatives, such as the green shipping corridors being developed by the European Union and countries such as Australia and Singapore, and the European Union's new carbon price on shipping, this is likely to rapidly accelerate investment in new, low-emission technologies.

For sectors of the industry this will cause considerable pain: as recently as mid-2022 only a third of the world's major shipping companies had announced policies to meet the grossly inadequate targets the IMO set in 2021, let alone to align with the far stronger cuts necessary to meet the targets agreed in 2023. Elsewhere change is already underway. In 2018, Danish shipping giant Maersk committed to reaching net zero by 2050; in 2022 the company brought that target forward to 2040, as well as establishing firm targets for 2030. As of mid-2023 Maersk has also ordered nineteen 'zero-emission' ships capable of carrying up to 17,000 containers each and of running on green methanol, the first of which should be in the water in 2024. The French shipping giant CMA CGM has also placed orders for eighteen container ships capable of running on methanol; China's COSCO has a dozen ultra-large methanol-ready container ships on order; and Pacific Basin Shipping, one of the world's largest carriers of dry bulk cargo, has announced it intends to transition its fleet to methanol.

Methanol has several attractions as an alternative to HFO. Conventional engines can be converted to run on methanol relatively easily and cheaply, and most major ports already have infrastructure capable of storing and supplying it. Burning it produces markedly less nitric oxide and almost entirely eliminates sulphur-dioxide emissions and the toxic particulates produced by burning HFO. Unlike oil it also dissolves in water, making it less harmful to marine environments (an important factor when one considers that ships leak or dump millions of tonnes of oil into the ocean every year).

Nevertheless the large-scale embrace of methanol should not be seen as a silver bullet. Currently most methanol is produced using fossil fuels, meaning its use will do little or nothing to reduce emissions. Moving to zero emissions requires a shift to green methanol in the form of bio-methanol, which is produced using biomass and renewable energy, or e-methanol, which is created using green hydrogen and captured carbon dioxide. Sourcing green methanol is likely to be a significant challenge, especially in the short term: in 2023 less than a million tonnes was produced worldwide, and while production is growing rapidly, just powering the ships Maersk already has on order is likely to require well over half a million tonnes annually, and the company itself estimates it would take 20 million tonnes a year to run its entire fleet.

Increased production of bio-methanol also has potentially serious implications for food and water security and biodiversity. In Europe, an area larger than Greece has already been turned over to growing biofuels, while in Africa, Asia and South America, rainforest and wetlands are being cleared and land once used for farming is being bought up and used to produce feedstock, leading to destitution and hunger for local people and siphoning wealth to companies in Europe, America and elsewhere. Some of these effects may be ameliorated by incorporating the use of waste, but even optimistic assessments suggest that by 2050 biofuels will require three times the land they require today.

Methanol is not the only low-emission fuel being explored: many shipping companies are also betting on ammonia. Although around seventy per cent of the ammonia currently produced is used as fertiliser in food production, ammonia has also been used as a fuel for more than a century. At present most ammonia is produced using industrial processes that create huge volumes of greenhouse gases (ammonia production is responsible for around 1.8 per cent of the world's emissions), but new technologies allowing it to be created more sustainably using renewable energy or in conjunction with carbon capture and storage are coming online.

Using ammonia as a fuel presents considerable challenges however. Although it does not produce carbon dioxide when combusted, burning ammonia releases nitrous oxide, a greenhouse gas more than 250 times as powerful as carbon dioxide, as well as nitrogen oxides and other pollutants, meaning its use requires technologies capable of capturing these emissions. Unlike methanol, which can be stored at room temperature, ammonia is highly toxic, and must be stored under pressure or refrigerated. Its energy density is also only around a third that of HFO, meaning fuel storage takes up three times as much space. And, as with biofuels, demand for ammonia to meet the demands of transport will have an effect on the supply of ammonia for other purposes, in particular agriculture.

Hydrogen offers another alternative, although again it is not without its drawbacks. While hydrogen is clean, and does not pose the safety risks ammonia presents, it still needs to be stored under pressure or at extremely low temperatures, and is significantly less energetically dense than either HFO or ammonia. More problematically, the vast bulk of hydrogen production still relies upon fossil fuels, and despite investment in hydrogen facilities that employ wind and solar power, the high cost and relative scarcity of green hydrogen mean its capacity to reduce emissions in the short to medium term is limited. Nonetheless its potential in the longer term is considerable: one 2020 study found that almost half the voyages along the crucial route between China and the United States in 2015 could have been powered by hydrogen fuel cells with no changes to operations or fuel capacity, and that with minor adjustments to fuel storage and routes, fully 99 per cent of voyages in that period could have been powered by hydrogen.

MEANWHILE OTHERS ARE LOOKING towards much older sources of energy. On a freezing morning in late February I board a train to Den Helder, in North Holland. Den Helder has been a naval base since Napoleon took a shine to the harbour after France annexed the Batavian Republic in 1810, and a port for far longer. These days it is home to the Royal Netherlands Navy, whose grey ships lie at anchor near the new naval headquarters on the Nieuwe Haven, but I am not here to see them. Instead I make my way through the quiet streets to the Willemsoord. The navy's former base of operations, it's now the gateway to a maze of old buildings and a dry dock clustered along the side of several canals and an artificial harbour, where a host of old vessels in various stages of repair are moored, ranging in size from canal boats to three-masted clippers.

I am here to see Captain Jorne Langelaan, the founder of EcoClipper – a venture that aims to build a fleet of sail cargo vessels. Langelaan is waiting outside the red brick building that houses a cooperative based in the harbour. He is tall, and his calf-high black boots and shoulder-length blond hair wedged under a black fisherman's cap make him look like he has stepped out of a Conrad novel, but there is something endearingly open and youthful about his manner. After he greets me we head into the building. Inside the air smells of sawdust and timber, and somewhere nearby people are hammering and sawing. Langelaan leads me through rooms crowded with tools and equipment; in one an old boat about 10 metres long is up on blocks in the midst of being restored. Pausing beside it, Langelaan tells me it is a fishing boat that was built more than a century ago. 'They would have taken her out on the Zuiderzee, which is now the IJsselmeer, and out into North Sea,' he says. 'But with no engine, only sail.'

Langelaan comes from a family of sailors and shipowners and has been working on boats since he was a teenager. After leaving school he attended the Enkhuizen Nautical College, which specialises in training officers for large sailing ships. This led to him serving on a range of vessels, including a sail cargo vessel that was working routes in the Caribbean, and the *Europa*, a steel-hulled barque that takes paying crew and others on cruises and races. It was on the *Europa* that he met his future business partners, Andreas Lackner and Arjen van der Veen, and later, while serving as the ship's artist, travelled to Antarctica.

Antarctica had a profound effect upon Langelaan. Newly alive to the fragility of the planet's systems he returned to the Netherlands and, along with Lackner and van der Veen, began to explore the idea of a new approach to shipping. Setting up a company called Fairtransport, the three refitted an old ship and named her the *Tres Hombres*, after the captain of the *Europa*'s name for the three of them. But when they went looking for investors they were greeted with ridicule. 'They thought we were crazy,' says Langelaan, with a shrug. Unable to find customers they volunteered their ship to help deliver aid to Haiti after the 2010 earthquake, and later began hauling rum.

Although Langelaan remains a shareholder in Fairtransport his energies are now tied up in EcoClipper, which is currently refitting *De Tukker* – a 40-metre-long, 70-tonne vessel that was built in 1912 and spent several decades working as a coastal trader – and working on plans to build the first *EcoClipper500*, a 500-tonne, steel-hulled sailing vessel capable of working major ocean routes such as the trans-pacific and transatlantic.

Langelaan based the design of the *EcoClipper500* on the *Noach*, a three-masted, square-rigged clipper built in Kinderdijk in 1857, and still regarded as one of the fastest Dutch sailing vessels of all time. His decision to look back to the age of sail for inspiration is not simply nostalgia, but about relying upon proven technology. 'The last generation of sail cargo ships had the highest possible innovations,' he says. 'Of course sailing has evolved since then. But the technology you see on the big racing yachts relies upon weight reduction, and only works because those ships are as light as possible. As soon as you want to start carrying cargo, you're working with different natural laws.' Langelaan also believes these older designs reduce emissions more effectively. 'A lot of modern sail cargo vessel designs use high technology, which still requires a lot of energy, or is based upon propelling ships that weigh almost nothing. And that means that although you get a little bit of thrust from the sails, you still rely mainly upon the engines.'

Langelaan's vision of a return to schooners and clipper ships might seem quixotic in the face of the dizzying scale and complexity of the contemporary shipping industry. Certainly it is difficult to imagine them replacing the 400-metre-long container ships and bulk carriers that plough up and down in the waters off Den Helder and Rotterdam. But the significance of what Langelaan and others like him are doing is more than just about practicality.

Christiaan De Beukelaer is an academic at the University of Melbourne. Originally from Belgium, he trained as a musicologist before spending time studying the intersection of development and cultural policy in Burkina Faso, Ghana and elsewhere. Half a decade ago he decided to refocus his work on the climate crisis, and after taking up sailing became fascinated by the challenges of decarbonising the shipping industry. In February 2020 he joined the crew of the *Avontuur*, a century-old schooner that transports rum and coffee and other goods between Europe and America, for a three-week trip from the Canary Islands to the French Antilles. But as countries around the world closed their borders, De Beukelaer and his crewmates found themselves stuck at sea for almost five months.

De Beukelaer cautions against focusing on the relatively small scale of ventures like EcoClipper and the *Avontuur*. 'It's important to remember that the sail cargo ships that are currently in operation are not representative of the types of ship that were being used in the late nineteenth century. In that era there were really sizable ships that carried 5000 tonnes of cargo. So what's in use now is one thing, but what's possible is something else altogether.'

He argues the significance of ventures like EcoClipper and the *Avontuur* lies in their capacity to enact a way of imagining our relationships with the planet and each other that is built upon ideas of collaboration and sustainability. 'Traditional sail cargo ships are experimental laboratories for radically different futures,' he says. 'But they also open the door to more ambitious approaches capable of offering practical solutions to the problem of decarbonisation.'

A GLIMPSE OF SOME OF THESE APPROACHES is already coming into view. In Portsmouth, on England's south coast, BAR Technologies is about to begin fitting the first set of WindWings, a 40-metre-high rigid sail system that can be retrofitted onto existing ships. Effectively vertical aeroplane wings and constructed out of steel and glass fibre similar to that used in wind turbines, WindWings are part of a fully integrated system that employs software and global positioning to optimise the sails' operation by orienting them to catch the wind most effectively, and reading weather maps to find the most efficient route. Meanwhile a second British company, Anemoi Marine Technologies, has developed a sail system that relies upon Flettner rotors, a century-old technology developed in the 1920s in which the wind spins a tall cylinder. Resembling tall white smokestacks that can be folded away, Flettner rotors have been operating successfully on the bulk carrier MV *Afros* since 2018, and Anemoi is moving ahead with installation of rotor sails on ships operated by Berge Bulk and others. And as I write this in 2023, sails are being fitted to the *Canopée*, a 121-metre-long low-emission vessel built around four 363-square-metre windsails that will be used to transport Ariane rockets to the European spaceport in French Guiana.

While BAR is already working on new hull designs that will allow the ships of the future to take full advantage of the power of the wind, for now at least, technologies such as the WindWings and Anemoi's rotors only function as an auxiliary power source, capable of reducing the fuel required by up to a third. Others have even more ambitious plans. Danielle Southcott is a Canadian sailor and entrepreneur. Originally from Ontario, she sailed as chief mate on the first transatlantic voyage of the *Tres Hombres*, an experience that inspired her to set up Sailcargo in 2014 and led to Sailcargo's ongoing project to build a fully sustainable, wooden container ship in Costa Rica. In 2021 Southcott left Sailcargo to set up Veer Voyage, which aims to deliver the first zero-emission, wind-powered container ship.

Southcott founded Veer out of a frustration with the small scale of many sail cargo projects, and their focus on anti-capitalist principles. 'So many of these projects say people should just stop buying things. And I'm not against that. I'm an environmentalist to the core. But I also want to effect change at scale. Little wooden cargo ships are gorgeous and satisfy a lot of things. But companies like Coca-Cola can't ship with them. They're not big enough, they don't work for the infrastructure. And so I decided to put a solution in their face that's so good, so perfect, that they can't not do it. So with Veer I'm working super-hard to provide solutions that are industrial, easily scalable, competitively priced – all the things that are the capitalist dream. But that's not because I'm deeply driven by capitalism. I have a list of problems as long as my arm with capitalism. It's because I want to create a solution so good that they can't say no.'

Veer's first prototype, the *Design No1*, should be under construction by the end of 2023. At 157 metres long and capable of carrying

157 containers, it will be powered by a DynaRig sail system and green hydrogen engine with a range of more than 2200 kilometres. But Southcott is clear the *Design No1* is not the end of what she wants Veer to do. 'Apple don't just make iPhones. They offer ways to interact with data. I see Veer the same way: we don't have a commitment to one design, or even wind power and hydrogen power. Instead we're about the delivery of clean energy solutions.'

Southcott is not alone in her focus on seeing zero-emission ships on the water as soon as possible. In France, Windcoop, a collaboration between maritime company Zéphyr and Borée, the designers of the *Canopée,* and spice importer Arcadie, has plans for a sail-powered container ship capable of carrying 60 to 68 containers that will draw 90 per cent of its energy from the wind and the remaining 10 per cent from a biofuel engine: their goal is to have the vessel working the route between Madagascar and Marseille by 2025. And in Sweden, shipping giant Wallenius Wilhelmsen is partnering with marine technology providers Alfa Laval to develop low-emission ships: their first prototype, the *Oceanbird,* a 220-metre vessel capable of carrying 7000 cars and designed to draw up to 90 per cent of its energy from the wind, should be on the water by 2026.

For Southcott this rush of innovation and investment demonstrates the industry is able to change. 'I get so frustrated. People act like decarbonisation is too hard, that nothing will change. But the shipping industry can do it if they want to. Have you seen the scale of the ships and canals and cranes? These guys can make anything happen.'

WHILE THE DIFFICULTIES ASSOCIATED with decarbonising shipping are not insurmountable, that should not distract from the scale of the task ahead. There are currently more than 50,000 ships moving across the ocean: replacing them or modifying their engines in a matter of decades will require the investment of hundreds of billions of dollars.

Yet decarbonisation is only a subset of a far more profound challenge. The ecological footprint of humans already outstrips the planet's capacity by 70 per cent. The shipping industry has played a central role in accelerating the patterns of consumption and extraction and imbalances of wealth that have enabled that overshoot. But while decarbonising shipping will reduce the emissions produced by ships moving goods and raw materials across the oceans, it will do little to change the larger problem of the unsustainability of the burden we place on the planet. 'Is a $50 air fryer really a good idea?' asks De Beukelaer. 'Because for that money it won't be a good machine, so it will probably break quickly. And then you've wasted so much in terms of energy, materials, the time of the unfortunate people stuck in factories producing this shit. I'm not saying that you shouldn't be able to get an air fryer if you really want to, but I think we have to start asking how much do we actually need to produce and need to have?'

Langelaan also contends that the problem demands more than simply flicking the switch to renewable energy – it requires a fundamental rethinking of the global economy. 'Change is happening, but most people are still at the point where they think we don't really have to make big adjustments, that we can just switch to a different fuel and everything else will stay the same. But that's not the case. It's going to be a totally different world in a hundred years. We are going to move in a direction nobody wants to think about, especially in shipping. And part of that has to be a return to a more localised world. I know that's against the predictions of the big economic think tanks and so on, but there's just no way around it. International trade is going to decline.'

There seems little question Langelaan is correct that the shipping industry is headed for major change: if nothing else the continuing expansion predicted by the OECD and others seems incompatible with a shift to a decarbonised world. Perhaps, though, shipping might also offer a way of seeing beyond the horizon. For just as its present is intimately intertwined with the historical processes that have shaped the world we now inhabit, its future contains the possibility of building a better and more sustainable world. 'Every sail cargo initiative offers a vision for the future,' says De Beukelaer. 'Choosing which one should be pursued touches upon some of the most fundamental issues we face in contemplating different climate futures: what exactly is the future we want? And who is "we" in this context?'

Staring at the vast machinery of Rotterdam I am reminded of my initial impression these machines might keep working without us, steered by some inhuman agency. It is tempting to see these processes as unstoppable, natural forces akin to gravity. Yet they are not. They are the products of very specific historical circumstances. The imbalances and processes of capitalist accumulation that have driven the planet to its current predicament were not inevitable and are not unchangeable. And neither is the course from here.

9

TRACES

'Let no one say the past is dead. The past is all about us and within.'
Oodgeroo Noonuccal, 'The Past'

THE PLANE'S APPROACH to the Cocos Islands is surprisingly steep – a wide bank past towering, geological columns of tropical cloud, followed by a rapid descent. For a few moments before we land it is possible to see the entirety of West Island, floating in the ocean like an inverted letter J, the scattering of other islands that form the broken circle of the atoll visible in the distance. The cleared airstrip follows the line of the island, which is barely 500 metres wide. On its eastern side a narrow border of coconut palms separates the runway from the turquoise lagoon; to the west a road marks a division between the low grass and a scattering of houses and trees, and beyond them, the darker blue of the Indian Ocean.

I am here with my friend Dr Jennifer Lavers. Originally from Canada, Lavers moved to Tasmania in the mid-2000s and has spent much of the time since studying the effects of plastics on birds and other animals. Her work has taken her to some of the most remote islands on the planet.

On the face of it the Cocos Islands might seem a curious choice for a study of plastic pollution. Officially an external territory of Australia, the twenty-seven islands that make up Cocos and the neighbouring Keeling Islands are sparsely populated and extremely isolated. Situated roughly midway between Broome and Sri Lanka, they are separated from the Australian landmass by more than 2000 kilometres of open ocean, and from Indonesia to the north by almost 1100 kilometres. Even their closest neighbour, Christmas Island, is 1000 kilometres away. Like most of the islands of the Indian Ocean, Cocos has no Indigenous population. The Cocos-Malays, who make up around three quarters of the 600 or so people who live permanently on the islands, were brought here as labourers during the nineteenth century, gradually developing their own dialect and customs. Today the Cocos-Malay community is centred on Home Island, or Pulau Selma, on the eastern side of the lagoon; the remainder of Cocos' residents – mostly mainland-born, mostly White – live on West Island, or Pulau Panjang.

Despite its isolation Cocos has drawn attention for the impact of ocean plastics on its ecology. In 2018, Lavers surveyed beaches on seven of the islands. What she found astonished her. Her data suggested there were at least 414 million pieces – or 238 tonnes – of plastic waste on Cocos's beaches, most of it concentrated in 'debris hotspots'. And this figure is probably conservative, since difficulties with transportation meant she and her team could not reach several of the worst-affected islands.

Lavers' findings on Cocos and elsewhere lay bare the scale of the plastic problem, but they also offer a reminder that our understanding of the effect of plastics on ocean environments is constrained by gaps in our knowledge of exactly where the more than eleven million tonnes of plastics washing into the ocean every year actually goes. Most of it just sinks; much of the rest breaks up into fragments of microplastic, which either mingle with the estimated 170 trillion particles of microplastic that are believed to fill the ocean's surface layers, percolate downwards to gather in sediments on the sea floor, or accumulate in the bodies of animals like some invisible, metastasising dark matter. But the sheer size of the ocean and the difficulty of studying the deep ocean means our understanding of exactly where and how plastics are distributed remains worryingly incomplete, and dependent upon sampling technologies that are slow and expensive. This problem is particularly acute when it comes to measuring the amount of plastic on beaches, which requires scientists to carry out manual surveys during which they laboriously count and record the amount of plastic they find.

We have come to Cocos because doctoral student Jenna Guffogg believes it is possible to accelerate this process. Together with a team of scientists from the Remote Sensing Research Group at the Royal Melbourne Institute of Technology, Guffogg intends to deploy drones to take high-resolution photographs of beaches. Artificial intelligence systems will then be trained to identify and count plastic items on the sand. If successful her system will allow scientists to survey entire beaches in minutes rather than days. Guffogg also hopes the system might be able to be combined with satellite imagery, an idea she hopes to test during this trip by using photos taken by the International Space Station (ISS) on one of its orbits overhead. This could make it possible to survey beaches from space, thus granting scientists access to data on plastics from even the most remote locations without the need to travel there. Even better, the principles underpinning the process might be able to be used to quantify the amount of plastic floating in the surface layers of the ocean. Such an exponential increase in the amount of data available to scientists has the potential to revolutionise our understanding of the quantity and distribution of ocean plastics.

THE FIRST RECORDED SIGHTING of the Cocos and Keeling Islands was in 1609, when Captain William Keeling of the East India Company noted their existence on his way back to England from the Indies. Europeans do not mention the islands again until 1749, when Swedish captain Carl Gustaf Ekeberg sketched an image of coconut palms on North Keeling, and it was not until the 1780s that the islands began to appear on maps and charts.

There is evidence that a ship from Mauritius was wrecked on the islands at some point and some of the crew survived for a time, but the first step towards settlement took place in 1825, when Scottish merchant seaman Captain John Clunies-Ross anchored off the atoll while travelling home from India. Conceiving a plan to establish a trading post on the islands, Clunies-Ross went ashore and nailed the Union Jack to a tree before returning to Britain to convince his wife to accompany him back to the islands with their young family.

Meanwhile Clunies-Ross's former employer, Alexander Hare, had also set his sights on the islands. Born the son of a London watchmaker in 1775, by the 1800s Hare had established himself in Malacca as a merchant. While there he came into the orbit of Stamford Raffles, who appointed him commissioner of Borneo. Hare used his time as commissioner to acquire more than 3500 square kilometres of land from the Sultan of Bandjermasin, on which he established his own independent state, Maluka.

Now styling himself the Rajah of Maluka, Hare set about building his own little kingdom, complete with its own flag and coinage. Like many wealthy Europeans in Malacca, Hare already owned a number of enslaved people, some of whom he had bought, others he had been gifted by local rulers and others. But in Maluka he expanded his household considerably and established the harem of Malay women for which he would later become notorious. It was also in Maluka that he became involved with Clunies-Ross, who acted as his overseer as well as supervising the transport of several thousand convicts from Java for use as forced labour.

In 1816 Hare fled Maluka for Java, taking with him a group of ninety or so slaves; some were domestic servants and labourers, others were part of his harem. Over the next decade he and his retinue spent time in Sumatra and South Africa, but after his insistence on keeping concubines placed him at odds with polite society in Cape Town, he began to cast around for somewhere he might resume the life he had known in Maluka. And so, in 1826, he hired a ship – commanded, as luck would have it, by John Clunies-Ross's younger brother, Robert – and took up residence on Home Island, which is where Clunies-Ross found him when he returned to the islands with his wife and children a few months later.

Although the two men had been friends in Maluka relations between the two settlements quickly deteriorated, especially when Hare's women began to desert him for the sailors who had accompanied Clunies-Ross. Finally, in 1831, Hare left the islands for Sumatra, where he died three years later.

With Hare gone, Clunies-Ross turned his attentions to expanding copra production on the islands. With no Indigenous population to put to work, he once again began to import workers from Java, relying for the most part on convicts who could be used as indentured labour. As the population grew larger and the family expanded and grew richer, the Clunies-Ross family began to style themselves as feudal rulers, making and enforcing the law and coining their own currency with which to pay their labourers. This arrangement was formalised in 1886, when Queen Victoria granted the islands to the Clunies-Ross family in perpetuity.

The Clunies-Ross ran the islands as a private kingdom for the best part of a century and a half. They controlled every aspect of life there. There was no legal system, or basic rights; instead the Clunies-Rosses relied upon a specially appointed council of headmen, or *juro*, who were rewarded with more pay and greater freedoms. Education was minimal and did not proceed beyond the primary level. The islanders were not even paid in legal tender but instead with plastic tokens that could only be redeemed at the company store.

Calisha Bennett's grandparents were born on Cocos in the 1930s, by which time John Sidney Clunies-Ross – the great-great-grandson of John Clunies-Ross – was the island's chief magistrate. In the 1950s her grandfather broke into a locked shed where food was stored. 'They used to control people's access to food as a way of making them behave,' says Bennett, who now runs a mentoring and education service for Islamic girls in Sydney's west. 'People were almost starving, so my grandfather took a traditional machete, or *parang*, and got food for the people. When they found out, Clunies-Ross put a rifle in his face and told him he was going to shoot him. After that he got an ulti-matum: he could either go to Sabah or to Christmas Island. He chose Christmas Island, and along with my grandmother and their eight children they were loaded onto a ship and banished.'

This state of affairs persisted until well after Australia took control of the islands in 1955, but by the 1970s the Australian Government was becoming increasingly uncomfortable with the family's style of management. In a report for the Department of External Territories in 1971, Commonwealth public servant Alan Kerr said he 'could not help but liken the situation to what I imagine life would be like for slaves on the estate of a benevolent slave owner in southern America'. Two years later the newly elected Whitlam Government invited the United Nations to visit the islands, then presided over by John Cecil Clunies-Ross, the son of John Sidney and the fifth Clunies-Ross to style himself as the ruler of the islands.

The language of the UN report makes it clear their represent-atives were deeply disturbed by what they found. It describes the Cocos-Malay workers as 'forced, underpaid labour', and is extremely critical of the 'anachronistic, feudal relationship' between John Cecil Clunies-Ross and the Cocos-Malay community and his 'complete control . . . over the life of the community'. It also notes that the people they spoke to were 'shy or not eager to provide information to the Mission', perhaps because of 'the presence of Mr Clunies-Ross and his estate manager', during the interviews, and observes that the community's extreme isolation and lack of political education had created a situation in which the local people were unaware of their fundamental rights. Clunies-Ross had previously told Australian officials that when it came to the Cocos-Malay, he 'did their thinking for them'.

The Clunies-Ross family's domination of the islands finally came to an end in 1978, when the Australian Government forced John Cecil Clunies-Ross to sell it the islands for $6.25 million, an arrangement that was followed in 1984 by a referendum on the islands' future in which the population voted overwhelmingly for formal integration with Australia. Yet the story of the Clunies-Rosses' domination of the islands is not a relic of the distant past – it is disturbingly proximate to the present. I was already at school by the time the UN visited the islands in 1974, and John Cecil Clunies-Ross himself only died in 2021, aged ninety-two.

ON OUR FIRST MORNING we wake early. The plan is to catch the early ferry across the lagoon to Home Island to carry out surveys at Cemetery Beach, one of the plastic hotspots Lavers surveyed for her 2018 study. Cemetery Beach is on the windward side of the island's northern tip, where the Indian South Equatorial Current sweeps warm water down from Indonesia, bringing plastic and other refuse with it.

Because we have not hired a car, the only way to get the gear to the ferry is on the bus that runs up and down the island every hour or two. This is easier said than done: the big drone Guffogg hopes to use to photograph the beach is contained in a case weighing almost 60 kilograms that requires four people to carry it any distance. We must also transport multiple bags of equipment, but with a little effort we get it all to the bus stop and then onto the bus. On the way we chat to the driver, who turns out to be a teacher at the local school who shares responsibility for ferrying people up and down the island a few days a month with several other members of the local community. When I ask the driver whether the local children cross the lagoon for school every day she tells me some do, although many leave in later years to attend high school on the Australian mainland.

At the dock on Home Island we carry the drone down onto the wharf. On the ride over Lavers has fallen into conversation with a woman she met on an earlier trip; seeing the drone the woman offers to drive us to the beach, and so we load into her van and bump out of the settlement and down the road.

She deposits us in the cemetery that gives the beach its name, and we make our way through the undergrowth separating us from the beach. Some sense of what lies ahead is apparent from the sea of plastic scattered among the low trees and shrubs. But it is only when we step out onto the beach itself that the sheer scale of the problem reveals itself. There is plastic everywhere – scattered on the sand, floating in the water, piled in banks against the trees and rocks.

Although I have seen photos and videos of pollution of this sort, encountering it in real life is deeply shocking. For a minute or two I stand motionless, taking stock of our surroundings and trying to make sense of what I am seeing. I am not alone in this: around me the others are also visibly shocked. Only Lavers seems unaffected, her face set and grim. I am familiar with Lavers' ability to withdraw into herself, which I have always assumed is connected at least in part to her traumatic childhood and adolescence: a member of the Métis Nation, she left her family as a teenager, and found herself homeless. When I ask her about her reaction later, she shrugs. 'After a while you lose the ability to feel anything,' she tells me. 'That human bit of you that feels, it just dies somehow.'

As we set up Guffogg keeps looking up at the wind smashing into coconut trees behind us, worried it will prevent the drone from taking off safely, rendering useless any data we collect manually. Her concerns turn out to be well-founded – we will not get the drone aloft today – but for the moment we decide we will proceed anyway and begin marking the beach out for our survey.

The survey process is surprisingly simple. To begin with the beach is divided up into quadrants, a grid of boxes 1 metre square. Then a number of these are selected at random and examined in detail. This latter process requires two people, one to lean over the quadrant and catalogue what it contains, the other to check and record results. Once the quadrants have been catalogued, the results of this manual survey will be able to be compared with the results from the drone for these same quadrants.

Lavers and I cross to our quadrant and begin. At first I do the counting, scratching across the sand. The sun is hot, and crouched down over the sand the glare is blinding. To be effective the survey must be extremely accurate, which means not just counting but also classifying scores of objects into various categories. Sometimes that is easy – a pen is obviously a pen – but for every easily identifiable object there are a dozen scraps and fragments, all of which must be catalogued by size and colour.

Patterns quickly become evident: among the most common objects are single-serve water cups and their pull-off lids; together with the tube-like wrappers of ice sticks and jello shots, their flimsy forms lie in banks against the trees and beneath the sand. But there are also huge numbers of other objects: plastic bottles, bottle tops, shoes, thongs, toothbrushes, rope and scraps of fishing net, building waste in the form of empty tubes of filler and glue, coffee scoops and ice-cream spoons (mostly Nestlé). Among the rubble of smaller detritus lie other, larger objects: fishing buoys, broken laundry baskets, polystyrene boxes and children's toys. Beneath a bush at the edge of the sand a soccer ball lies half-buried in broken shards of cups, its pale, bleached dome like a grotesque parody of a skull or an egg. And everywhere tiny scraps and shards, pieces so small or damaged as to be unidentifiable.

THE COCOS ISLANDS would also have a lasting impact upon the work of Charles Darwin, who visited them in 1836 on the homeward leg of his journey aboard the *Beagle*. In his account of his journey Darwin complained of the 'oppressive' glare created by the stark white sand of the islands' 'calcareous beaches', but like many early visitors found himself charmed by their beauty, declaring that 'nothing could be more elegant, than the manner in which the young and full-grown cocoa-nut trees, without destroying each other's symmetry, were mingled into one wood'. In so doing he also reveals the degree to which the islands had already been transformed by a decade of settlement, noting the native hardwood trees were almost entirely gone, felled for building and replaced by coconut palms.

Darwin also provides an account of the flora and fauna of the islands, noting it was 'exceedingly limited', and comprised of some twenty species of plants, including half-a-dozen species of trees, plus an unknown number of mosses, lichens and fungi, and those plants brought to the islands more recently by the settlers. He notes the presence of rats, 'smaller and much more brightly coloured' than 'the English kind', and believed to have been deposited by the ship from Mauritius that was wrecked upon the islands before the arrival of Hare and the Clunies-Rosses, a handful of birds, including a snipe and a rail, one species of lizard, a limited number of insects and spiders, and a great variety of hermit crabs.

Yet Darwin reserved his greatest wonder for the marine environment of the islands. On 6 April he and the *Beagle*'s Captain Robert FitzRoy travelled to Cocos's South Island, winding their way through a narrow and intricate channel bordered by 'fields of delicately branched corals'. Upon reaching the island they crossed it, and found 'a great surf breaking on the windward coast'. Moved, Darwin observed:

> There is to my mind a considerable degree of grandeur in the view of the outer shores of these lagoon islands. There is a simplicity in the barrier-like beach, the margin of green bushes and tall cocoa-nuts, the solid flat of coral beach, strewed here and there with great fragments, and the line of furious breakers, all rounding away toward either hand.

Darwin's appreciation was not purely aesthetic, however. For as his journals make clear, he had for some time been considering the question of how coral atolls such as Cocos form in the first place.

This subject was a matter of intense debate among scientists of the time, and had, perhaps not coincidentally, been identified as one of the questions the *Beagle*'s voyage might address. Although scientific understanding of the biology of coral was still limited, scientists were aware reefs are created by the slow accumulation of coral skeletons over time. Yet if corals only grow in relatively shallow water, how could they form atolls, which often rise from waters many kilometres deep?

Scottish geologist Charles Lyell had already proposed the answer might lie in volcanic activity, noting many atolls resemble the craters of submerged volcanos. But Darwin soon discarded this idea as inadequate. Instead, after observing land uplifted after an earthquake in Chile in 1835, he speculated that as areas of land rose, regions of the seabed might fall. If that happened gradually enough, corals growing on the seamounts formed by volcanoes might keep pace, growing upwards towards the light.

Cocos offered Darwin an opportunity to test his ideas. At Darwin's request, FitzRoy took soundings on the drop-off on the outer perimeter of the island, collecting evidence that as the water grew deeper the coral and sand present closer to the surface disappeared, presumably replaced by some other form of stone. In his diary Darwin wrote:

Capt. FitzRoy at the distance of but little more than a mile from the shore sounded with a line 7200 feet long, & found no bottom. Hence we must consider this Isld as the summit of a lofty mountain; to how great a depth or thickness the work of the Coral animal extends is quite uncertain . . . Under this view, we must look at a Lagoon Isd as a monument raised by myriads

of tiny architects, to mark the spot where a former land lies
buried in the depths of the ocean.

The result of Darwin's investigations – *The Structure and Distribution of Coral Reefs* – was published in 1842, and was to become
his first major scientific contribution. Reading it now, it is impossible not to be struck by how much its clarity, closely argued reasoning
and inexorable marshalling of evidence resemble that of *On the Origin
of Species*.

Lyell was delighted when Darwin wrote to him to explain his
theory, but cautioned the younger man not to 'flatter yourself that you
will be believed, till you are growing bald, like me, with hard work &
vexation at the incredulity of the world'. Unfortunately for Darwin,
Lyell's words would prove prophetic: many did not accept his ideas,
preferring the rival theories of Scottish oceanographer John Murray,
who postulated that reefs occurred when accumulations of marine
sediments rose high enough to allow corals to grow.

It would be 1953 before Darwin was proved right. In the aftermath of the American nuclear tests in the Marshall Islands, the
US Government sent scientists to assess the damage to Bikini and
Enewetak atolls. Determined to establish the geological structure of
the islands, the scientists drilled down into Enewetak. For weeks they
found nothing but coral, until finally, more than 1400 metres below
the surface, they struck volcanic rock.

The discovery of a volcano beneath Enewetak was also significant because it offered a glimpse of the dynamism of the Earth's crust.
Over the next few decades this realisation would transform scientific
understanding of the Earth's inner structure. By mapping the seismic
rumbles of nuclear blasts as they reverberated back and forth through
the planet, scientists discovered the Earth is not composed of simple
concentric layers; instead it is a far more complex and dynamic place,
in which continental plates buckle and dive beneath each other as they

collide, plumes of magma snake up through the crust, and huge hunks
of denser material such as the blobs some believe are the remnants of
Theia that were buried there after the collision that created the Moon
lie concealed far beneath our feet.

As the British scientist Steve Jones has pointed out, this convergence of military interests and scientific research did not just help
revolutionise scientific understanding of the structure of the planet,
it also helped lay the foundation for the Cold War military-scientific
complex that led to the space race, the development of the microchip
and even the internet. Simultaneously, though, the blasts on Bikini,
Enewetak and elsewhere speak to the callousness and violence that
lies at the heart of the colonial project. Between 1946 and 1958,
the United States detonated sixty-seven nuclear devices in the
Marshall Islands. These devices had a total yield equivalent to almost
110 megatonnes of TNT, more than 7200 times the yield of the
bomb that was dropped on Hiroshima, and a hundred times the yield
of the atmospheric tests the United States carried out on its own soil.
The unimaginable destructive force of the blasts vaporised entire
islands, boiled fish in the sea and cooked seabirds in the air up to
20 kilometres away. Radioactive fallout from the blasts coated nearby
islands in a layer up to 2 centimetres deep, contaminating the water
supply, poisoning fishing grounds and in the case of Rongelap Atoll,
which was exposed to fallout from a test in 1954, causing severe
radiation sickness in the inhabitants. The US claimed the incident
on Rongelap was an accident, but there is no question they understood the risk when they undertook the test. And when inhabitants of
the island wanted to return to their home, US scientists deliberately
withheld information about the risks because it would allow them
to study the long-term effects of radiation on human beings. As one
Atomic Energy Commission official told a meeting of the agency's
Biology and Medicine Advisory Committee in 1956, 'Data of this
type has never been available. While it is true that these people do not

live the way that Westerners do, civilised people, it is nonetheless also true that they are more like us than the mice.'

More than seventy years later the Marshallese are still grappling with the toxic legacy of the nuclear violence unleashed on their island. Those exposed to the radiation on Rongelap have suffered elevated levels of miscarriage and cancer, and their children have suffered birth deformities, thyroid problems and leukemia. People relocated from Bikini and Enewetak before the tests are still unable to return, and on several of the islands levels of radiation remain higher than those around Chernobyl. And as waters warm and sea levels rise, there are growing fears the ageing concrete dome known as 'the Tomb' that the US Army constructed on Runit Island in the late 1970s over a pit containing more than 70,000 cubic metres of plutonium is at risk of cracking and leaking irradiated material into the ocean.

The tests in the Marshall Islands would also provide one other link to Cocos and Darwin. In the aftermath of the blasts, scientists took note of the dispersal patterns of the fallout. Tracking the radio-active material as it travelled on the jet streams high above and spread outwards through the water to settle in sediments and permeate coral reefs thousands of kilometres away revealed hitherto unknown information about the speed and direction of the winds and currents that circulate through the atmosphere and ocean. And in a piece of synchronicity so perfect I am not sure whether it is beautiful or sinister or both, one of these currents bears this radioactivity west through the islands of Indonesia into the Indian Ocean, to deposit it on the reefs of Cocos, thus closing a circle between the origin of Darwin's theory about the formation of coral atolls and its violent proof on Enewetak more than a century later.

IN RECENT YEARS much of Lavers' research has focused on the effects of plastics on seabirds. Plastic is now believed to be present in the bodies of more than 90 per cent of all pelagic birds, and more than a million die every year from ingesting it. Many of these deaths are the result of malnutrition, as birds mistake brightly coloured pieces of plastic for the fish and krill and squid upon which they feed – or worse yet, feed it to their chicks, which then either starve or are so weighed down by the plastic in their bellies that they cannot fly. Lavers has witnessed these effects firsthand: on Lord Howe Island, off Australia's east coast, where the number of living shear-water chicks has fallen substantially, she recently removed more than 400 pieces of plastic from the bodies of some baby birds.

But Lavers' work on Cocos also illustrates the effect of plastics on other, less charismatic animals. Like many remote islands Cocos is home to multiple species of crabs, a number of which have diversified to fill ecological niches usually occupied by mammals and birds. In Darwin's time giant robber crabs up to a metre across were common on Cocos, and even today one cannot walk more than a few steps without glimpsing horn-eyed ghost crabs, whose armoured eyestalks bear a disconcerting resemblance to former Australian prime minister Robert Menzies' eyebrows, skittering across the sand, or hearing heavily armoured, purple-eyed land crabs rustling through the leaf litter between the palms.

Even more abundant are the hermit crabs. Hermit crabs, which take their name from their habit of relying upon scavenged seashells to protect themselves from predators, are everywhere on Cocos, crawling across the sandy beaches or through the palms behind them. Some are tiny; others, like the huge purple variety that wears coconut shells in place of seashells, are as big as baseballs.

Despite their ubiquity there is anecdotal evidence that hermit crab numbers on Cocos have declined in recent years. Until recently it was assumed that was due to local fishers using them for bait, but in 2018 Lavers came across evidence that suggested another reason altogether. During a survey on Home Island she came across a plastic bottle filled with dead hermit crabs. At the time she thought little of it, but shortly afterwards she found another container full of dead crabs, and then another, and another. As the bottles and containers multiplied, Lavers and her team realised what was happening. Like many scavengers, hermit crabs are naturally curious and, upon encountering a plastic container, will often crawl inside. If a crab climbs into a container that is angled downwards it is easy enough for it to scramble out. But if the container is angled upwards, the crab's claws cannot grip the slippery surface, making it impossible for it to escape. Trapped inside the container, the crab quickly dies of heat and dehydration.

There are a lot of bottles and plastic containers on the beaches of Cocos, but even if every one of them killed a crab, it would be unlikely to have a significant impact on crab populations. Unfortunately, hermit crabs have another peculiar adaptation: upon dying, their bodies release a pheromone that alerts other crabs nearby that their shell is now available. As a result, as one crab becomes stuck and dies, others quickly arrive and crawl in after it, only to find themselves trapped as well. As they die they too release the pheromone, attracting yet more crabs who also become trapped and die. And so on, and so on.

Although Lavers has explained the phenomenon to me before, it is not until we come across a container on Cemetery Beach filled with the dead bodies of the crabs that its awful reality becomes clear. Emptying the container – a black plastic jar that once held 320 doses of a weightlifting supplement ('micronized aminos!' trumpets the remnants of the label) – Lavers and I find a mess of decaying limbs and claws and a dozen empty shells, ranging from a couple of centimetres in diameter to four or five times that size. In another container shortly afterwards I find another eight, although Lavers tells me this is nothing: she has found containers containing as many as 526 dead crabs.

Over the weeks following the discovery of that first container in 2018, Lavers and her team counted hundreds of containers filled with dead crabs on multiple beaches. Extrapolating from their results they calculated more than half a million crabs were dying in plastic containers every year on Cocos's beaches. And the problem is not confined to Cocos: a similar survey on Henderson Island in the Pacific calculated 61,000 crabs were dying annually there as well, suggesting it is likely to be happening on beaches around the world, and that the total number of crabs dying every year must be far, far higher.

Lavers has recently stumbled on another, potentially even more disturbing side effect of the proliferation of plastic on the world's beaches. 'I was standing on Cemetery Beach and I had this absolute lightbulb moment. I was looking at the layer of plastic that has accumulated there, which wasn't just an object or a scrap here or there, it was an actual biofilm, so wherever you step your foot sinks down into more plastic. And I thought, hang on, there's no way this is innocuous and harmless. It's not just that it's unsightly, it's not that simple. It's literally creating a barrier. I didn't know what that barrier was doing, but I knew it had to be doing something.'

Together with her colleague Alex Bond, at the Natural History Museum in London, Lavers began to study the sand itself, tracking temperature, conductivity, permeability and other measures of soil condition in an attempt to establish whether they too were affected by the presence of plastics. What they found astonished even them. 'We were so utterly stunned by the results that we thought we must have made a mistake somewhere, so we sat there for two days trying to work out where we'd gone wrong, because the results couldn't possibly be as bad as what we had found. But they were. Our work shows the presence of plastic in these sorts of quantities on beaches is raising the temperature of beach sediment by upwards of ten degrees, and often substantially upwards.'

Exactly what effect such a dramatic rise in temperature might have on beach ecosystems is not yet known, but it seems safe to assume it won't be positive. As well as larger organisms such as worms, crabs and seashells, the sand is home to a teeming array of minute creatures – tiny crustaceans, molluscs, echinoderms, nematodes and other animals, many only fractions of a millimetre in size. These organisms are known collectively as meiofauna – a catch-all term for organisms small enough to pass through a 1-millimetre mesh but large enough to be caught by a 45-micrometre, or 0.045-millimetre mesh. And they are vital to the health of beach environments. 'The organisms in the surface layers are so critical,' says Lavers. 'All those teeny-tiny little things – the minute invertebrates and worms and bugs and so on – are fundamental to what we call the turbation or turning over of the soil, or as food for other animals.'

Hotter sand also has the potential to affect larger animals. The sex of green sea turtles is determined by the temperature of the sand in which their eggs are incubated; on some islands in the Great Barrier Reef sex ratios have already been so drastically affected that close to 100 per cent of all hatchlings are female. Lavers and Bond's work suggests the presence of plastics in the sand may exacerbate the effects of global heating upon the species' future.

Declines in meiofauna populations are also likely to affect species that rely upon them for food. This has especially serious implications for birds. As Lavers observes, 'One of the reasons you get hotspots for migratory birds is because those beaches have these meiofauna living in the surface layer.' If the meiofauna is reduced or removed, so are the resources for the arriving birds.

While the effects of the process require further study, Lavers is already concerned by what such research will uncover. 'You throw plastic in the mix and you start changing the temperature and humidity of the whole environment, so it's no longer conducive to the same diversity of life. Basically that means you have a beach that's no longer alive.'

*

THESE THOUGHTS ARE VERY MUCH on my mind on our second morning, when we load our gear into outrigger canoes and set off across the lagoon towards South Island.

Known as Pulau Atas – 'Upwind Island' in Malay – South Island is the second-largest of Cocos's twenty-seven islands, but has remained uninhabited because the lagoon around it is so shallow that even now, close to high tide, the bottom is less than a metre below us. As we shoot along I gaze down into the turquoise water. When Darwin visited these islands 200 years earlier he marvelled at the fields of branching corals that filled this lagoon. Most of that coral is gone now, courtesy of deoxygenation events in the 1980s and 1990s, and while there are still fish here and there, and once or twice a turtle darts away at startling velocity, for the most part it is pale sand that the shadows of the canoes dance across.

On the island we make camp beside the lagoon. After checking the gear has made it across unscathed, we set off to begin the next set of surveys. The beach we are heading for is on the ocean side of the island, and to get to it we must load the gear back into the canoes and travel along the channel that connects the lagoon to the ocean. Once the channel becomes too shallow the canoes have to be unloaded again, and the bags of food and equipment and the 60-kilogram case containing the drone must be carried several kilometres along a beach covered in shifting chunks of coral until finally we reach the beach skirting the seaward edge of the atoll.

When we finally arrive we put down the drone and rest for a few minutes. Breakers are striking a reef several hundred metres offshore; the power of their impact as they detonate against the coral reverberates through the sand beneath our feet, but otherwise it is almost silent, the only sound the movement of the coconut trees in the wind, which blows hard and steady off the ocean. After the disaster of Cemetery Beach, the wind is not good news: if it continues we will not be able to get the drone aloft, and while that will not mean the expedition is a complete failure – there will, if the weather holds, at least be the images from the ISS to work with – it will render a lot of the work so far essentially pointless.

Picking up the drone again we begin to hike along the beach, and as we do it quickly becomes clear why Lavers has brought us here. As on Cemetery Beach, there is plastic everywhere, a sea of refuse that spreads from the water's edge up into the sandhills behind the beach.

After another kilometre or so we stop again, and, unpacking the equipment, mark out quadrants and transects and begin the manual surveys. It is difficult work. Although I am better prepared than the day before, the sheer volume of plastic on the sand is still overwhelming. During a break from the surveys I walk along a strip of beach just past the survey area and make a list of what I can see: bottles, thongs, rope, a child's shoe with the word 'happy' printed on it, dozens of thong cut-outs clearly discarded from a factory, pieces of industrial waste, toothpaste containers, toothbrushes, bottles, tyres, plastic buoys, children's toys, sections of coolant bottles, cigarette lighters, cutlery, inner-sole inserts, as well as the now familiar bottle tops and fragments.

Seeing it spread out like this I am suddenly and forcibly aware of my own implication in this catastrophe, of the fact that almost every object I touch or use – my razor, my car, the soles in my shoes – is part of the same unending torrent of refuse that is choking the planet. Just as the currents that bear the plastic and fallout through the oceans connect the world into one vast network, so too my life is part of the disaster that is unfolding here.

Over the next few days we return to the beach several times, conducting more surveys in more quadrants while we wait for the wind to drop far enough to allow the drone to be used. As we gather data I make my own notes, and stare seaward. This is the beach Darwin came to, the place whose 'grandeur' and 'simplicity' he rhapsodised, and despite the plastic it remains astonishingly, breathtakingly beautiful.

Seated on the sandhills it is possible to see fish moving in the water just offshore, sometimes as silvering schools, sometimes as fugitive shadows that dart by at surprising speed as they follow the line of the beach. One morning I see huge humphead wrasse, a metre long or more, fighting, their blue and green bodies twisting and flashing as they break the water's surface. Further around, on the lagoon side, reef sharks splash and play in the shallows, often only half-submerged, seemingly unconcerned by our proximity.

Finally, on our last morning, the wind drops enough to get the drone aloft. Removed from its case and unfurled it is surprisingly large, its multiple limbs and carapace oddly insectile in appearance. As it lifts off a pair of noddies appear from over the hills behind us, floating on the wind as they watch the drone. Although up close they are a surprisingly dainty bird, the pale plumage around their head and eyes soft and delicate, in flight they are almost sculptural, the outline of their wide wings and long beaks like the strokes of an engraver's evocation of a bird.

After a few moments they glide closer, and then wheel in opposite directions around the drone and disappear back over the hill. Thirty seconds later an entire flock of noddies appears over the hill, as if summoned by the first two. Dipping and surfing on the wind they surround the drone and follow it along the beach, seemingly wary of this strange interloper in their territory.

THIS WARINESS OF INTRUDERS is not limited to the birds. That same afternoon several of us walk over the ridge behind the beach into the trees. Down among the palms we come across a group of air force personnel out on a hike. They are off duty, and dressed causally in shorts and singlets. We stop to speak to them – it turns out they were aboard an RAAF Orion I saw land on the airstrip on West Island the night after we arrived from Australia. But when we ask them why they are here they become evasive and unfriendly. 'Are you out here on some kind of exercise?' I ask one of them. She looks almost physically uncomfortable. 'Something like that,' she says, staring away into the middle distance.

The presence of air force personnel on Cocos is a reminder of the islands' continuing strategic significance. In 1901 a telegraph station was established on Direction Island, in the north of the atoll; this station was to be vital to Allied communications across two world wars, and was the target of the German cruiser SMS *Emden*, which was overpowered by the HMAS *Sydney* off North Keeling Island in November 1914. Thirty years later, during World War II, the islands provided an important base for British aircraft involved in the defeat of Japanese forces in Southeast Asia. In May 1942 they were also the backdrop to a little-known episode that offers a reminder of the larger context of World War II, when a group of Ceylonese soldiers mutinied and attempted to hand the islands over to the Japanese. Their leader, Gratien Fernando, a Sinhalese Marxist who had signed up to fight fascism, saw the action as a way of resisting British colonial power. Three months later he and two of his conspirators became the only Commonwealth soldiers to be executed for mutiny during World War II; Fernando's final words were 'Loyalty to a country under the heel of a White man is disloyalty.'

The airstrip and cable station also played an important part in the considerations surrounding Cocos's transfer of the islands to Australian control in 1955, and while the cable station ceased operations decades ago, the islands have continued to feature in the calculations of military leaders, especially as the focus of American military planning has shifted towards the Indo-Pacific region. A decade ago there was talk of using the islands as a base for US drones capable of monitoring the South China Sea, and more recently the Australian Government has announced upgrades to the runway on West Island that will allow modern surveillance aircraft to land.

The role of the islands as an outpost of empire extends beyond surveillance and communications. In late August 2001, the Norwegian freighter MV *Tampa* answered a distress call from an Indonesian fishing vessel 140 kilometres north of Christmas Island. On board were 433 refugees, most of whom were Hazaras fleeing the Taliban in Afghanistan. The *Tampa*'s captain, Arne Rinnan, took the refugees and five crew members on board and requested permission to land on Christmas Island to unload the refugees and crew. The Howard Government – facing an election and behind in the polls – refused, and when Rinnan eventually did enter Australian waters dispatched the SAS to board the *Tampa*, and prevent it from docking.

In the midst of this standoff the Howard Government put in place what it dubbed the Pacific Solution. First it excised Christmas Island, Cocos and several other Australian external territories from Australia's migration zone, placing anybody who arrived on them without authorisation outside the jurisdiction of Australian law. Then it entered into agreements with Papua New Guinea and Nauru that would see unauthorised arrivals transferred to detention facilities there while their cases were considered.

Played out against the backdrop of the 2001 federal election and al-Qaeda's devastating attacks on the World Trade Center and Washington on September 11, the incident and its aftermath convulsed Australia. While refugee advocates decried the Howard Government's actions, public anxiety about border control and paranoia about refugees and Muslims helped underpin John Howard's unlikely victory in the election that November, and drove a long-term realignment of Australian politics.

Whatever its morality, there is no question the Pacific Solution was brutally effective. Over the next seven years, unauthorised boat arrivals slowed to a trickle. In 2008, the newly elected Rudd Government relaxed the provisions. 'Labor rejects the notion that dehumanising and punishing unauthorised arrivals with long-term detention is an effective or civilised response,' said Immigration Minister Chris Evans at the time.

Over the next few years, boat arrivals increased sharply, partly due to the intensification of the civil war in Sri Lanka. In 2012 alone, seventy-three boats were intercepted on or near Cocos, carrying several thousand people.

For many of the inhabitants of Cocos these were unsettling and upsetting times. 'It was all people knew about us,' one resident tells me. 'And that meant the only people who came here were journalists.' Worse yet was the knowledge that people were dying. 'For every two or three boats that made it there was one that didn't,' says another, describing finding a child's backpack lying on a beach amid fragments of wreckage. She looks away before continuing. 'One day I found a lifejacket with 50,000 rial tucked into it on a beach on South Island. Later thirty-eight more lifejackets from the same boat washed ashore.'

In response to the ballooning number of arrivals the Pacific Solution was reinstated, this time with a proviso that not only would unauthorised arrivals be denied the protection of Australian law, they would never be permitted to set foot on Australian soil. The human cost of that decision has been incalculable.

The cruelty and inhumanity of these policies remains a source of anger and distress for many in Australia more than twenty years after the *Tampa*. The wars and violence from which the people aboard the boats that foundered off Cocos and other islands are, almost to a one, legacies of colonialism, or wars driven by the desire of the great powers to secure their energy supplies. Similarly, the preparedness of Australian politicians to use the suffering of these people – almost exclusively people of colour – as political capital, and of Australia as a nation to avert its gaze from the cruelty of the camps on Nauru and Manus, are an expression of those same processes and the logic that underpins them, a logic that also lies at the heart of the climate crisis and its unequal effects.

ON MY FINAL DAY we cross the lagoon back towards West Island. To the east huge thunderclouds mass, dark against the cerulean blue of the sky, marking the edge of a cyclonic low that has been building between Cocos and Christmas Island over the last week.

As the canoes glide across the shallow water the mood is buoyant. After the successful deployment of the drone the day before it feels as if the team's luck is turning. While the others laugh and call between the canoes I stare at the islands in the distance. From here they look pristine, untouched. But they are not, and not just because of the plastic strewn on their beaches.

When Darwin visited almost 200 years ago he observed that the trees were crowded with 'gannets, frigate-birds and terns'. Some of that birdlife remains: as well as the noddies on the seaward side of South Island and the rails that dart here and there on the airstrip on West Island, I have seen five frigatebirds moving slowly along one of the beaches, and a solitary tropicbird, its magnificent tail trailing behind it. But for the most part birds have been noticeable by their absence during my time on the islands.

The reasons for this are both complex and simple. Unlike the birds on the uninhabited Keeling Islands to the north, the birds on Cocos have suffered from the destruction of the native vegetation and the harvesting of their eggs by human beings. As a National Parks officer says to me at one point, 'Once the nesting sites are lost, they're just gone.'

This sense the world is empty of birds is curiously dislocating, a silence one cannot ignore once made aware of it. But it is also suggestive of a larger absence. Prior to the arrival of humans the islands seem to have been covered with a mixture of coconut palms and several species of softwood and hardwood trees. These other species were mostly cleared to make way for the coconut plantations that were the basis of the islands' economy when the Clunies-Ross family ruled here. Since the copra business collapsed the plantations have reverted to forest, although on satellite images the rows in which the palms were planted are still faintly visible, spectral traceries against the riotous green.

THESE GHOSTLY TRACES of the past are all around us and within us. As I write this the Anthropocene Working Group of the International Union of Geological Sciences is inching towards a formal declaration that the planet has entered the new geological era dubbed the Anthropocene, in which human activity has become the defining force shaping Earth's processes and systems.

Less than a century ago the idea humans might have been able to alter the planet so profoundly would have seemed risible. Now, though, we know better. The oil, coal and gas extracted from the Earth's crust have altered the makeup of the atmosphere, raising global temperatures, changing the weather and altering the motion of the currents. The mass of the objects made over the past century and a bit outweighs the biomass of every living thing on Earth combined. Human waste is heaped beneath the surface in land-fills, drifts down into the ocean's sediments and layers itself into the shrinking ice sheets and glaciers, and industrial agriculture has disrupted the planet's phosphorus cycle. Chemicals released by industry have eaten away at the ozone layer that shields the planet from the Sun's ultraviolet radiation. Humans are even the most powerful geological force on Earth, moving twenty-four times more sediment than all the world's rivers, and extracting so much sand to make concrete that it would be possible to cover the entire planet in a 2-millimetre-thick shell.

For the designation of a new epoch to become official, it is necessary to identify a specific site at which these transformations are given physical form. Such locations are marked with a bronze plaque, or 'golden spike', on which a line marks out the boundary. In the case of prior chapters in our planet's history, these transitions are relatively easy to identify: the transition from the Precambrian to the Cambrian is marked at a site in Newfoundland where it is possible to see a fossil of a complex organism; similarly the transition from the end of the Cretaceous to the Paleogene is marked by the presence of iridium connected to the impact that wiped out the dinosaurs in a reddish layer of clay in Tunisia.

Identifying a marker for the beginning of the Anthropocene has been more complex. Many have argued the obvious choice is the unmistakable fingerprint of the radioactive isotopes produced by nuclear blasts that appears in sediments and ice cores and corals in the years after 1945, or the sudden spike in levels caused by testing in the Pacific in the early 1950s. This has the advantage of coinciding with the rapid acceleration of the impact of human activity on the planet's systems in the years after World War II. Others have suggested the sudden proliferation of plastic, or phosphates, or a particular kind of ash known as spheroidal carbonaceous particles that is produced by burning oil and coal at high temperatures, and which begin to appear in large quantities in places as far apart as Greenland and Antarctica during the post-war industrial expansion. Still others have proposed the bones of the tens of billions of chickens that humans consume every year.

In mid-2023 the Anthropocene Working Group selected a sediment core from Crawford Lake, in Ontario, Canada, as its choice for the location of the Anthropocene golden spike. A sinkhole lake, Crawford's sediments contain traces of plutonium from nuclear testing and specks of fly ash produced by the burning of fossil fuels over the past three quarters of a century.

There is no question both these markers capture moments of transition. But they also obscure as much as they reveal. The Anthropocene did not begin with the first nuclear blasts, or even the rapid uptake of fossil fuels in the nineteenth and twentieth centuries. Instead its roots lie far deeper, in the shift in the relationship with the natural world that began in Europe hundreds of years ago, and which gave birth to the process of violent exploitation and extraction and movement of resources and people across the oceans that has driven the transformation of the planet.

It is for this reason that the scientists Simon L. Lewis and Mark A. Maslin have suggested the sudden collapse in atmospheric carbon dioxide levels caused by the deaths of tens of millions of Native Americans following the arrival of Europeans in the Americas just over 500 years ago as an alternative marker. They argue that 'placing the Anthropocene at this time highlights the idea that colonialism, global trade and the desire for wealth and profits began driving Earth towards a new state'.

For others even the term is problematic, suggesting responsibility for this process is shared equally by some abstract 'Anthropos'. In its place researchers have offered a rapidly growing list of alternatives, ranging from the Capitalocene to the Thanatocene, the Plasticocene and the Eremocene, or Age of Loneliness. Yet among the most suggestive is one coined by the philosopher Donna Haraway and the anthropologist Anna Tsing in 2016: Plantationcene. In an echo of Raj Patel and Jason W. Moore's argument that capitalism was not born on the factory floor but on the plantation, Haraway argues that emphasising the role of the plantation allows us to see the broader processes at work. 'When I think about the question, *what is a plantation*, some combination of these things seems to me to be pretty much always present across a 500-year period: radical simplification; substitution of peoples, crops, microbes, and life forms; forced labour; and, crucially, the disordering of times of generation across species, including human beings.'

Whatever we call it, though, the reality is that the traces of this transformation are not just etched into the planet. A fine haze of microplastics now fills the atmosphere and oceans, drifting on the winds and carpeting the ocean floor. We inhale and ingest tens of thousands of these particles every year; once in our bodies they disperse, leaching into our blood and organs, crossing the membrane that is supposed to protect our brains from poisons, even gathering in breast milk to be passed into the bodies of the newborn. Like the radiation that permeates the bodies of the Marshallese, this toxic accumulation is not just biological – it contains within it the history of violence and dispossession that has shaped our world. The marker, in other words, is within us, written into the very fabric of our bodies, as surely as the burning of fossil fuels is recorded in the rain of ash onto the ice sheets or the vertiginous rise in global temperatures.

These thoughts come back to me as the plane ascends and the islands disappear into the immense blue of the ocean, and along with them the memory of a moment on my second day. Exhausted by the heat and the sickening sea of plastic, I took a break from the survey on Cemetery Beach and walked up into the graveyard behind it. Walking the hundred metres or so to the other side of the island I gazed out over the lagoon. The sand was gleaming white, and the water lapping against the shore was crystal clear, but the shore itself had been replaced with a wall of sandbags almost 2 metres high to slow down erosion. Just offshore, a huge orange and white Border Force ship sat at anchor, a reminder of the flow of people from across the water.

Turning back inland I made my way through the graves. Most were of local Cocos-Malay people, each of them a small, fenced plot about a metre wide and a metre and a half long, within which sat one or two small wooden grave markers. Where there were two, the husband's sat at the front, with the wife's behind it encircled with a headscarf – as if to ensure modesty in the afterlife. But beyond them, at the island's northernmost point, I came upon another group of graves bearing the names of members of the Clunies-Ross family reaching back almost two centuries. Carved of basalt and granite that must have been shipped in from elsewhere, they had once been grand things but were now slowly collapsing into the sandy soil. I stood there for a long time, listening to the sound of the wind in the palms, the waves on the reef. And at the confluence of those lonely graves, the baleful bulk of that ship offshore and those two beaches, one choked with plastic, the other slowly collapsing before the inexorable power of the rising sea, I was suddenly, inescapably aware not just that the past has not ended, but of the way its presence shadows the future that is already taking shape around us.

10

REEF

'. . . the future
takes shape
too quickly.'
Jorie Graham, 'Sea Change'

IN MARCH 2016, Professor Justin Marshall was working at the Lizard Island Research Station on Jiigurru 250 kilometres north of Cairns. A neurobiologist specialising in the visual ecology of marine animals, he was there with a group of students as part of a project exploring the role of colour and vision in shaping the inner worlds of marine animals.

The area around Jiigurru is one of the jewels of the Great Barrier Reef, a place surrounded by vibrant and thrillingly diverse reefs, each of them home to complex communities of coral, fish and hundreds of other species. It is also a place Marshall knows well and cares deeply about, having spent decades diving and snorkelling there.

As March began, Marshall was apprehensive about what the weeks ahead might hold. For months, water temperatures in the Pacific had been at an all-time high, and reefs in Hawaii and elsewhere had already experienced significant bleaching events. Meanwhile along the Queensland coast, where ocean temperatures had shattered records dating back more than a century, signs corals weren't coping had been visible since November the year before.

'You could feel it in the water,' Marshall says. 'It was so hot, and the water was unnaturally still.' Marshall had seen the effects of elevated ocean temperatures before and was familiar with the effects of bleaching, but what he saw when he went into the water on 16 March was like nothing he had seen before. 'It wasn't just that the coral bleached. Coral can come back from that. Instead, it just boiled.' Eight years after that hot summer, Marshall takes a breath, trying to compose himself. 'The polyps just sloughed off, so the water was like a fluorescent yellow soup of dying coral. Everywhere you looked there were dead bodies. It was like swimming in pus.'

The catastrophe Marshall witnessed that day wasn't confined to the waters off Jiigurru. Over the weeks that followed more than 90 per cent of reefs spread along the 2300-kilometre length of the Barrier Reef bleached. The northern part of the Reef, stretching from Port Douglas to New Guinea, which had previously been regarded as the most pristine, was hit particularly hard: by the time the heat finally began to dissipate more than 80 per cent of reefs along its 700 kilometre length had been severely bleached, and more than two thirds of the corals were dead. The extent and severity of the event left the Reef transformed, permanently altering its structure and the mix of species in many areas. 'We've never seen anything like this scale of bleaching before,' said Professor Terry Hughes, then head of the National Coral Bleaching Taskforce. 'It's like ten cyclones have come ashore at once.'

Marshall's voice still shakes when he describes the experience. 'It was awful. When you got out of the water it was all over you, so you reeked of death, and the Reef stank of death, and the fish that were still there looked stunned and confused.' Later, still deranged by grief, he declared to reporters, 'This is the most devastating, gut-wrenching fuck-up.'

ONLY FORTY YEARS AGO, the events of 2016 would have been unimaginable. Coral reefs are the most complex ecosystems on Earth, fantastically diverse communities in which a profusion of species thrive. Diving on one is an extraordinary experience: life and energy ripple through their structure, crowding into every space and at every scale. Where the reefs rise from the ocean floor their surfaces are crowded with the massing shapes of corals and sponges, some plate-like, some like boulders, some a profusion of branches. Swim close to them and you will find shrimp, crabs and other crustaceans crowded into the nooks and crannies between the corals alongside worms and molluscs and countless other invertebrates; drift further out and gaze across the Reef and you will see anemones and urchins and other even more peculiar invertebrates clinging to the walls and slopes while clouds of vividly coloured fish hunt and graze above them.

The cornerstone of this extraordinary system is the corals themselves. Although to our terrestrial eyes corals look like plants, they are in fact colonies of tiny animals, or polyps, that are related to anemones and, more distantly, jellyfish. Although their design varies from species to species, individual polyps resemble their anemone cousins. Their bodies are rooted at the bottom and open at the top into circular mouths surrounded by tiny tentacles. These tentacles are covered in stinging nematocytes, which the corals use to catch zooplankton and any other suitable animal that makes the mistake of wandering within reach (some corals will happily dine on small fish).

A few species of coral rely entirely on the food they catch with their tentacles for subsistence. But most survive by virtue of a symbiotic relationship with photosynthetic microalgae, or zooxanthellae, which the polyps secrete in their tissues, and which give the corals their distinctive colours. The polyps supply the zooxanthellae with the carbon dioxide and nitrogen they need to fuel their photosynthesis; in return the zooxanthellae produce food in the form of glucose and other sugars and amino acids. This process is startlingly efficient: coral reefs are among the most productive ecosystems on Earth, and healthy corals produce energy at many times the rate of most terrestrial plants.

The zooxanthellae also provide the corals with calcium carbonate (the same material that molluscs and other animals use to create their shells). The corals secrete this to create the hard skeletons that give them their distinctive shapes; as these accrete over thousands of generations, they gradually create reefs and islands.

Coral bleaching occurs when the polyps become stressed and expel the zooxanthellae. It is not immediately fatal, but because the polyps depend upon the zooxanthellae for food, bleached corals are at grave risk of starvation and disease.

Although a variety of factors can cause corals to react in this way, the most common is extended exposure to elevated water temperatures. Until the 1970s, such conditions were rare, and so while there were occasional episodes here and there, they were highly localised and transitory. But as ocean temperatures started to tick upwards as a result of global heating, larger bleaching events began to occur. One of the first of these was in 1980, when reefs in the Caribbean and parts of the Pacific suffered significant bleaching. This was followed by a second major bleaching event on the Reef in early 1982, and smaller events in 1992 and 1994.

Then, across 1997 and 1998, something entirely new occurred. As an intense El Niño raised ocean temperatures, reefs around the world began to bleach. On the Great Barrier Reef 14 per cent of outer reefs and 67 per cent of inshore reefs suffered severe bleaching, and a quarter suffered extreme levels of bleaching: more than 60 per cent of corals were affected. Elsewhere the effects were even worse: reefs in the Maldives, Sri Lanka, Singapore and Tanzania were almost wiped out, and severe bleaching was recorded on other reefs from Kenya to Brazil. By the time the event was over almost a quarter of the world's coral was dead.

This pattern was repeated in 2002 when mass bleaching again struck the Great Barrier Reef. This time, though, there was a new and even more frightening twist. In 1998 bleaching had been the result of an El Niño supercharging the background warming. But by 2002 the additional energy of El Niño was no longer needed: the water was just hot. And if 1998 and 2002 were bad, 2016 was next level. By the time it was over, more than 20 per cent of the Great Barrier Reef was dead. And whereas in previous bleaching events there had been a spectrum of winners and losers, this time the heat was so extreme that on many reefs it killed even the hardiest corals.

Then in 2017 it happened again. This time it was the reefs in the central section of the Great Barrier Reef that bore the brunt of the event. By the end of this bleaching, it was clear the Reef was now in desperate trouble, leading some to describe it as 'terminal'. Across the summers of 2018 and 2019, researchers nervously monitored weather reports, terrified of what another event might mean. Thankfully the Reef got lucky, but in 2020, as the first wave of the COVID-19 pandemic washed across the world, it happened again. Although it attracted significantly less attention because of the pandemic, 2020 was second only to 2016 in severity, and left 25 per cent of reefs up and down the entire 2200-kilometre expanse severely affected. In 2022, the Reef was struck by its fourth mass bleaching event in just seven years.

That year felt different, and not just because it took place in a La Niña year, which would usually be associated with cooler ocean temperatures and should have offered the Reef a respite. Instead its proximity to the host of other climate-related disasters that began in the early 2020s made it feel as if a breaking wave had swept across the world, bearing the certainties of the past away with it. More disquietingly again, though, it made clear what should have been apparent all along, which is that the escalating crisis on coral reefs is a terrifying preview of the future that's rapidly overtaking the planet as a whole.

OVE HOEGH-GULDBERG was one of the first to recognise the scale of the threat rising temperatures posed to coral reefs. These days Hoegh-Guldberg is Professor of Marine Studies at the University of Queensland and chief scientist for the Great Barrier Reef Foundation, but back in 1998 when the first major bleaching event occurred, he was a researcher at the University of Sydney. Concerned by the potential for more bleaching events, he began trying to model what might happen if the planet continued to warm in the ways that were then predicted. What he found shocked him. 'At first I didn't believe the results I was getting, and assumed I must have made some kind of mistake. But when I went back and recalculated I got the same results. When you took the potential trajectories and applied them to multiple places on the Great Barrier Reef and multiple places around the planet you always came to the same conclusion. By 2040 or 2050 reefs would be suffering annual, back-to-back bleaching events, caused by marine heatwaves hotter than anything we see now. And that would devastate reefs.'

When Hoegh-Guldberg published his results in 1999, some in the coral research community accused him of being alarmist. 'A lot of my colleagues piled on me for being doom and gloom. There was a real attitude that I must have made a mistake. But my response was to ask they show me where that mistake might be. Because it's a very simple story. Corals can only survive to certain limits. If temperatures rise past those limits then we'll lose them.'

The two-and-a-bit decades since Hoegh-Guldberg's paper have made it clear that his predictions were not alarmist – if anything, they were too conservative. 'What we're realising is that the climate's sensitivity to increased levels of greenhouse gases is much greater than we expected, so things are warming faster than we thought they would back in 1999. Over the past six years we've had four years in which we saw mass coral bleaching and mortality. So what's happening today looks very much like the kind of rolling, back-to-back events I originally predicted we wouldn't see until 2040 or 2050.'

Charlie Veron is one of the world's leading experts on coral reefs. Now in his late seventies, he is responsible for identifying more than 20 per cent of all known coral species. Warm, generous and possessed of the peculiar combination of curiosity and fearlessness that characterises many of the most original scientific minds, his passion for coral reefs – and sense of urgency about their plight – comes through with every word he speaks about them.

Veron first encountered mass bleaching in Japan in the 1990s. 'It was so dramatic that I thought somebody must have tipped a whole lot of chemicals into the water. But I was working somewhere remote where that couldn't possibly have happened.' Up until then he'd been focused on the threats posed by pollution and mismanagement of reefs, and hadn't seen rising temperatures as a big problem. But the reef in Japan changed that. 'Alarm bells rang, because I could see what was happening – and it was serious.'

Despite his focus on the challenges they face, there is delight in Veron's voice when he describes a living reef: 'There's nothing remotely like it anywhere else. There are things happening absolutely every-where. There are fish all over the place, the corals and their colours, but also all sorts of molluscs and crustaceans, and at night there are just masses of little critters and weirdo things that are attracted to your torchlight. And it's so noisy! All these animals have something to say to the others. If you stop breathing for a moment, you can hear the racket all around you.'

His voice slows: 'That's a hell of a contrast to what you see after a bleaching event. All those animals, they have their food and their shelter disrupted.' He pauses. 'A freshly bleached reef is really beau-tiful,' he says. 'It's like a surreal sculpture where everything is white, or bluish-white. But that beauty only lasts for a few weeks. Then the macroalgae begin to take over and the corals start falling apart, and what was a vibrant coral reef becomes a bed of algae. It's silent. It's dead. It's boring.'

Death and destruction are a natural part of life on a coral reef. Storms and cyclones and abrupt changes in salinity can all damage them severely. Healthy reefs can recover from these sorts of distur-bances because most corals reproduce by spawning. On the Great Barrier Reef this takes place in late spring, usually just after the full moon in late October or early November. In a massive, synchronised event across a single night, billions of corals up and down the Reef's 2300-kilometre length release bundles of eggs and sperm, filling the water with a milky soup. As it drifts upwards this soup spreads outwards on the currents, dispersing the larval polyps so they can colonise new locations and regenerate damaged areas.

But while the effects of disturbances such as cyclones tend to be relatively localised, mass bleaching events are spread across hundreds or even thousands of kilometres, making it harder for corals to repopulate. Furthermore bleaching events tend to affect reefs in a more uniform way, meaning it is less likely there will be patches of reef that survive relatively unscathed scattered amid the destruction: after Cyclone Yasi hit the Reef in 2011, the average distance between relatively undisturbed areas and the severely damaged sections of the Reef was only 17 kilometres. After the 2017 bleaching event the average distance between severely bleached reefs and only mildly affected reefs was more than 100 kilometres. This significantly affected the Reef's ability to regenerate: in the aftermath of the 2016 and 2017 bleaching events the rate of coral recruitment – or the number of larval corals that succeeded in attaching themselves to reefs – fell by almost 90 per cent.

The devastation wrought by the mass bleaching events since 2016 also reflects a change in the way heat is affecting the Reef. In earlier bleaching events corals tended to die slowly from starvation after they expelled their zooxanthellae. But in 2016 and subsequent events the water was so hot it simply killed the polyps. 'On the hottest reefs the corals didn't die of starvation, they just cooked,' says Terry Hughes, who believes the tendency to focus on the immediate effects of bleaching can obscure the longer-term devastation. 'Mortality during heatwaves is much more complex. You get instant deaths because the coral tissue itself melts away. And you get slower death if the algae don't come back. But you also find that even if the coral survive the initial event they're much weaker, which means the incidence of coral disease goes through the roof. And you get much higher predation on surviving corals because crown-of-thorns starfish and coral-eating snails like *Drupella* are much less affected by the heat than coral. So if you're down to 10 per cent of corals but the number of predators is still the same, the surviving corals are subject to ten times the predation. And these processes all have different timelines. While the cooking might take a couple of weeks, the starvation might play out over three to six months, and the disease and predation over a year or more. So the mortality from these bleaching events just keeps rolling.'

But as Hoegh-Guldberg realised in 1999, the real threat to reefs isn't isolated incidents of bleaching but the cumulative impact of multiple events that progressively erode the diversity and resilience of the system as a whole. This process is already visible on the Great Barrier Reef – even before the 2022 event only 2 per cent of the Reef had escaped being bleached, and coral recruitment had declined by more than 70 per cent.

Charlie Veron puts it this way: 'I've seen some reefs go through three cycles of bleaching and recovery. But others don't recover. They go into a phase of ecological collapse. It's like burning a piece of rainforest. Before it burns it will be a fabulous place, full of life. If it burns, you'll lose some of that, but some will come back. If you burn it again, though, it won't recover as well. And if you burn it another time it won't recover at all. It just becomes a weed patch.'

Dr Mike Emslie, who heads up the Australian Institute of Marine Science's Long Term Monitoring Program, compares the Reef to a prizefighter. 'You can get slugged and knocked to the ground and keep getting back to your feet, but at some point you've been knocked down too many times, and you can't get up again.'

*

AGAINST THIS BACKDROP the outlook for coral reefs appears bleak. More than half the reefs around the world have already been lost, wiped out by human activity or devastated by bleaching and disease. And as the chance of holding global heating to 1.5 degrees slips away, that rate of change looks set to accelerate rapidly: one recent study found that at just 1.5 degrees of warming, 99 per cent of reefs worldwide will be exposed to frequent thermal stress, while the IPCC predicts that at 2 degrees reefs will almost entirely disappear, with devastating implications for the oceans as a whole.

Nonetheless there are those who believe it may be possible to find a way to help coral reefs survive. Nestled amid humped granite hills and tropical forest, a bit over half an hour's drive from Townsville, is the headquarters of the Australian Institute of Marine Science (AIMS). At its heart is the National Sea Simulator, or SeaSim – a purpose-built suite of seawater experimental research aquaria spread across 1500 square metres. Capable of pumping around 1.5 million litres of seawater a day, its industrially calibrated systems allow researchers to delineate water temperature, salinity, acidity and other factors with razor-sharp precision. This makes it possible to simulate conditions on the Reef both now and in a potentially much hotter future, and – more importantly – to explore strategies that might help coral and other marine organisms survive in a radically altered world.

On the spring morning when I visit it is hot and bright, the air outside full of the cries of friarbirds and other tropical species. Inside it is cool and quiet, save for the occasional shriek of power tools from construction work at the far end of the facility. The manager of the SeaSim, marine biologist Craig Humphrey, meets me at the entrance. After a quick tour of a couple of smaller labs and one of the plant rooms responsible for regulating conditions in the tanks, he opens a sealed door and we step into the SeaSim's nerve centre.

We are standing on a platform above a large space in which dozens of shallow, open-topped tanks are arranged in rows. The light is low and diffuse, and the air is full of the sound of gurgling water and the low thrum of pumps. Humphrey leads me down towards ground level, and I stop in astonishment. At the bottom of the stairs a huge aquarium is built into the wall, its glass front almost two metres high. Inside is an artificial reef made up of hundreds of corals. At the front clownfish dart between the purple-tinged tentacles of a magnificent sea anemone; behind it, cobalt-blue surgeonfish and orange damsel-fish shoot here and there among the brainlike forms of *Platygyra daedalea* and convoluted plates of *Montipora aequituberculata* that rise up the back of the tank in a wall, interspersed here and there with the delicate shapes of branching corals. Rather like the experience of diving on a reef for the first time, the effect is almost overwhelming, a riot of impossibly vivid colour and movement.

Next to me Humphrey laughs. 'There are about sixty species of hard coral in the Caribbean. Last time I counted this tank had sixty-three species. So this tank is as biodiverse as the entire Caribbean in terms of coral species. And what's in there is just a fraction of what you see out on the Reef.'

When I finally tear my eyes away I turn to the first of several smaller tanks to the left. In one a handful of living corals sit in an algae-choked wasteland, lonely outcrops of colour in a mass of scurfy green. The contrast to the main tank is stark. 'The Reef is going to continue to exist in some form,' says Humphrey. 'But it certainly won't look like it does today. This is one version of the future, one where you've seen a phase shift from a coral-dominated ecosystem to an ecosystem dominated by macroalgae.'

Leaving behind the display area Humphrey leads me towards the white-bottomed experimental tanks that hold serried rows of branch-ing corals or racks containing multiple fragments of coral, some bathed in bright white light, others glowing in the purplish gloom of ultraviolet. In one, several stonelike *Goniastrea* have extruded dozens of translucent tentacles that dance about them in the water. 'They use the tentacles to sting other corals, so we try to keep them separated from other species,' Humphrey says. In the next tank a pair of yellow surgeonfish swim among the racks of coral: they are used to help control algae. 'They're probably our worst-paid staff,' Humphrey laughs. 'But also our most efficient.'

For the surgeonfish the physical conditions in the tank are likely to be almost indistinguishable from those they would experience in their natural habitat: the SeaSim's systems are designed to emulate not just environmental factors such as temperature and salinity, but also natural sunlight conditions and the rhythmic daily fluctuations in acidity that occur on the Reef. This ability to mimic real-world conditions while also exercising total control over other factors results in higher-quality data. But it also makes it possible to undertake longer, multi-year studies that operate on generational time-scales. 'Climate change is a gradual process, so the corals which will encounter the changed conditions aren't the corals of today, but their offspring. That means the ability to look at multiple generations and see how they will respond is really important.'

Evolutionary ecologist Dr Line Bay is one of the researchers utilising the SeaSim's capabilities. Originally from Denmark, Bay fell in love with the Reef when she visited Australia as a backpacker in the 1990s. Now one of the team leaders on the multi-institution Reef Recovery, Restoration and Adaptation Program, the world's largest program to future-proof coral reefs, Bay's research is concentrated on trying to understand the genetic and environmental factors that underpin heat tolerance of corals. She is particularly focused on the question of why corals from one part of the Reef are able to survive temperatures that would kill corals of the same species that live in cooler waters. 'We know that corals are adapted to the temperature regime of the region they come from. So corals from warmer reefs, like inshore reefs or the reefs in the northern part of the Great Barrier Reef, have a higher heat tolerance than corals from cooler reefs because they're adapted to that environment. That means that if you could take corals from warmer reefs and move them to cooler but warming reefs, then it might be possible to speed up those natural adaptive processes.'

This kind of genetic flow already occurs naturally across large sections of the Reef as larval corals released during annual spawning events spread out to colonise new areas. But because the prevailing currents on the southern and northern sections of the Reef flow in opposite directions, there is little transfer between north and south – meaning the corals adapted to the warmer conditions that prevail in the north don't tend to spread south.

Bay has been investigating the possibility of artificially assisting this process, either by physically transplanting corals from northern regions to the south, or by using the SeaSim to produce large numbers of heat-tolerant larvae that can be released into cooler waters. The results have been encouraging. In 2017 AIMS scientists collected corals that had survived the bleaching events of 2016 and 2017 from the far north of the Reef and transported them 1000 kilometres south to Townsville, where they were placed in the SeaSim and bred with corals from cooler waters that had also survived earlier bleaching events. This procedure not only produced healthy offspring, but those with at least one parent from the hotter waters to the north were thirteen to twenty-six times more likely to survive when subjected to elevated temperatures. The team later placed hundreds of these selectively bred juveniles on the Reef itself, and found they thrived in the cooler waters of the central reef.

Together with new techniques for identifying heat-tolerant corals and increasing success in encouraging corals to spawn in nurseries, results like these suggest that assisted gene flow has the potential to speed up the natural rate of adaptation, effectively buying the Reef some of the time it so desperately needs to adapt to a rapidly heating world. But it also opens the door to other, more direct interventions. Madeleine van Oppen is one of the pioneers of this growing field. An ecological geneticist, she became fascinated by the use of genetics in ecology after using early genetic sequencing technologies to understand the phylogeny of marine algae as a student in the Netherlands in the 1990s. 'It made me realise the power of genetic data, especially in combination with other ecological data.' Now a professor at the University of Melbourne and a senior principal research scientist at AIMS, van Oppen and her team have been investigating the techniques to accelerate natural evolutionary processes.

One avenue van Oppen and her team are exploring involves hybridising different species of corals to produce more resilient varieties, in the same way farmers crossbreed strains of wheat and other crops. Although this work is still at an early stage, experimental crossbreeding of different species of *Acropora* yielded young that exhibited higher growth rates, fertility and survival rates when exposed to conditions simulating likely future temperatures and ocean acidity on the Reef.

An even more promising scheme does not involve the coral polyps but their algal symbionts. One of the ways wild corals respond to rising temperatures is by changing the mix of symbiont species they hold within their tissues, or by taking up entirely new species of zooxanthellae. Known as 'shuffling' or 'switching', this process can have a significant effect on the heat tolerance of the host corals.

With this in mind, van Oppen and her colleagues inoculated corals in their crossbreeding study with three different species of microalgae. This had a significant effect on the subsequent ability of the corals to survive in warmer water, with those inoculated with one particular species – *Durusdinium trenchii* – bleaching considerably less when exposed to warmer water, especially when the host coral had at least one parent from the northern section of the Reef.

Experiments in which the naturally occurring symbionts are replaced with strains of microalgae that have been 'hardened', by removing them from their hosts and subjecting them to high temperatures over multiple generations, have also shown extremely promising results.

Other possibilities van Oppen has been exploring focus on the use of bacterial probiotics to boost the health and resilience of corals. But work in this area has been slower, partly because our understanding of the coral microbiome remains limited. 'The bacterial communities in corals are highly diverse,' says van Oppen. 'And while we know those communities play an important part in nutrient cycling and protecting the corals from microbial pathogens, we don't have a good insight into what individual species of bacteria are doing. That makes it difficult to pick species for manipulation.' While there have been some encouraging results, there are also significant hurdles, not the least of which is the challenge of stably integrating the probiotic bacteria into the coral microbiome. 'If you have to reinoculate the coral every few days that's never going to be economically feasible in the field. It's just too expensive.'

Whether such interventions will work outside the lab is also still unknown, although small-scale tests of the modified symbionts are underway. Nonetheless some in the scientific community question their applicability in the real world. 'The Great Barrier Reef is the size of Italy,' says Hoegh-Guldberg. 'So while there are many beautiful projects growing corals, once you do the numbers they really fall short of the scale that you need to significantly impact the future of the Reef.'

Terry Hughes is also sceptical. 'Let's say you raise 3000 heat-adapted corals in a tank. That would probably take five years and cost millions of dollars, because you'd need a huge facility and multiple technicians. So where do you put those corals? Do you put one each on 3000 reefs? Or do you put all 3000 in a 100-square-metre area on a single reef? And either way, do you think that's going to change the gene pool of the wild coral population, which is orders of magnitude larger?'

Hughes' objections aren't just logistical. Although he concedes the chances of it happening are low, he is also worried about the potential for engineered corals to supplant wild corals. 'There's a nightmare scenario where we produce a weedy coral or something similar and it gets out of control.'

For Hughes this is part of a larger concern about the ethics of such projects, and in particular about who gets to make the decisions about which corals should be manipulated and where they should be placed. 'Who gets to decide the future mix of the Great Barrier Reef? The minister for the environment? Traditional owners? Some scientist who works for CSIRO? And what are the governance structures around that process?

'People love the concept that clever scientists are going to save the Great Barrier Reef, but it's an idea we have to be really careful of, because it can be used as a smokescreen, a way of convincing the public that they don't need to worry about global warming because we can regrow the Reef. The media are a big problem, because they serve up stories about brightening clouds and breeding super-corals, but without ever talking about the risks or the governance issues. But fossil fuel companies love coral-reef restoration as well, because they can use it in their greenwashing campaigns. You know, "BHP is saving the turtles" and that sort of nonsense.'

Van Oppen believes such objections misrepresent the nature of the work she and her colleagues are doing. 'The experiments we do in the lab are really small-scale and focused on the nitty-gritty, because they're about providing proof of concept. But once we're confident that crossing the comparatively heat-tolerant individuals from a reef with other individuals will produce offspring that also have an enhanced tolerance, then we can do that in bulk. We don't have to separate sperm and eggs from each colony and do individual crosses,

and so on. We can just bring in these corals once, use the rapid pheno-
typing that Line Bay does to assess what their relative tolerance is,
and then just let them spawn.'

Bay also believes it is possible these processes will be ready to
be deployed at scale sooner than many understand. 'We're making
advances in the technology we use to breed corals in captivity and to
settle them onto the Reef in a cost-effective manner. That's now at a
point where we're getting ready to do much larger field trials, placing
tens of thousands of heat-tolerant corals onto the Reef. And that's just
the start. We'd like to be able to put between one and ten million corals
onto the Reef every year.' Veron agrees. 'The process of artificially
introducing larvae has been proved beyond any doubt. So the notion
you can't do it at scale is just bullshit. Because you might not be able to
repopulate the whole thing in a year, but perhaps you can in ten.'

Bay also sees her work and that of her colleagues at AIMS as
part of a larger process. 'In other parts of the world they've shown
that even at local scales restoration and adaptation can have massive
ecological, social and financial benefits, big impacts on local communities and tourism and coastal protection and so on. So I don't think
that just because you can't do this stuff at scale over the entire Great
Barrier Reef you should just pack up and go home. I think we'll get
there over time.'

Yet as the pressures on the Reef build, some are looking to even
more radical solutions to buy its ecosystems time to adapt to a hotter
planet. A 2019 report commissioned by the Australian Government
explored the possibility of protecting sections of the Reef from
extreme temperatures by constructing shade structures and pumping
cold water from deeper in the ocean, as well as developing techniques
to stabilise the physical structure of damaged reefs to promote coral
recruitment. Another proposal involves using fleets of ships equipped
with mist machines to create plumes of vaporised water that would
brighten clouds and increase their ability to reflect sunlight back

upwards into the atmosphere. A trial of this scheme in 2021 produced encouraging results but also attracted considerable controversy, not least because of fears the use of such geoengineering technologies might undermine action on emissions.

Meanwhile, others are moving ahead with the creation of coral arks capable of preserving living corals and their genetic material. This opens the possibility of using the preserved material to repopulate coral reefs in the future once global temperatures stabilise. It also creates a repository of genetic material capable of supporting the work of Bay and van Oppen and others like them if reefs continue to collapse.

There is something difficult to accommodate about this idea, a sense in which its recognition of the scale of the transformation taking place around us dissolves the scaffolding of denial that usually surrounds our thinking about the climate crisis. But it also requires us to think differently about time and our relationship with the future. As Charlie Veron, who has been instrumental in helping set up one of the most important repositories – the Great Barrier Reef Legacy's Living Coral Biobank – says, 'If corals go completely extinct, then it's all over for them. But if they're kept alive in the biobank, then we're able to support the kind of work they're doing at AIMS, and other work in the future. Because if clever people like Madeleine [van Oppen] can come up with a way to create a temperature-resistant cultivar, then it's a short step to breeding up millions of larvae, and once you've got that you've got a path to creating something really positive. But you've got to have that initial stock there, otherwise you've got no hope.'

CORAL REEFS AREN'T JUST VITAL to the health of the oceans, they are also agents of planetary change. Given enough time – and corals have had hundreds of millions of years – the tiny polyps will create towering structures not just in the form of living reefs,

like the 2300 kilometres of the Great Barrier Reef, but the limestone deposits found in many parts of the world. As Darwin observed,
'We feel surprise when travellers tell us of the vast dimensions of the
Pyramids and other great ruins, but how utterly insignificant are the
greatest of these, when compared to these mountains of stone accumulated by the agency of various minute and tender animals!' Now
these same structures – or at least the living communities that create
them – are at risk because the changes humans have unleashed are
outpacing their capacity to adapt.

I find myself thinking about this idea of time – and the mismatch
between the speed of transformation and the larger timescales needed
for adaptation – when I visit Dr Neal Cantin in his lab at AIMS.
Cantin is an ecophysiologist with a particular interest in using the
long-term history of the Reef to inform our understanding of its
response to the rapidly multiplying effects of climate change.

Much of Cantin's work depends upon cores extracted from the
skeletons of massive, long-lived corals such as *Porites*. As corals grow,
their polyps lay down layers of calcium carbonate. Like the growth
rings in trees, these layers – or density bands – act as time capsules
containing not just the history of the coral's life, but vital information
about the climate and other environmental conditions stretching back
hundreds of years.

Cantin is a softly-spoken Canadian with sandy hair and a gentle
but methodical manner, who seems to weigh each word carefully
before speaking. When I arrive he shows me the suite of labs where
he works. We start with the drills themselves, the largest of which is
a silver-and-black machine with four wheels that looks a little like the
floor polishers used by cleaners. Like a lot of underwater machinery,
it depends upon technology originally developed for the oil industry.
'Which is ironic,' Cantin says with a sad smile. After the drills he leads
me through to the rooms where AIMS' coral core library is housed.
Opening a drawer he reveals a selection of about a dozen cores, each

40 to 80 centimetres in length, many of which have been divided in half down the middle. The flat sections are smooth, like cut stone, and subtle gradations in the bleached white of the calcium carbonate form lines across them, like strata in stone. Cantin picks one up and passes it to me. I turn it in my hands. Although it looks porous it is surprisingly heavy, weighing about the same as a similarly-sized piece of stone. Cantin hands me a second section, telling me the coral it came from was several hundred years old, meaning the larvae from which it grew first affixed itself to the Reef before Captain Cook sailed past in 1770. Finally he hands me a third. This one dates back an astonishing 500 years, meaning the coral was already growing when Henry VIII was on the throne in England. I run my hand across its surface, wondering at the idea that an animal able to survive for 500 years is now at risk of being wiped out by human activity.

Cantin replaces the cores in their drawers and leads me through to an adjoining lab where a series of large machines are set up. He positions a pair of sectioned cores on a shallow tray in one of them and flicks a switch, bathing them in ultraviolet light. No longer pale, the cores instead glow Klein blue like cyanotypes against the dark background of the scanner, their density bands marked out like waves on the ocean's surface. Cantin points to the most luminous of the bands, telling me they indicate flood years – 'the brighter they glow, the bigger the flood event' – and contrasting them with the darker lines that correspond to cycles of drought.

In combination with x-rays and chemical analysis, these bands mean the cores contain a detailed record of ocean temperature, rainfall, salinity and sediment. No less importantly, they provide snapshots of other impacts: corals bear traces of fallout from nuclear tests, as well as registering shifts in industry such as the growth in aluminium production in Gladstone. ('You see a spike in the amount of aluminium in the ocean in the 1980s, when the smelter opened,' says Cantin, 'and there's another signal that pops out when they expanded production during the 1990s.') This data has allowed AIMS scientists to extract information about weather patterns and water temperatures stretching back to the eighteenth century, while examinations of cores of dead corals have provided evidence about river flows and rainfall up to 6000 years ago.

These findings offer a frightening window onto the scale of human impact on the climate. 'The coral cores give us a very accurate picture of the variation in temperature over the past 450 to 500 years. And what we see is a warming signal on both decadal and century time-scales. At first that warming enhances coral growth, because warmer water makes enzymes more efficient, at least up to the point where the temperature exceeds the thermal tolerance of the coral. Then we see a bleaching event, which is captured in the core as a high-density stress band, which marks an impact on the growth of the coral that lasts three or four years.' Cantin pauses. 'In the collection we have corals that let us look back more than 400 years. And we don't see negative growth anomalies or stress bands until the big El Niño that caused the first global bleaching event in 1983. This tells us that what is happening now is not normal.'

Cores from living corals are not our only source of information about the Reef's past. By extracting cores from the ancient reefs on the ocean floor, geoscientists have mapped the ways in which the Reef has responded to changing conditions over the past 30,000 years. This process revealed that the Reef has suffered several near-death events in response to changes in the Earth's climate across that time. The first of these events occurred about 30,000 years ago as falling global temperatures caused sea levels to drop – a process that continued until the Last Glacial Maximum 21,000 years ago, when sea levels were almost 120 metres lower than they are today. During this period shallow-water reefs were exposed and died, which caused the Reef to migrate many kilometres eastward into deeper water. This process was reversed about 17,000 years ago, when the ice sheets began to melt again and the Reef was forced to migrate inwards once more as sea levels rose. This culminated in a final near-death experience approximately 10,000 years ago when the ocean flooded the land along the coast and released large amounts of sediments.

While the story of the Reef's survival over tens of thousands of years speaks to the remarkable capacity of coral reefs to adapt to changing circumstances, the lesson for today is quite different. 'This ability to respond to temperature changes of three to four degrees by migrating backwards and forwards from the coastline is pretty amazing,' says University of Sydney geoscientist Professor Jody Webster, who headed up the study. 'But the critical thing is the rate of change. Those changes took place over 10,000 years; what we're seeing now is a degree or more in just a century.'

The effects of this unhinging of temporality are all around us, visible in the hastening wave of extreme temperatures and natural disasters, and the accelerating collapse in biodiversity. 'Coral reefs are very good historians,' says Charlie Veron, who has compared climate change to a snowball rolling down a hill, gathering size and speed until it becomes unstoppable. 'They keep track of changes in the environment better than any other group of organisms. And when you look back at the mass extinctions that have taken place in the past, coral reefs are always the first to go. And then everything else follows. I've never seen an exception.'

WHAT THEN IS THE FUTURE of the Reef in a rapidly heating world? Veron's assessment is blunt. 'It doesn't have a future. If you look at the rate of decline on the Great Barrier Reef and almost every other coral reef around the world, you'd have to be on drugs to believe these ecosystems aren't plunging into disaster. Of course they are. It's obvious. The Reef is going to continue its downward spiral. That's just the reality. The water is going to keep getting warmer, and that will trigger mass bleaching more and more often, and ocean acidification will finish off whatever's left.'

Justin Marshall offers a similar assessment. 'The Reef will be mostly gone within my lifetime. There will be remnants here and there, but I think the best thing we can do is what botanists are doing with seed banks and botanical gardens, which is to try and hold on to as much as we can and hope we can reseed at some later point. So although we should be celebrating the resilience that's still there in the system, it won't be enough, because the reality is we've been so slow to take action on climate change that we're probably heading for more than two degrees. That means a lot of shit is going to go down, and some of that shit, unfortunately, is going to be that the Reef will mostly die.'

For Terry Hughes, the answer is more qualified. 'It's too late to save the Reef in its current configuration, but not too late to save it in a new configuration. What the Reef is going through is one hell of a series of natural selection events, but we now know that when a bleaching event occurs you can have two individuals of the same species side by side: one will bleach and die, the other one won't. So we know there are tougher, more heat-adapted genotypes out there, and they're being favoured by natural selection. Reefs are going to have to keep running that gauntlet until temperatures stabilise. So we can expect coral cover and the mix of species to keep declining in the short to medium term. There will be spikes upwards in cover in between bleaching events, but the window for those recovery phases is going to keep shrinking. Once sea temperatures stop rising and things begin to re-equilibrate we'll see a period of rebuilding, where the heat-tolerant species and genotypes will increase in cover and you'll see a new configuration.'

These are confronting scenarios, especially for people whose lives have been dedicated to studying and preserving coral reefs. For Marshall, the cost has been particularly high: '2016 was awful,' he says, his voice shaking in a reminder of the way trauma collapses time.

'I was swimming around in the rotting bodies of billions of animals that I love. It's left me with definite PTSD symptoms where I burst into tears or go blank when I try to talk about it. What's surprising to me is that I always thought I was fairly robust psychologically, but you encounter something like that and you realise you're not. Because I'm fucked, and I think I'll probably remain fucked for the rest of my life.'

Although Hoegh-Guldberg is more circumspect, he also concedes his work has come at considerable personal cost. 'Sometimes I feel like I've lost the capacity to relate to it emotionally. It's so traumatic that you begin to protect yourself by not looking at it, so it just becomes an academic exercise about exactly how soon reefs are going to die.' He hesitates. 'I was obsessed with coral reefs as a kid. All I wanted was to go and dive on them. I've had to somehow dislocate myself from that little boy who loved reefs and wanted to study them for the rest of his life.'

Bay also speaks about the difficulty of managing negative emotions. 'I think you find ways of not thinking about it, of putting it aside, but then you have moments of tiredness or darkness when it all comes back.' Yet despite that, she finds solace in what remains. 'There's always beauty to be found on the Reef. It's just incredible. From the microscopic level up, there's just so much there. The colours, the diversity, the complexity. We're so lucky to get to experience it. So we just can't give up, no matter how depressing it seems.'

Curiously few admit to feeling anger. 'I don't feel angry because I've been part of the problem,' says Hoegh-Guldberg. 'We've all been burning fossil fuels, so we're all to blame.' But he also admits it hasn't always been easy. 'There was a point when I was getting death threats and people egged my house because they wanted to send me a very specific message. I still get a twinge of anger when I think about that, or when I get attacked by climate deniers in the media. But eventually you just get numb to it.'

All, however, express their frustration with the dishonesty and disinformation peddled by right-wing politicians and their enablers in the media. 'How can you be in a public position and say these things?' says Cantin.

Hoegh-Guldberg believes the influence of denialists is waning, but he also worries about the ability of often well-intentioned ideas to distract from the central task. 'You see so much bullshit published, stuff that's about people giving each other false hope. And that concerns me, because it's slowing down our ability to respond in the way we need to. I sometimes think that if humans have a lethal flaw it's our unbridled optimism, and that's part of what drives the false hope. But I also don't feel like I can criticise my colleagues – because shit, it's a scary thing.'

Van Oppen says something similar. 'I think humans always wait until the last minute before they act. Scientists have been warning about climate change for decades, and governments have been ignoring it. But with all these floods and droughts and other extreme-weather events, things are changing. So although things are going to get worse over coming decades, and we're going to keep losing diversity, I'm hopeful we will be able to do something before we lose everything on the Reef. If we don't stop climate change, then there's no hope for coral reefs. Because while we can push the limits a little with the sorts of interventions we're working on, we can't achieve an endless increase in thermal tolerance.'

Veron says he is pleased scientists are finally becoming vocal about the scale of what is taking place – and he doesn't hold back in his disgust about many involved in the IPCC process, whom he believes have soft-pedalled reality for too long. 'But at long, long last climatologists and glaciologists are speaking out in public forums.'

Hughes, too, notes the different ways the stories might be told. 'It's incredibly important that we don't just give up on reefs. The media often talk about coral-reef collapse and the disappearance of coral. That's not language a scientist would ever use, because there will always be a tropical ecosystem. We could see temperature rises of five degrees and there would still be something growing out there. There would probably be a lot more seaweed and almost no coral, but there would still be something. But I don't think we'll get to that extreme. And the reason I say that is that the cyclones and droughts and floods and fires caused by 1.2 degrees of warming have already got the attention of most people, and as we see another half a degree of warming over the next decade the rest are going to wake up as well.'

Yet their words are also a reminder that although we often talk about the uncertainty of our historical moment, about the sense that we have stepped out of a predictable world and into the unknown, much of what is difficult about life in the shadow of climate catastrophe is in fact how much *is* certain, and the amount of change that is now unstoppable.

Cantin led many of the surveys of the Reef during the 2016 and 2017 bleaching events, and the experience had a profound effect on him. 'When I got in the air it was so obvious. There was just reef after reef that was completely white and fluorescent.' In the aftermath of this, the focus of his work shifted. 'Bleaching has always been a central part of my research. But after 2016 I started to focus more on ecological responses, and how we can minimise the mortality from the future bleaching events we know are coming.' When I tell him this sounds more like a strategy for an emergency than a change in research focus, he looks away before he nods.

'Yeah,' he says in a quiet voice.

For Marshall, 2016 also ushered in a profound shift in the emphasis of his work, one that saw him leave academia altogether and move into advocacy and public education. 'After 2016 I couldn't in good conscience continue doing what I was doing. I think there are quite a few scientists in the same position. Because you can't keep saying, "Oh yes, there's a problem, but I'm going to go and discover a few new corals or keep researching the vision of this particular fish." I'm already in the position where some of the fish I used to study aren't there any more. I can't just stand by and watch the thing I love most disappear.' Marshall's main regret is that he didn't change his focus earlier. 'When he was seven my son asked me whether the Reef was dying. He's nineteen now, and I feel terminally guilty I didn't do more sooner.'

For all of them, the urgency of the situation only confirms the importance of the work they are doing. 'The honest truth is we don't know if these approaches are going to work,' says Bay. 'We don't know exactly how we can implement them. But that's precisely why we're doing the research, so we can find answers to those questions. And that's why I'm involved in this work. Because I want to make sure: if we're putting reef restoration and adaptation forward as a credible solution to keep our reefs as healthy as we can while we get our act together on temperatures, then I want to make sure that approach is backed by the best science.' Bay pauses for a moment, then continues. 'I feel a moral and professional obligation to engage in the work I'm doing. This is where I can apply my expertise to provide the best solutions.'

Even Veron refuses to give way to despair. 'The biggest thing we have at the moment is hope. If people decide things are hopeless they'll just give up. And we can't let that happen, because it would be like turning off the artificial life support when that's the only thing keeping us going.'

Marshall believes the growth of broad-based social movements means momentum for change is growing, but worries people are still too prepared to cede control over the process to governments. 'We need to recognise that we're individually responsible. Politicians are basically lubrication: they can help things happen, but, really, it's up to us.'

For Marshall this process is also inseparable from a larger societal shift in the way those in developed nations imagine their relationship with the natural world. 'We tend to imagine ourselves as these super-beings that are somehow managing nature. That's a mistake. We're not separate from nature or above it. We're part of it.' For him, this also underlines the degree to which hope is not an abstract state but instead a personal and political practice, a refusal to give in. 'Don't get hung up on whether you should be doing things at a consumer level or pushing for change at a governmental level. Do both. Do everything. And don't wait for the scientists to fix it. Do something yourself. Not tomorrow. Today.'

11

NET

'Let me not return with an empty boat.'
Senegalese fisher's prayer
Quoted in Anna Badkhen, *Fisherman's Blues*

IT IS JUST AFTER DAWN in Bonthe, on Sherbro Island in Sierra Leone, when I finally make contact with Amara Kalone. He is standing on the jetty: behind him on the foreshore the square, white shape of the town's clock tower is visible, red-roofed buildings behind it. In the water in front of the jetty a line of fishing boats are moored along the stone seawall, their names written in low bands of white below the gunwales of their blue hulls.

Kalone and I have been bouncing messages back and forth for days, trying to find a time to talk. The son of a fisher, he grew up in a village not far from Bonthe before going to school in Bo, Sierra Leone's second-biggest city. During Sierra Leone's brutal civil war he returned home to his community; once the war ended he worked in tourism before becoming involved with the Environmental Justice Foundation. He now works as a community organiser, helping fishers and others advocate for better environmental management.

As a boy, Kalone worked with his father on his boat. 'If you grow up in the coastal area everybody fishes, especially if you're from a poor background,' he says. 'When I came back from school for the holidays, I had to join my dad to do some fishing.'

Fishing is the principal industry in Bonthe, and along much of the Sierra Leonean coast. Upwelling of nutrient-rich water from the deep ocean sustains schools of sardines, sardinella and mackerel, as well as larger fish such as skipjack, bigeye, bonga and bonito. For centuries this abundance has fed the people of Sierra Leone, and even today, fishing employs around half a million of the country's 8.5 million people and provides four-fifths of the country's protein needs. Yet in recent decades, catches for local artisanal fishers have dropped precipitously. Kalone shows me a basket sitting on the dock. A few dozen small, silvery fish gleam in the dawn light. 'They are fishing the whole night and they only catch forty or fifty small fish,' he says.

Kalone says the reason is simple. The schools of fish that once supported Bonthe and communities like it are being targeted by industrial trawlers capable of catching more than 100 tonnes of fish a day. The overwhelming majority of these industrial vessels are owned and operated by foreigners: some operate under licences issued by the Sierra Leonean Government; others fish without proper licences, or in the inshore exclusion zone that is supposed to be reserved for artisanal fishers. Many employ highly destructive practices such as bottom trawling, which involves dragging weighted nets across the sea floor, smashing sponges and corals and destroying benthic ecosystems. Others simply under-report catches, or employ illegal or oversized nets with mesh so fine it catches juvenile fish. 'These fishermen, they are not honest to themselves,' says Kalone. 'They get a licence to do a particular sort of fishing, but when they go to sea they do a different kind. So, for example, they have a licence to catch groupers, but they use nets to catch a different species.'

The problem of illegal, unreported and unregulated (or IUU) fishing by foreign fleets is not confined to Sierra Leone. In Ghana, where fishing supports around 10 per cent of the population, over-fishing has pushed the country's once-bountiful fisheries to the point of collapse. The total catch by the artisanal fleet declined by almost 40 per cent between 1996 and 2016, and by almost 15 per cent in 2018 alone, with small pelagic fish such as sardinella, which are the principal catch for much of the artisanal fleet, falling the fastest. Incomes in fishing communities have dropped, and fishers speak of being unable to catch enough to feed their families, and of going for days on end without food. Similarly in Senegal, where as many as one in six people work in industries connected to fishing and fish makes up well over half the protein consumed, fish populations are in steep decline as a result of intense competition between artisanal fishers and industrial vessels, a situation that is repeated in The Gambia, Mauritania, Guinea, Guinea-Bissau, Liberia, Côte d'Ivoire and elsewhere. Some estimates suggest that almost half the fish caught off West Africa are caught illegally – or, to put that another way, for every fish that is caught legally, another is caught illegally.

The effects of this on communities like Kalone's are immense. Illegal fishing siphons an estimated US$9.4 billion out of West Africa every year. It also places local communities under considerable pressure, affecting the ability of both fishers and the large networks of mostly women who process and sell the fish to earn a living and feed their families. And, as fishers travel further to catch fish their fuel costs increase, further eroding their ability to make a living – as does the risk they will be lost at sea.

While things have improved in inshore zones in recent years, competition between artisanal fishers and industrial vessels also generates conflict. Kalone says there are frequent confrontations – his brother had his nets destroyed by a trawler, and stories of local fishers having their nets cut or run down or their boats swamped or sunk by trawlers are common, as are accounts of attacks upon local fishers by the trawler crews. 'What mostly happens is that the trawlers will enter into the fishing nets, and the fishermen will try to defend themselves by going closer to the trawler and saying, "You guys have damaged my net." Then somebody will come and take a stick and hit that person to move them on from the trawler. If you're not fortunate you can fall in the water and drown. So people are losing their equipment out there, but they're also losing their lives.'

Those unable to make a living are increasingly abandoning their communities. Kalone tells me of men who have left their wives and children behind and moved to the city, while others head north to Europe, often making the fateful decision to attempt the passage out across the Atlantic to the Canaries. For those who make it, an uncertain life awaits; for the families of the thousands every year who disappear along the way, there is only grief and uncertainty.

ALTHOUGH WEST AFRICA is one of the world's illegal fishing hotspots, accounting for as much as 40 per cent of all IUU fishing worldwide, the situation there is just one part of a much larger problem. Some studies suggest that over the past seventy years populations of large fish around the world have been reduced by about 90 per cent. According to the United Nations Food and Agriculture Organization (FAO), close to 95 per cent of the world's fish stocks are being fished to their limits or beyond, and the number that are being fished unsustainably is increasing, rising from just 10 per cent in 1974 to 35.4 per cent in 2019. On the other side of the ledger, the number of stocks that aren't being fished to their limits has steadily decreased over the same period, falling from almost 40 per cent in 1974 to 7.2 per cent in 2019. Other authorities suggest even steeper declines: in 2022 the Global Fishing Index, produced by Australia's Minderoo Foundation, found that almost half of the 1439 stocks it assessed were overfished, and almost one in ten were on the brink of collapse. Sharks and rays have been particularly hard-hit: more than half of pelagic ocean shark species and nearly two-thirds of coral reef sharks are at risk of extinction, while sawfish have already vanished entirely from half the waters they once frequented as a result of overfishing, and run the risk of complete extinction without an immediate effort to stop their exploitation. 'These sharks and rays have evolved over 450 million years and survived six mass extinctions, but they can't deal with this fishing pressure,' said the author of one recent study documenting the precipitous collapse in shark numbers.

These declines are placing pressure on communities in many countries. Billions of people in developing countries rely upon fish as a source of protein, and in addition to the 40 million people employed as fishers, another 560 million are dependent upon fisheries and aquaculture for their livelihoods.

Daniel Pauly is a University Killam Professor at the University of British Columbia and chief investigator at the Sea Around Us, an international research initiative focused on measuring the true impact of fisheries. Now in his late seventies, his research and ideas have reshaped the study of fisheries and our understanding of marine ecosystems. Pauly has compared the impact of fishing on marine life to the impact of the meteorite that wiped out the dinosaurs on terrestrial life, and describes fisheries as 'a global disaster . . . There are too many boats, they don't earn enough money and are kept afloat with subsidies.' He argues this catastrophe began in the second half of the nineteenth century. Prior to that, the impact of humans on fish populations was constrained by the scale of the ocean and the difficulties of keeping fish fresh. But the advent of railways capable of efficiently transporting large quantities of fish inland in the middle of the nineteenth century, and the introduction of steam-powered trawlers in Britain not long afterwards, saw catches rise sharply. 'Before then the fish were mostly okay,' says Pauly. 'But fossil fuels plus giant boats, and you're talking about a wave of extermination that washed out across the world like a steamroller.'

The advent of steam-powered trawlers also drove a rapid expansion of bottom trawling. Within a few decades industrial trawlers were sweeping weighted nets across the sea floor off Australia, New Zealand, Canada and the United States, reducing vast tracts of the seabed to a lifeless desert. Not coincidentally, this period coincides with rapid falls in fish populations in affected areas: in the North Sea, where the first intensive trawling took place, catches dropped by two-thirds or more between the 1860s and the beginning of the twentieth century. But it wasn't just fish that were affected. The spread of trawling also resulted in what marine biologist Callum Roberts describes as 'the greatest human transformation of marine

habitats ever seen, before or since', and the cause of 'the shift from
the biologically rich, complex, and productive environments to the
immense expanses of gravel, sand and mud that predominate today'.

This process accelerated drastically in the years after World War II
as developed nations expanded and modernised their fleets, often
drawing upon technological advances that had their origins in the
recently-ended conflict. Steel hulls replaced wooden ones, allowing
ships to stay at sea longer, sonar started to be used to locate fish, and
mechanical refrigeration began to become more commonplace.

As fleets grew larger they also began to fish further from shore.
From 1950 to 1980 the reach of industrial fisheries expanded by more
than a million square kilometres a year; by the 1980s this had accelerated to between 2.8 and almost 3.9 million square kilometres a year.
As the reach of industrial fishing fleets expanded, so did the number
of fish being caught: in 1950 reported landings were just under
17 million tonnes a year; by their peak, in the 1990s this had increased
almost fivefold, to more than 85 million tonnes.

Simultaneously, fish populations started to suffer sudden collapses.
In the late 1960s the billions of massing sardines that sustained
Monterey's Cannery Row, the place immortalised by John Steinbeck
as 'a poem, a stink, a grating noise, a quality of light, a tone, a habit, a
nostalgia, a dream', almost entirely disappeared; a few years later the
same thing happened with the anchoveta in Peru, while catches of
other fish, such as haddock and rosefish, also fell rapidly. But perhaps
the most shocking of these collapses concerned the Atlantic cod. Cod
once filled the waters of the North Atlantic in almost unimaginable numbers: English sailors in the fifteenth and sixteenth centuries
said lines and nets were not necessary, instead they could be caught
by simply dropping a basket overboard; others describe them being
so thick in the water it was difficult to row a boat through them. As
fishing pressure on the cod increased, however, their numbers began
to decline, before collapsing entirely in 1992 in a calamity that saw

thousands of job losses and whole communities thrown into disarray. Expressed in raw numbers, the collapse is even more shocking: historical reconstructions suggest that in the sixteenth century the total biomass of the cod in the North Atlantic was around seven million tonnes; in 1992 it was barely 22,000 tonnes, or a third of 1 per cent of the original population.

Pauly, who spent much of his early career working in the Philippines and other developing countries, witnessed this process of expansion, over-exploitation and collapse firsthand. 'In 1975 I was working in Indonesia. Trawling was already happening in the Gulf of Thailand and the Malacca Strait, but it had started to spill out into the Java Sea. Within a decade the resource was essentially gone. In the Gulf of Thailand the catch per hour of a trawler fell from 400 to 500 kilos to around 5 per cent of that. Today it's approximately 20 kilos an hour, and that's all little fish and crap.'

As falling numbers of fish cut into the profits that had resulted from geographical expansion, industrial fishers turned their attention to new populations, fishing deeper and deeper for previously unexploited species such as the orange roughy, and further and further down the food web. Yet the benefits of this process could not hide the fact that the once seemingly inexhaustible bounty of the ocean was becoming increasingly depleted, so much so that some reconstructions suggest the number of fish in the ocean is now less than a tenth of what it was a century ago, and many places that once teemed with fish, sharks and other species are now largely empty of life.

This crisis has been masked by several factors. One is systematic under-reporting. Although the FAO provides biennial assessments of the state of the world's fisheries, the data they rely upon is often sketchy and unreliable. Many poorer nations do not have the resources to monitor catches; others only record commercial catches, thus ignoring the fish caught by artisanal and small-scale fishers. Others manipulate and falsify data: through the 1990s Chinese authorities provided false figures that made it appear catches were increasing when in fact they were declining (even today fisheries data exists in two forms – with and without China).

An awareness of this under-reporting was one of the motivations behind the Sea Around Us, which uses multiple data sources and statistical techniques to construct a more accurate picture of the state of world fisheries. These reconstructions suggest that between 1970 and 2010 the actual catch was approximately 50 per cent higher than the FAO data suggests. And while the Sea Around Us's reconstructions also show a peak around 1996, they suggest that since then the number of fish caught has been dropping three times faster than the official figures suggest.

In some cases this is a sign of better management: over the past few decades many wealthier nations have imposed catch limits to allow fish populations to recover. But the blurriness of the idea of sustainability means that even where this has happened, populations are far smaller than they used to be. 'The whole concept of sustainability is just completely crazy,' says Pauly. 'If I put you in a basement and come and beat you within an inch of your life every day that's sustainable, because as long as I also feed you, you're not going to die. Maintaining an overfished stock in an overfished state is just the same. If you have a devastated system, what you should be doing is rebuilding it, not sustaining it in its devastated state.' These problems are exacerbated by what Pauly dubs shifting baseline syndrome, a form of historical amnesia in which environmental change becomes normalised so that each new generation regards the environment they knew when young as normal. In other words the real question isn't whether a population is sustainable, but 'what the population used to be before fishing really got started'.

Simultaneously the effort required to catch these fish continues to increase. The total distance travelled by industrial fishers has doubled since 1950, largely as a result of the large distant-water fleets operated by China, Japan, Taiwan, South Korea and Spain, while improvements in technology such as GPS, echo sounders and fish finders have effectively doubled the capacity of these fleets to find and catch fish over the past thirty-five years. In other words, fishing fleets are working harder and harder but catches are still declining, in many cases significantly: in the Indian Ocean, for instance, the catch per unit effort fell 78 per cent between 1950 and 2010.

As profitability has fallen, fishers have been kept afloat by government subsidies. In 2018 the total value of these subsidies, which range from support with the cost of fuel, to tax structures designed to encourage investment in ships and equipment, to investment in ports and processing plants, was estimated to be in the order of US$35.4 billion, of which US$22.2 billion was directed at increasing the already-unsustainable capacity of fishing fleets. The role of these subsidies in disguising the unprofitability of fisheries is starkly evident in the distant-water fleets that operate on the world's high seas. In 2014 high-seas fishing fleets globally received subsidies amounting to almost US$4.2 billion, more than half of which flowed to the fishing fleets of just four nations: Japan, Spain, China and South Korea. Yet even with this injection of support, many high-seas fisheries are barely profitable; indeed in 2014 total subsidies were around eight times as high as the total profits of the global distant-water fleet.

These subsidies do more than disguise the unprofitability of industrial fisheries, they encourage fishers to continue exploiting fisheries well beyond their natural limits. 'Many ocean fisheries could not operate if there were no subsidies,' says Pauly. 'Fishers don't produce fish the way that farmers produce a crop. They don't sow, they don't fertilise, they don't do anything towards the production. All they do is collect. And if you can't do that profitably you shouldn't be doing it. Because that lack of profit is Nature telling you to leave me alone so that I can recover. Subsidies mean you don't have to pay attention to that signal.'

The lack of profitability also feeds into the problem of IUU fishing. As fish populations have declined, heavily subsidised and uneconomic fishing fleets have increasingly sought to prop up their bottom lines by fishing illegally, under-reporting their catch or targeting fish populations on the high seas, where regulation is patchy and largely unenforced. Illegal fishing can take various forms, ranging from fishing outside of authorised areas and targeting of protected species, to the use of illegal equipment. Although West Africa remains the hotspot for these sorts of activities, there is also a serious problem with illegal and unregulated fishing on the high seas, where patchy regulation and a lack of monitoring and enforcement allow illegal and unregulated fishing to run rampant. Of particular concern in recent years has been the expansion of fishing activity just outside the territorial waters of Oman and Yemen in the northwestern Indian Ocean. Between 2015 and 2019 the number of ships detected operating in the area rose from just thirty to nearly 300; across the same period the number of refrigerated cargo vessels, or reefers, capable of transporting the catch to processing facilities on land detected in the area rose from just three to thirty-four, suggesting a massive rise in the volume of squid being caught. Likewise in the Pacific, where island nations have worked collaboratively to create intergovernmental systems to license and manage fishing in their territorial waters, industrial fishing in international waters is placing increasing pressure on squid and tuna populations, as well as sweeping up large numbers of endangered sharks, dolphins and other marine animals.

Most of the vessels engaged in this plunder are Chinese. Prior to the 1980s, China did not have a distant-water fleet, but as fish numbers in Chinese waters began to collapse under intensive fishing pressure, it started to look further afield for fish to feed its massive population. Today China fields the world's largest distant-water fishing fleet – estimates of its numbers range from around 3000 to almost 17,000 vessels. The uncertainty is partly due to the fact many Chinese ships do not operate under the Chinese flag. Instead they are flagged to other nations, especially African nations, and their ownership is hidden behind complex and opaque business structures designed to make regulation and enforcement difficult. Many also seem to operate without the satellite automatic identification system (AIS) used to track ships, rendering them effectively invisible.

Some of these ships are trawlers, but many are huge factory ships capable of catching more than 100 tonnes of fish a day, often supported by sophisticated reefers and motherships capable of taking on board fish and shifting them to processing facilities onshore. These vessels frequently work together to strip whole regions of the ocean: in 2020 the Ecuadorian Government called in the United States Coast Guard after an armada of close to 300 Chinese fishing boats was detected massing just outside the biologically rich and diverse waters of the World Heritage–listed Galápagos Marine Reserve. After the US deployed a Coast Guard cutter, the Chinese fleet moved south into Peruvian and then Chilean waters, provoking outrage across South America. Concern was sharpened by previous encounters with Chinese fishing vessels in the region: in 2017 the Ecuadorian Navy captured a Chinese trawler in the Galápagos Marine Reserve with 300 tonnes of illegally caught sharks, including endangered species such as whale sharks and hammerheads, on board. There have been confrontations between Chilean and Peruvian naval vessels and Chinese fishing vessels as well. These ships often operate with the support of armed Chinese Coast Guard vessels; even when they are without an escort, they are notorious for behaving aggressively towards anybody they regard as a threat to their activities. According to the Brookings Institution the Chinese fishing fleets in the Pacific 'do double-duty as maritime militias that threaten and intimidate the fishers from neighbouring nations'.

While the Chinese are by far the largest offenders, they are not alone. Taiwanese, Russian, South Korean, French and Spanish vessels also have large distant-water fleets that engage in similar illegal and semi-legal activities. But illegal and unregulated fishing is only one part of a broader spectrum of fishing activity that exists outside the regulations and agreements designed to protect dwindling fish populations. Peter Horn heads up the Pew Charitable Trusts' International Fisheries project. Formerly an officer in the British Royal Navy, he commanded the HMS *Middleton* and led deployments in the Atlantic, Middle East and Asia, as well as working in strategy and intelligence. He says the focus on illegal fishing can sometimes overshadow the significance of allied issues associated with under-reporting. 'People often focus on illegal fishing, but illegal fishing is only a small subset of IUU. In fact, unreported fishing is the biggest part of the problem.'

Unreported fishing is made possible by the murkiness of industrial fishing supply chains, which often see fish transferred from ship to ship at sea or in port and by weak inspection regimes. 'You can have vessels that are authorised to go into an area because they're licensed to fish there, but they can be 100 per cent non-compliant in their reporting of what they're actually taking from that area,' says Horn. It can also involve sizable takes by artisanal fishers – in West African nations, where the artisanal fleet has increased more than fourfold since the 1950s, much of the catch by local fishers is not reported – as well as practices such as *saiko*, in which foreign-owned industrial trawlers sell frozen blocks of small pelagic fish such as sardinella to West African fishers for sale in local markets. Originally a way for trawlers to exchange bycatch for fresh food and water, as larger fish have been fished out and local communities have found it increasingly difficult to catch enough to eat, trawlers have begun specifically targeting smaller fish to sell to local fishers, helping push their populations closer and closer to complete collapse.

By helping undermine attempts to regulate fishing, IUU fishing places fish populations under even greater pressure, intensifying the ongoing degradation of marine ecosystems. But as its effects in Sierra Leone and elsewhere make clear, its impacts are not just environmental. The total value lost to IUU fishing is estimated to be as high as US$50 billion a year. Most of this loss is borne by developing countries, including some of the poorest nations in the world, denying them not just the capacity to exploit their own resources, but also revenue in the form of taxes and licence fees. In this respect IUU fishing is simply another manifestation of a larger asymmetry: vessels flagged to higher-income countries are responsible for 97 per cent of industrial fishing and 78 per cent of fishing within the national waters of lower-income countries, meaning the benefit of the industrial fishing that is stripping waters in West Africa and elsewhere flows overwhelmingly to wealthier countries. This leads to problems with governance. 'The distant-water fleet operators often use their bigger financial muscle to sweeten the pie for governments and government officials in other ways,' says Horn. 'I'm not saying that's corruption, but there may be a *quid pro quo* on support in other areas. So we've got to help those countries find their way there and then make the right decision for their peoples.'

These pressures add to existing problems, destabilising communities and creating issues with security. Horn argues the piracy problem off the coast of Somalia is an example of this. During the civil war in the 1990s, foreign fishers swept in and pillaged Somalia's waters, driving fish populations to the brink of collapse and destroying the marine environment by trawling in sensitive areas. Unable to feed their families, some Somali fishers started pirating Indian fishing boats and quickly escalated to attacks on cargo and other vessels. 'So one of the contributing factors to that whole maritime security challenge was fishermen who could no longer feed their families or pay for their livelihoods.'

IUU fishing is also part of a larger web of illegality and exploitation. Duncan Copeland is executive director and one of the founders of Stiftelse Trygg Mat Tracking (TMT), a Norwegian non-profit that focuses on assisting governments to identify and investigate illegal fishing. Copeland first became aware of the scale of the problem of IUU fishing in West Africa in 2005, when he crewed aboard a yacht that was being sailed from Guinea-Bissau to Brazil. When darkness fell on the first night, the ocean ahead of them was lit up by a blaze of light that spread to the horizon in both directions. As it drew closer Copeland realised it was coming from dozens of huge industrial trawlers. 'It was like a city out at sea,' he says. 'And behind me was one of the poorest countries in the world where you could barely see a candlelight.'

For the next two days the yacht Copeland was on zigzagged its way through scores of fishing vessels. The bottom trawlers they encountered only 20 kilometres offshore gave way to enormous mid-water trawlers targeting sardinella and other small pelagic fish. For Copeland part of what made the experience shocking was the juxtaposition between the sheer scale of the industrial fleet and the dugout canoes being used by the Guinea-Bissauan fishers closer to shore. 'It was clearly not a level playing field,' he says.

Copeland says the operators of distant-water fleets often feel a degree of impunity. 'It's an operational environment that lends itself to openings for people to commit crimes. People who are working on the other side of the world also often feel very far away from the long arm of the law in their own countries. And we see this repeatedly – also that people are more likely to take legal risks in another country than they are in their own.'

Copeland argues illegality connected to fishing falls into three interconnected categories. 'There is violation of fisheries laws such as where licensing conditions aren't being met, or fishers are using illegal equipment or exceeding quotas. Then there's a middle ground which is about violation of other laws to facilitate the profits from illegal fishing. This can be using indentured labour or cutting corners on safety, or document fraud or forgery or tax avoidance. But it's still about making money from fishing. The third category is where the fishing operation is being used as a cover for other crimes like smuggling arms or drugs or wildlife, or even piracy.'

Recent years have seen a long line of incidents involving slavery and labour abuse on board fishing vessels. In 2017 journalist Ian Urbina published video of Taiwanese fishers executing at least four men. In the footage the fishers laugh and shout as they fire shot after shot at several men floating in the water; afterwards they pose for selfies, grinning. Multiple incidents involving men and children being used as slave labour or beaten and even killed on Thai, South Korean and Chinese vessels have also been documented: 'I thought I was going to die,' said one Cambodian. 'They kept me chained up, they didn't care about me or give me any food . . . They sold us like animals, but we are not animals – we are human beings.' Other studies have found more than half the men rescued from slavery have seen people murdered in front of them. 'The worst was when I saw one of my co-workers fall into the sea. We were ordered to continue to work and prevented from helping him . . . He drowned,' said one. 'We were beat frequently by the Thai crew, on the back of the head and across the back. The captain had a gun . . . We saw a Thai captain decapitate a Vietnamese fisherman, and another Thai captain decapitate a Thai fisherman,' said a second.

The causes of these cultures of exploitation and abuse are complex, but they are intimately entwined with overfishing. In Thailand, where studies suggest three quarters of the more than 100,000 mostly Cambodian, Laotian and Myanmarese migrants employed in the fishing industry have been held in debt bondage, close to 20 per cent have experienced conditions of slavery, and almost two fifths have been the victims of human trafficking, the catch per unit effort has declined by 97 per cent since the 1960s.

Professor Jessica Meeuwig is director of the Centre for Marine Futures at the University of Western Australia, and an expert on the social and biological impacts of overfishing. In 2019 Meeuwig and her colleague marine ecologist David Tickler developed a model to assess the risk of slavery in different countries. They found the countries in which the risk of slavery was highest shared certain characteristics. These included high levels of vessel and fuel subsidies and low catch value per crew member; high levels of unreported catch, suggesting poor governance of fisheries; and a high level of dependence upon distant-water fishing. 'If you're not profitable, you're going to have to cut your input costs,' Meeuwig says. 'And in fishing there are two basic input costs. One is vessel plus fuel, and the other is labour. And the only way to drive down labour costs is by exploiting people.'

Meeuwig argues these problems are being exacerbated by the growing dependence of fishing fleets upon factory ships and the practice of transshipment, in which catches are moved between ships at sea, making regulation of what and how much is being caught effectively impossible. 'One of the major shifts in industrial fisheries in recent years is the development of factory ships that can stay offshore pretty much permanently,' says Meeuwig. 'Smaller and even medium-sized ships used to go out and fish and then come back to port, and when they came to shore the crew were potentially visible. But if you've got a factory ship that never comes to shore, the crew on that boat are effectively invisible.'

Daniel Pauly agrees. 'Slavery on board ships is not a thing specific to Asia – it's a thing specific to overfishing. Why would you not pay your crew? Why do you have to shanghai them? Why do you have to bribe them or get them drunk to get them on a boat? Because you don't catch enough fish, so you have to reduce your costs.' Or in the words of the Environmental Justice Foundation's CEO, Steve Trent, 'Ecosystem decline and slavery exist in a vicious cycle. People are trafficked as a result of environmental crises and forced to endure terrible human rights abuses while working in industries that also harm the environment. Unrestricted industrial exploitation damages ecosystems and exposes vulnerable populations to trafficking and abuse. Overfishing exacerbates pirate fishing, which further drives slavery and environmental degradation.'

UNRAVELLING THESE PROBLEMS poses huge challenges for both individual nations and the international community. The scale of the ocean makes monitoring the activities of fishers difficult, especially on the high seas, and many of the countries most affected by overfishing lack the resources to effectively police their own waters. Likewise, the migratory nature of many fish populations – and indeed the transnational nature of much of the crime associated with fishing – means effective management depends upon inter-governmental cooperation across a range of fronts. And, perhaps most importantly, many of the imbalances of power and wealth that make nations vulnerable to overfishing are legacies of long histories of colonial exploitation.

Combating these issues therefore requires coordinated action on a wide range of fronts. Better monitoring is key, especially when it comes to the actions of industrial fishing fleets. Continuing to expand the number of large fishing vessels fitted with AIS, that allow their movements to be tracked and analysed is vital on this front, as are initiatives such as the Combined IUU Vessel List, which allows multiple intergovernmental organisations to share information on vessels that are suspected of engaging in IUU fishing. Technology also has an important role to play. Global Fishing Watch integrates satellite monitoring, AISs and other forms of tracking data with shipping registry information to track and analyse fishing activity around the world. Researchers there recently developed a machine learning system capable of using satellite data about average voyage time, distance travelled and patterns of port visits to identify fishing vessels on which crew were at high risk of forced labour. In collaboration with the International Monitoring, Control and Surveillance Network, Global Fishing Watch and TMT have also created the Joint Analytical Cell (JAC), a cooperation initiative between non-state organisations that is improving the ability of governments and communities in developing countries to monitor and control fishing activity in their waters by providing them with high-quality intelligence about fishing activity and the ships and companies involved. The JAC's capacity has recently been extended by the addition of satellite tracking systems and beneficial ownership identification capability.

Progress also depends upon better cooperation between governments. Again there is movement on this front: in the Pacific, where illegal tuna fishing costs island nations upwards of US$200 million a year, the Pacific Islands Forum Fisheries Agency monitors the movement of fishing vessels within the economic zones of all seventeen member countries, and the intergovernmental Western and Central Pacific Fisheries Commission is charged with conserving and managing tuna stocks in international waters. Similarly in the Indian Ocean FISH-i Africa, a taskforce to combat IUU fishing that relies upon information sharing and regional coordination between Comoros, Kenya, Madagascar, Mauritius, Mozambique, Seychelles, Somalia and Tanzania, has successfully uncovered multiple instances of fraud and illegal fishing. And in the Atlantic, Liberia, Côte d'Ivoire, Ghana, Togo, Benin, and Nigeria's Fisheries Committee for the West Central Gulf of Guinea (FCWC) have established the West Africa Task Force, which with technical support from TMT and Stop Illegal Fishing has resulted in more effective sharing of information between member nations, increased collaboration on operations, and improved enforcement. Meanwhile, in a reminder of the degree to which fisheries are bound up in security concerns, in 2022 the Quad security grouping of the United States, Australia, India and Japan launched the Indo-Pacific Partnership for Maritime Domain Awareness, a satellite-based tracking service that will supply intelligence about the movement of vessels in real time, as well as using radio and radar monitoring to identify ships that have switched off their AIS because they are engaged in illegal fishing a move that was widely seen as designed to help control illegal fishing by Chinese vessels.

Combating IUU fishing depends upon more than simply controlling the activities of fishing vessels, however. Fishing supply chains are notoriously complex and opaque, and frequently involve transshipment at port or in sea. This makes tracking the movement of fish and fish products difficult. This murkiness does not just obscure illegality and exploitation of poor and vulnerable people. It also means consumers in wealthy countries, far from the abuses in question, are directly implicated in the web of human-rights abuse and illegal fishing.

As part of their research, Meeuwig and Tickler modelled flows of seafood through global supply chains. They found the importation of seafood in countries such as the US and Europe can increase the exposure of consumers to potentially slave-caught or processed seafood as much as eightfold. Similarly in West Africa, where mostly Chinese-owned and -operated fishmeal factories now produce hundreds of thousands of tonnes of fish oil and fishmeal a year, mostly for use as feed in fish farms and piggeries in China, Türkiye and Vietnam, lack of transparency in supply chains means farmed fish and pork fed on West African fishmeal is being sold by major retailers in Europe and the UK. In other words, fish that could be used to feed people in West Africa is instead being fed to animals for consumption in wealthy countries.

In all cases successful implementation of these solutions depends upon improving compliance and enforcement in ways that drive positive change. Professor Rashid Sumaila is an economist at the Institute for the Oceans and Fisheries and the School of Public Policy and Global Affairs at the University of British Columbia. As a boy in Nigeria his grandfather advised him to walk as if the Earth felt pain, a maxim that has helped shape his thinking about ocean ecosystems and their role in sustaining human communities. In 2023, he and Daniel Pauly were jointly awarded the Tyler Prize for Environmental Achievement – an honour sometimes described as the Nobel prize for the environment – for their work on ocean fisheries. 'The reason that people fish illegally is mainly because they make money out of it. So the solution is to make it uneconomic for them. And how do you do that? One way is through monitoring, so that if people are fishing illegally you can apprehend them, but you also have to have a penalty system that acts as a deterrent.'

Copeland agrees, and argues that in some respects at least, the situation is gradually improving. 'When I started working in West Africa you used to be able to sit on the beach and watch trawlers going up and down just offshore. We still have problems with trawlers coming into those inshore zones but it's not so common any more, because most of the countries have got vessel monitoring systems in place – or there are very active local fishermen reporting on them.'

Copeland argues these changes have been complemented by better systems for identifying the real owners of boats, as well as improved controls in ports and elsewhere. Nonetheless he is quick to point out these improvements are only one side of the story. 'In general, the responsibility of distant-water nations is slowly moving in a positive direction in terms of improving controls of vessels and monitoring the vessels. But on the flip side, it's not just about illegal fishing. We're still seeing an increase in terms of the number of vessels being built, and countries are licensing too many boats. And while the big problem used to be that vessels didn't have licences, now almost every vessel will have a licence or some kind of authorisation, which they use to give themselves a veneer of legality. But behind that they may be using illegal mesh in their nets, or transshipping, or targeting the wrong species, as well as having issues with labour and vessel safety and so on.'

Increased monitoring and compliance can only do so much without tackling the overfishing that lies at the heart of the problem and the subsidies that support it. 'Subsidising fisheries disconnects fishing from the amount of fish that's available,' says Meeuwig. 'Because if we didn't subsidise, people would only fish where and when there were enough fish to maintain their operation. But subsidies mean they can keep going until the fish are completely gone.'

Sumaila agrees. 'Subsidies mean you get into a vicious cycle where you get negative feedback from people to nature, and if you do that long enough you start to get negative feedback from nature to people. But what we need to do is find ways to reinforce positive feedback from people to nature and nature to people. If you take away the subsidies you can do that, because then economics can serve as the boundary.'

Hopes of an end to harmful subsidies were dealt a serious blow in mid-2022 when, after years of campaigning, the World Trade Organization finally reached agreement on a set of rules ostensibly designed to tackle overfishing and IUU fishing. Although the new rules improve transparency and accountability, they specifically avoid direct controls on the capacity-enhancing subsidies that encourage overfishing by keeping uneconomic fishing fleets afloat.

Pauly describes the deal as 'a massive disappointment'. Sumaila, who has spent much of the past decade involved in the push to bring the matter before the WTO, is more circumspect. 'When you look around the world it often seems that countries can't agree on anything, so the mere fact these 164 countries managed to agree upon something is a positive, because it shows we can get there . . . But for those of us who care about sustainability and the fish in the water, it doesn't deal a fatal blow to harmful subsidies. So it's a big failure in that regard.'

In the meantime other more radical plans are being mooted, in particular a total ban on fishing on the high seas. Only a few years ago such an idea might have seemed fanciful, but in a move Sumaila describes as a 'big deal', the 2023 United Nations High Seas Treaty declared 30 per cent of the high seas should be set aside as marine reserves, suggesting a more complete ban is not beyond the realm of possibility.

Pauly and Sumaila argue a total ban wouldn't just give marine ecosystems the fighting chance they need. Over time it would actually increase the size of the catch in coastal waters as rising populations began to spill over from protected areas. 'Fish really is the one thing where if you do less you get more,' says Sumaila.

No less importantly, a ban would lay the groundwork for a more equitable distribution of the wealth generated by fisheries. 'Closing the high seas would result in no loss in total global catch, just a more equitable distribution,' says Pauly. 'Most commercially fished species move back and forth between the high seas and coastal areas, where they can be caught in a country's exclusive economic zone.' Meeuwig agrees. 'Instead of all the economic benefit flowing to Spanish, Taiwanese and Korean distant-water fishing fleets, countries like Indonesia and nations in West and East Africa would be able to take advantage of their domestic fisheries.'

IN THE ABSENCE OF CHANGE, fisheries will only continue to decline, pushing communities such as Kalone's closer and closer to the brink. 'Humans are running an experiment called global industrialised fisheries,' says Meeuwig, 'and that experiment is transforming the oceans. We're stripping it of its biomass but we're concentrating on the top predators, which is like knocking off all the lions and tigers on land. The problem is we have no idea what the ecological consequences might be of a diminished, depleted, exhausted ocean. And when you remember that the ocean produces 50 per cent of the oxygen we breathe, and throw in climate change on top of that, it means we're altering a system that covers 70 per cent of our planet surface and 90 per cent of its inhabitable volume.'

'Essentially we're managing misery,' says Pauly. 'Because the big abundances are gone, what we're doing is very, very precise stuff. And you need that precision because we're balancing on the edge of a precipice. And you have to watch your step when you do that.'

As Meeuwig notes, these problems are being exacerbated by the growing impact of global heating. As the oceans grow warmer, fish are shifting their ranges. In places such as West Africa this is creating tensions between communities dependent upon already-depleted fish stocks, but it is also affecting fish populations more directly. 'Most of the inshore fish that the small-scale sector in West Africa rely upon spend some part of their life cycle in the river estuaries,' says Copeland. 'But heavier rainfall combined with logging and development upstream means there's more silt coming down, and combined with sea-level rise that means that breeding areas that used to be known for good catches are only half as deep, or no longer there at all.'

'A system that is depleted will react more strongly to disruption,' Pauly says. 'When we had several million tonnes of cod in Canada there was enormous genetic variance, and some of that would have included fish that were pre-adapted to warmer temperatures. With so many fewer fish you have less genetic variability, so if the population comes under pressure from rising temperatures they have less capacity to resist. In other words, on top of everything else we have made their population more vulnerable to global warming.'

Meeuwig argues these processes are transforming the ocean in fundamental ways that will have profound ramifications for human society. 'We're seeing a complete simplification of ocean ecosystems,' she says. 'We already have food security crises in developing countries that rely on seafood. That's only going to increase.'

Pauly says the same. 'People ask when the catastrophe is going to happen. It has already happened. There isn't a cliff we're going to suddenly fall off at a particular date in the future. Instead everything has been eroded away. And unless there's change, that's going to continue. We'll lose more fish, and more fish. My colleagues and I will write papers about this one and that one, and eventually we'll be writing about how the bacteria are in danger. We'll be saying save the bacteria because that's all that's left.'

Nonetheless, none of these people are without hope that change is possible if the pressure of overfishing can be reduced. 'These are very rich waters in West Africa,' says Copeland. 'Given a little bit of time and space they can respond quite quickly. Small things like local communities deciding not to fish in certain creeks for three months of the year because they know fish are breeding there then, or making sure people don't use mosquito nets to catch fingerlings, can have really positive results in terms of the next year's catch or the year after that. At a government level we're also seeing action such as the joint closed season that Ghana and Côte d'Ivoire have agreed for 2023, which Togo and Benin will join in 2024. If fishers and government then see tangible results in terms of catches and income, we can head in a better direction.'

Sumaila argues that fisheries are an infinitely renewable resource that has the capacity not just to feed the world forever, but to help build prosperity for hundreds of millions of people in developing countries if they are managed in the interests of all. 'If we're going to have a world that works environmentally, socially and economically then I say we have a responsibility to make life liveable for people all over the world,' he says, and then laughs. 'But I'm a dreamer, aren't I?'

12

SOUTH

'All ice wants to be water.'
Jean Sprackland, 'Tilt'

IT IS A MUGGY, UNSEASONABLY WARM day in March and I am standing at the back of a building in the Australian Antarctic Division's (AAD) headquarters on the outskirts of Hobart. I am here with Dr So Kawaguchi and Rob King. An ecologist and one of the world's foremost experts on Antarctic krill, Kawaguchi leads AAD's krill research projects. King is a biologist and marine research facility specialist, and for the past eight years has been the biology lead in the design process for the division's new icebreaker, the RSV *Nuyina*. The two of them are about to show me one of the AAD's most prized possessions: living Antarctic krill.

After disinfecting our shoes we step through the door into a small lab. To my left and in front of me workbenches line the walls, the space above them plastered with photos of krill and ice. To my right a glass cabinet holds half-a-dozen plastic tubes about a metre and a half high and 20 centimetres in diameter that are filled with bubbling water and bright-green or dirty-orange algae. As we pass, Kawaguchi points to one and tells me it was bred from samples collected in a lake near Australia's Davis Station in the early 1980s and has been sustained as a continuous culture ever since.

At the far end of the lab, King opens a door and ushers me into a second, smaller space that contains three round tanks a metre or so across, their white shapes swaddled in black insulation. The area is suffused by the soft thrum of pumps and the sound of the water in the hoses that connect the tanks to the pipes on the wall; combined with the low, almost greenish illumination, specially designed to replicate the light in Antarctica, it lends the room a sense of calm.

We pause beside the middle and largest tank. The water – refrigerated to just half a degree above zero – is so cold you can feel the

chill rising off the surface, but despite the frigid temperature thousands of krill swarm within it. These are sub-adults: probably a year
or two in age, and slightly shorter than my little finger. They look a
little like prawns, although slimmer and more delicate, and save for a
translucent dusting of red along their bodies, which fades from a rusty
hue at the front to a brighter red along the segmented abdomen and
tail, they are mostly transparent, their most prominent features their
large, jet-black eyes and the digestive gland a little way behind them.
On those that have fed recently the latter is bright green, distended
with algae.

Despite their small size the krill do not dart or flicker in the manner
of schooling fish. Instead, they move like tiny boats or torpedoes, the
rippling motion of their back legs – or pleopods, to use the technical
term – pushing them through the water with surprising speed. Most
seem to be moving in formation, following the same path clockwise
around the tank, swimming steadily against the current created by the
pump; the remainder twist and circle, feeding on floating algae.

In the next tank are a smaller number of adult krill. At about
5 centimetres in length they are longer and larger than the sub-adults,
as well as less brightly coloured, the red bled through their bodies like
selenium in glass. In contrast to the krill in the first tank only a few
are swimming in formation, but nonetheless they move quickly and
purposefully through the water.

Beside me Kawaguchi and King are gazing into the tanks as well.
The two of them make an appealing, if mismatched pair: Kawaguchi
is slender and softly-spoken, while King, who is well over 2 metres
tall, possesses a generousness and bashful humour that belies his
towering frame. Their obvious affection for their submarine charges
is disarming. But as the minutes pass I find myself falling under the
spell of these tiny creatures as well. There is something hypnotic about
them, their beauty and their pressing numbers, the way they dance
beneath the water.

Yet the krill are not here because they are wondrous. They are here because there is no easy way to study them in the hostile and icy waters of the Antarctic. And it is imperative we understand them as best we can, because their world is changing in ways that have frightening implications not just for the krill, but for the Antarctic ecosystem as a whole.

ANTARCTIC KRILL ARE CRUSTACEANS, but despite their appearance they are not a variety of prawn or shrimp. Instead, along with the eighty-four other species of krill, they form their own order, the Euphausiacea, which is separated from prawns and shrimp and their relatives, lobsters and crabs, by 100 million years of evolution.

Unlike prawns and crabs, which tend to live on or near the sea floor, krill swim freely in the open ocean. But they are also distinguished from other crustaceans by the sheer scale of their population. Although estimates vary, there is little question Antarctic krill are the most abundant wild animals on the planet, with a total biomass of somewhere between 300 and 500 million tonnes. Even at the lower end of that range this is close to five times the biomass of every living wild mammal, bird and reptile combined. Only humans and cattle weigh more, with the eight billion people on the planet clocking in at around 600 million tonnes and the billion or so cattle at slightly more again. And if these figures are so large as to feel meaningless, the number of individual krill – somewhere in the region of 800 trillion, although again this varies from year to year and season to season – is even more so.

While this vast population is distributed throughout the Southern Ocean and at all depths (although krill spend most of their time in the upper layers of the ocean, they have also been sighted 4500 metres down on the sea floor), as much as half of it concentrates into huge swarms of billions of individuals that gather near the surface.

The largest of these swarms, which can extend for kilometres and contain as many as 60,000 krill per cubic metre, are so immense they are visible from space and so complex they are effectively a form of super-organism in their own right. Video taken within swarms shows the massing krill as clouds of innumerable individuals that swirl and dance, like moths around a lamp, the aggregation of bodies so dense it reduces the light from above to a dull red glow.

The krill in Kawaguchi and King's tanks were caught off the coast of East Antarctica, somewhere between Australia's Davis and Casey stations, but 70 per cent of the world's krill population is concentrated in the region below the southwest Atlantic, an expanse of ocean that takes in the ice-bound Weddell Sea and stretches from the Drake Passage in the west to Queen Maud Land in the east. These waters, and in particular those off the Antarctic Peninsula and around the string of islands scattered from its tip to the South Georgia and the South Sandwich islands, seem to be attractive to the krill because their geography suits their complex life cycle and its close synchronisation with the annual ebb and flow of the sea ice and the solar cycle.

This cycle begins in spring as the ice that has formed in the winter darkness begins to retreat, exposing the waters of the polar sea to the Sun. As the days grow longer the increase in the amount of solar radiation triggers great blooms of phytoplankton. These attract the krill, which arrive in huge numbers from their winter hiding places to gorge themselves on the plankton.

While the krill devour the plankton the female krill begin to prepare to breed, the body of each swelling with up to 10,000 eggs. As the eggs approach maturity the krill migrate away from the coast to mass above the transition zones where the sea floor drops away into the deep ocean in preparation for spawning.

Mating is a speedy affair. The males place a bundle of sperm on the genital opening of the females using their front pleopods;

this then fertilises the eggs as they are expelled. The fertilised eggs drift slowly downwards until they are somewhere between 700 to 1000 metres below the surface. It is only then, in the gloom of the mesopelagic, that they finally hatch, and the krill larvae emerge.

In their initial larval form the krill look nothing like their parents: instead they resemble some sort of mouthless four-legged flea. Yet unlike the eggs, which were at the mercy of gravity and the currents, they are able to move under their own power. And so they begin to ascend once more.

The journey back to the surface takes several weeks – long enough for the larval krill to pass through several stages of development along the way. With each phase they move closer to their adult form, acquiring more legs, increasingly complex organs and, finally, as they reach the surface layers of the ocean, a mouth.

The larval krill are sustained by nutrients from their egg. But by the time they reach the sunlit zone once more that supply of energy has been exhausted, making it critical they find food as soon as possible. Yet this presents a problem, for in the time it has taken the juvenile krill to make it to the surface the brief Antarctic summer has ended, and the polar waters are hastening into winter once more. The bountiful supply of phytoplankton that sustained the adult krill only a few months earlier has disappeared.

The krill's solution to this challenge is ingenious. Although we tend to associate the frozen wastes of Antarctica with stark, white cleanliness, in fact the underside of the sea ice is home to a thriving ecosystem of ice-algae and other phytoplankton and microbes, all drawing sustenance from the light leaking through from above. Hidden beneath the rapidly expanding sea ice, the larval krill feed on this film, moving across the underside in a zigzag motion as they rake organic matter from the ice with specially adapted bristles on the tips of their thoracic legs. 'They're like the combine harvesters of the ocean,' says King, before pointing out that despite their focus on algae they're also quite capable of gobbling up any small animal unfortunate enough to get in their way. 'If combine harvesters did the same that would make it very dangerous walking around a farm during harvest,' he laughs.

The juvenile krill pass their first winter in this strange, inverted world, protected from predators by the cracks and runnels in the ice's underside. But as the summer returns and the sea ice begins to break up, they are forced to abandon its protection and move back into open water.

This swarming is fundamental to krill behaviour, although the precise reasons behind it are a matter of debate. What is certain is that it's at least partly about safety in numbers: as the aggregations grow, the odds of any particular krill being eaten decline. But by swimming in formation the krill are also able to confuse the visual systems of their predators, who find it difficult to select a particular krill to attack when so many bodies are moving in unison. The first time I read about this strategy, I found it unlikely: after all, how could it be difficult to choose a target? Yet watching the krill in the tanks, I grasp how effective it is. The synchronised movement of so many bodies is oddly mesmerising, transforming their massing numbers into a shifting whole, and making it almost impossible to focus on any individual krill.

It will be another year before they are sexually mature – although determining their age is difficult, krill seem to live to be about ten – but after leaving the ice the juveniles join the adults, and begin to congregate into swarms. The juvenile krill spend their first summer with the adults feasting on the phytoplankton and growing larger. Then as the days grow shorter and the sea ice begins to expand again, they disappear once more.

The question of exactly where they vanish to remains a mystery. 'Where do 400 or 500 million tonnes of krill go each winter?' says King. 'Do they sit under the sea ice and eat the tiny amount of plankton they can find there? Do they dive down to the sea floor and eat the detritus from the summer blooms and the faecal pellets that have drifted down? Or do they attack the benthic community for whatever food they can get?'

It is possible a new project involving cameras moored on the sea floor across the Antarctic winter will solve some of these questions. Nonetheless one part of the answer may have already been uncovered by accident during a separate AAD study that attached cameras to fishing lines to assess the impact of longline fishing on animals on the sea floor. 'The other team came up to So's office with some of their footage and said, "I think we filmed some krill down at 700 metres,"' says King. 'So started watching the footage and his mouth fell open. Because almost immediately a couple of krill mated in front of the camera, which is something nobody had ever seen before.' King laughs. 'We wrote a whole paper on those eight seconds of video.'

BOTH BY VIRTUE OF THEIR UBIQUITY and their sheer weight of numbers, krill are the cornerstone of the Antarctic ecosystem. Penguins, seals and baleen whales, such as humpbacks and blue whales, all depend upon them for survival, as do the fish and squid that sustain predators such as sperm whales. Unlike more complex food chains, in which large amounts of energy are lost at every level, the simplicity of the Antarctic food chain is extremely efficient: the phytoplankton transform the energy of the Sun and the nutrients they absorb into chemical energy, which the krill in turn consume and convert into energy that is consumed by penguins, seals and whales.

This efficiency is visible in the bodies of the animals at its top, many of which are startlingly dense with stored energy: as much as 30 per cent of the body mass of some penguins can be fat, while in Weddell seals this percentage can be even higher. But nowhere is it

more evident than in the bodies and the life cycles of the great whales.
Blue whales are the largest animals that have ever lived, growing
to up to 30 metres in length and reaching a weight of more than
180 tonnes – up to half of which can be fat (a greater proportion than
any other mammal). A female blue whale only feeds for half the year
but produces around 180 litres of milk a day: this has nine times the
fat content of human milk, and provides such concentrated sustenance
that her calf can gain as much as 90 kilograms a day and reach sexual
maturity at just five years of age. The milk of humpbacks is even
richer – with a fat content of up to 50 per cent it is closer to cream in
consistency – and females produce as much as 350 litres a day.

It was this stored energy that lay behind the horrifying massacre
of Antarctic mammals and birds during the nineteenth and twentieth
centuries. When James Cook landed in the Kerguelen Islands in
1776, he found seals 'so fearless we killed as ma[n]y as we chose to
make oil for our lamps and other uses'. Within a decade, sealers had
also begun to take advantage of this fearlessness, initiating a blood-
bath that resulted in the death of at least seven million fur seals during
the first three decades of the nineteenth century.

Before long elephant seals were also being slaughtered in their
millions, the blubber in their huge bodies flensed away and boiled
down to make oil. Even penguins were not spared: in the 1880s
and 1890s vast numbers of king penguins were butchered on
Macquarie Island, their bodies converted into oil in specially designed
digesters.

Yet nothing compares to the carnage inflicted upon whales
and their societies. Although the difficulties of whaling in the
remote and hostile southern latitudes meant industrial whaling in
Antarctic and Subantarctic waters did not really commence until
the early twentieth century, that changed with the establishment of
a commercial whale-processing station on South Georgia in 1904.
Initially attention was concentrated on humpbacks, which were

relatively slow and easy to catch. But as diesel-powered ships began to replace sail, and explosive harpoons took over from barbed metal, whalers began to pursue larger and faster whales. This led to the slaughter of almost 30,000 blue whales a year by the end of the 1920s; as the population of blues began to collapse, whalers transferred their attentions to fin whales, leading to catches in excess of 25,000 a year by the 1950s, at which point whalers moved onto sei whales and then, finally, minke whales. In all, more than two million whales were killed in the Southern Hemisphere in the twentieth century.

In the eighteenth and nineteenth centuries whale oil was mostly burned in lamps and in lighthouses, where it was prized for burning clear and clean, although cheaper grades tended to produce a revolting odour. Later, it was incorporated into products ranging from soap to industrial lubricant, nitroglycerine and linoleum. Yet by the twentieth century, whale oil's most important use was in food. In 1929 chemists working for Britain's Lever Brothers and the Dutch company Margarine Unie discovered how to harden whale oil into fat – and thus manufacture margarine. A merger of the two companies created the company we now know as Unilever, which used its market power to control the oil market so successfully that by 1935 84 per cent of the world's whale oil was being used for margarine (Unilever used whale oil interchangeably with other oils, such as palm and groundnut oils, as it occasionally left a 'fishy taste'). The economic and cultural significance of margarine is difficult to overstate: in the build-up to World War II the British declared the spread a 'national defence commodity', while the Nazis sent a secret mission to Antarctica in the hope of securing Germany's supply of whale oil by taking control of the waters off Queen Maud Land.

The effects of this grotesque destruction were not limited to the whales themselves. Although prior to the advent of industrial whaling whale populations pre-whaling consumed huge amounts of krill – close to half a billion tonnes a year – the collapse in whale numbers did not lead to an explosion in krill. In fact the opposite occurred: the krill population actually fell by as much as 80 per cent. This happened because whales are not simply passive consumers. Instead, in a reminder of the intricate webs of ecological connection and energy exchange that underpin ecosystems, iron-rich whale excrement provides vital nutrients for the phytoplankton upon which the krill depend; without this input phytoplankton numbers crashed, and so did krill populations. Furthermore, because whales play a vital part in transporting nutrients between different layers of the ocean and different parts of the ocean, the collapse in their numbers also reduced the ability of the ocean to absorb carbon dioxide: a 2019 study by the International Monetary Fund calculated that allowing whales to return to their pre-whaling populations would allow the oceans to absorb 1.7 billion tonnes of carbon dioxide a year, slightly more than Russia's total emissions.

Even today, when whale numbers are a fraction of what they once were, krill play an important role in this process. Like all plants, algae and phytoplankton absorb carbon dioxide, and as the krill consume them they absorb this carbon in turn. Most of this carbon is fixed in the matter of their bodies; the rest is expelled in faecal pellets that sink into the deep, locking away the carbon they contain. In the area around the Antarctic Peninsula and the Scotia Sea, krill faeces and shell casings sequester some 23 million tonnes of carbon dioxide across the summer alone. Once the contribution of other regions, larval krill and the downward movement of krill carcasses is included, this figure is likely to be considerably higher.

ALTHOUGH THE SCALE OF THEIR POPULATION and their contribution to ecological and planetary processes makes it easy to imagine krill only through the lens of statistics, they are also animals in their own right. When I ask Kawaguchi whether he ever wonders about the lives of individual krill, or what the world must be like for the elegant creatures darting about in his tanks, his face lights up. 'Often!' he says, laughing delightedly.

In practical terms this desire to imagine the way the krill think has affected the design of the tanks in which they are housed. Early on King and Kawaguchi realised normal schooling behaviour was a strong indicator that the captive krill are doing well. But designing tanks that encourage schooling proved surprisingly challenging: not only did the krill not school when placed in black-sided tanks, they also refused to school in clear-sided tanks. It was not until it occurred to Kawaguchi and King to try a white tank with no direct light that the krill began to behave normally.

Kawaguchi believes this is because it resembles the diffuse light and white boundaries of the world beneath the ice. But it is also because in clear-sided tanks the glass becomes reflective from within: 'The krill are very visual animals and use their eyesight to form schools; because they're always watching each other, being able to see their reflection confuses them, so they just retreat to the corner and hide.'

For Kawaguchi these challenges are a reminder of the vital role the krill's sociability plays in its success as a species. 'One hypothesis is that the krill form schools because it's energy efficient, and swimming and turning in unison is hydrodynamically effective. But they also seem to be able to sense and respond to the movement of their neighbours in the swarm through changes in the water. What that shows is they're really social animals, and it's that sociability that has allowed them to become the key species in the Southern Ocean.'

This sociability is equally evident in the krill's tendency to become stressed when separated from each other: if isolated from their shoalmates their hearts beat faster, just as they do when exposed to threats such as whales. But the scale and complexity of the krill's swarming also means scientists are really studying two things: the behaviour of the individual krill and their aggregate behaviour, which exhibits a form of collective intelligence, rather like a beehive or an ant colony.

'The swarms are a whole other animal,' says King. 'And when you're trying to understand how they're interacting with their ecosystem, or how they'll respond to fishing, then understanding how krill behave as that superorganism is just as important as understanding the behaviour of the individual krill that underpins it.' This is especially obvious in interactions with predators such as whales. 'When a whale attacks a swarm you get a whale-sized hole in it – not because the whale has eaten all the individual krill, but because the krill are using tail flicks to attempt to escape, and those tail flicks warn neighbouring krill, who also attempt to escape and so on, and the process of those tail flicks radiating outwards means the super-organism changes shape to avoid the whale.'

These massive concentrations are not just attractive to predators such as whales and penguins. Over the past fifty years a growing fleet of fishing vessels has begun to target the krill directly. Most of the krill caught by these ships is processed into feed for aquaculture or used to meet the skyrocketing demand for krill oil as a dietary supplement.

For some years now the body charged with managing the krill fishery, the Commission for the Conservation of Antarctic Marine Living Resources (CCAMLR), has set the allowable catch in the southwest Atlantic sector at 5.6 million tonnes, or 1 to 2 per cent of the krill's total biomass, with further limits imposed in specific areas in order to prevent overfishing of particular regions. Until recently catches have remained well below these levels, but in the past few years they have been growing rapidly: in 2022 the reported catch was just over 415,000 tonnes, more than double the catch in 2010. And this upward trend looks likely to continue: South Korea and Russia are both committed to increasing the size of their krill industries, while

China, which has already more than doubled its catch since 2019, recently commissioned the world's largest krill trawler, and Norwegian company Aker BioMarine, which in 2020 accounted for nearly half the total krill catch, is also expanding its fleet. The predominance of a relatively small number of countries and companies is partly a reflection of the technical and economic challenges krill fishing poses: the ships that harvest the krill are floating factories that are able to operate for weeks at a time in the icy waters of the Antarctic. But it is also a reminder of the ongoing transfer of the value of ocean resources to richer nations at the expense of all, a situation that is underlined by studies showing the value of the carbon sequestered by the krill is at least sixty times higher than the value of the fishery.

Concerns about the effect of fishing have been sharpened by research suggesting krill populations in the southwest Atlantic sector have fallen by 70 to 80 per cent since the 1970s, probably due to changes in sea ice. Kawaguchi, who works closely with CCAMLR, is wary of such figures: while he acknowledges krill populations may have declined and the range of krill habitat may have shifted southwards across that period, he argues the data is not robust enough to be so definitive. Instead he believes the real threat of fishing is not to the krill but to the animal communities that depend upon them. 'When they're raising young, land-based predators such as penguins and seals are restricted to shorter journeys of a day or so. That means their breeding colonies are positioned very close to where the krill aggregate. The problem is that those regions of reliable production are also very attractive to fishing vessels, and if they keep returning to them there is the potential they will deprive the penguins and seals of the food they need to rear their young, which could cause a significant and irreversible effect on that particular population or ecosystem.'

Kawaguchi argues this problem can be managed effectively by ensuring fishing is avoided in vulnerable areas, or at times when

penguin and seal populations are particularly reliant upon an abundance of krill, and points to CCAMLR's continuing efforts to assess
these risks. Others are less sanguine: a 2022 report by Imperial
College London scientist Dr Emma Cavan found that the krill
fishery is becoming increasingly spatially concentrated, with ships
often returning to the same areas year after year, and that in recent
seasons several juvenile humpbacks have died after being caught in
trawling nets. The report also raised concerns about the accuracy
of the methods used to measure catches, as well as noting that despite
the strict reporting guidelines imposed by CCAMLR, over the past
five years catches have not only consistently exceeded catch limits, but
the amount by which they have exceeded them has been rising year on
year. Perhaps most ominously, Cavan's report notes that the current
catch limits do not account for environmental variability or the impact
of climate change.

IT IS THIS LAST QUESTION that looms largest in the minds
of many scientists. Every year as the winter descends, the surface of
the ocean around Antarctica begins to freeze, and the sea ice around
the continent expands outwards; every year as the summer returns
it retreats inwards once more. Because the krill's life cycle is closely
synchronised with this annual dance and the solar cycle that drives it,
they are acutely vulnerable to changes in it. Denied abundant supplies
of phytoplankton at the right moment, female krill grow at a slower
rate and produce fewer eggs, reducing their ability to spawn effectively.
Similarly larval krill must find sea ice, and therefore food, within ten
days of their return from deeper waters, or starve.

Until 2015 the sea ice seemed to be resisting the warming that
has affected the rest of the Antarctic. Yet since 2015 the total area
covered by sea ice around Antarctica has declined rapidly. In 2017
and 2018 both the winter maximum and summer minimum coverage

were the lowest on record. Coverage recovered slightly in subsequent years, but in the first months of 2022 the sea ice dropped below 2 million square kilometres for the first time. This was only the beginning. In 2023 the ice shrank even further, falling to a mere 1.79 million square kilometres during February. After previous record lows the ice had always recovered relatively normally across the winter months, but in 2023 its recovery was far slower. By July, when the ice would usually have expanded to around 16.4 million square kilometres, it had only reached 14.1 kilometres. This anomaly was so large that it is what statisticians dub a five-sigma event, or five deviations beyond the mean. The odds of a five-sigma event are so remote that it is a virtual certainty that what is taking place is not simply noise in the data or statistical fluctuation – speaking in July 2023, University of Tasmania oceanographer Edward Doddridge calculated the odds of the anomaly occurring naturally were about 7.5 million to one. Instead it seems the forces that once powered annual rhythm of the sea ice have been fatally disrupted, and the ice has shifted into a new state. In other words the record-breaking sea-ice anomalies in 2022 and 2023 herald a more profound transformation of the Antarctic environment, in which sea-ice extent will continue to decline.

The loss of sea ice is already having serious impacts on species such as emperor penguins that rely upon it: when sea ice in the Bellingshausen Sea broke up early in 2022 thousands of penguin chicks drowned, leading to what scientists describe as 'complete breeding failure' in all but one of the colonies in the region.

Changes in the behaviour of sea ice also have the potential to drastically affect krill populations. One 2016 study found that the combined effects of warming water and shrinking sea ice could lead to a reduction in krill habitat of as much as 80 per cent, while more recent work makes it clear that as ocean temperatures rise the krill's habitat will contract southwards and deteriorate significantly. There is also evidence that populations of salps – free-floating, gelatinous animals sometimes known as sea grapes or sea snot, which also feed on phytoplankton and form huge aggregations like krill – are increasing, raising the possibility they may supplant the krill's role in the Antarctic ecosystem.

These changes may already be affecting animals that rely on krill. Humpback whales spend the summer in the Antarctic, gorging themselves on krill before migrating thousands of kilometres northwards into warmer waters to calve and rear their young. Despite the huge distance they travel the whales do not eat during their migration north and south – they survive on the fat they build up during summer, sometimes losing up to half their body weight across the course of their journey. The demands of this process often push the whales to the edge of their physical endurance, and mean it is vital they are able to locate abundant krill easily and reliably during the summer.

It is significant, therefore, that in recent years whales arriving on Australia's east coast have been in poorer condition, with lower levels of body fat and higher concentrations of organic pollutants in their blubber. This was particularly pronounced in 2017, when low levels of sea ice and the break-up of the Larsen C Ice Shelf on the Antarctic Peninsula resulted in wide areas of open water where there would usually be ice, and in 2011 when an unusually strong La Niña caused sea ice to retreat further than normal.

Data about the effects of 2022's record low levels of sea ice on the whales is not yet available, but observations suggest their behaviour may have been affected. Griffith University's School of Environment and Science's Professor Susan Bengtson Nash heads up the Humpback Whale Sentinel Programme, which coordinates long-term, standardised biomonitoring of humpback whales to help track changes in Antarctic ecosystems. According to her, '2021 was certainly a very different year. Even before we went into the field, we had the tourist operators in Hervey Bay asking where the whales were, because they hadn't arrived. And although they did finally turn up, they then left early as well. And the reports from New Caledonia and Brazil were similar: the whales arrived late and then disappeared very quickly.' Disturbingly these difficult years also seem to be associated with sharp upticks in the number of whales stranding themselves: in 2021, 230 humpbacks stranded themselves in Brazil, almost double the previous record set in 2017. 'The question now is whether we're going to see these events occur with greater frequency,' says Bengtson Nash. 'And if we do, will they still be interspersed with bumper years that somehow counter the poorer years?'

Rising temperatures are not the only challenge facing the krill. As the ocean absorbs more carbon dioxide from the atmosphere, seawater is acidifying. This is especially true in polar regions, where the cooler water absorbs carbon dioxide more efficiently. For animals with calcium carbonate shells or internal structures – such as molluscs, crustaceans and corals – this can have serious effects on growth and development, or in extreme cases result in their shells simply dissolving.

Research suggests adult krill are mostly unaffected by the sorts of increases in ocean pH expected in coming decades. But experiments on embryonic krill exposed to the levels of acidity likely to be experienced in deeper waters by 2100 showed they failed to develop, prompting fears of a tipping point beyond which krill populations could almost entirely collapse. According to Kawaguchi and his colleagues this could have 'catastrophic consequences for dependent marine mammals and birds of the Southern Ocean'.

'The problem is that it isn't just warmer water, or changes in sea ice, or acidification or even organic pollutants,' says King. 'It's the synergistic effect of all of them at once.'

Kawaguchi agrees. 'The krill are experiencing an environment that they've never experienced before, and that's very difficult to comprehend.' But he believes that when change arrives, it may arrive abruptly. 'These ecosystems are operating in a very fine balance. That means that once things change, there might well be a major shift in the structure of the system.'

ELSEWHERE IN ANTARCTICA that shift in the structure of the system is already underway. In mid-March of 2022 Antarctica experienced an unprecedented heatwave. At France and Italy's Concordia Station, high up on the frozen plateau of East Antarctica, temperatures were more than 40 degrees above the March average; while at Vostok Station, some 560 kilometres away, the maximum temperature was 15 degrees above the previous record, and at Australia's Casey Station temperatures were close to what would normally be the midsummer maximum. Although part of a broader trend – over the past fifty years temperatures on the Antarctic Peninsula have risen by more than 3 degrees, triple the global average – this was far outside the bounds of anything experienced before.

'Antarctic climatology has been rewritten,' said one scientist, who described the temperature anomalies as 'unthinkable'. Another said it 'upended our expectations about the Antarctic climate system'. A week later East Antarctica's Conger Ice Shelf abruptly collapsed, its almost 1200-square-kilometre structure disintegrating without warning in a day or two. The cause of the collapse is not yet clear, but it seems likely the heatwave in combination with the effects of warming water beneath the shelf pushed it past a tipping point and its integrity simply gave way.

The Conger is not the first ice shelf to break up in recent years. Over the past two and a half decades large sections of the Larsen Ice Shelf – a vast structure that once extended out into the Weddell Sea from the Antarctic Peninsula's eastern flank – have collapsed. The northernmost section, Larsen A, broke up in 1995; this was followed by the much larger Larsen B in early 2002, a process that saw almost two thirds of the ice sheet – 3300 square kilometres of ice some 220 metres thick – break away into the sea. Finally, in January 2022, much of what remained of Larsen B broke up in a few days, releasing yet more ice into the ocean.

Meanwhile the even bigger section known as Larsen C has been growing increasingly unstable. The fourth-largest ice shelf remaining in Antarctica, Larsen C covered some 50,000 square kilometres – at least until 2016, when scientists observed a huge crack, more than 100 kilometres long, along its seaward side. Within months the entire section, which accounted for about a tenth of Larsen C's total area, had broken away to form an immense iceberg 175 kilometres long, 50 kilometres wide and 200 metres thick. What remains of Larsen C seems to be held in place by two distinct ice rises, or areas where the ice comes into contact with the sea floor. But unprecedentedly early melting in 2019 and 2020 that resulted in large bodies of meltwater on the shelf has given rise to concerns the entire structure might suddenly destabilise in the same way as Larsen B or the Conger.

The behaviour of Antarctica's ice is both extremely simple and incredibly complex. Its ice sheets, almost 5 kilometres thick in some places and containing 70 per cent of the world's fresh water, are constantly replenished by falling snow; as the snow builds up it gradually compacts into ice, which then flows downwards towards the ocean in the great rivers of ice we call glaciers. Where these meet the water they gradually melt from beneath, resulting in floating ice shelves like Larsen and the Conger, which project outwards over the ocean for tens or even hundreds of kilometres, before finally beginning to break up.

The speed at which glaciers flow depends upon a host of variables, ranging from the slope beneath them to the presence of obstacles and barriers such as ridges and valleys beneath the ice. Yet in Antarctica one of the most significant constraints on the movement of glaciers are the ice shelves. These slow down the loss of ice into the ocean by acting as barriers. Without them the glaciers behind them slip outwards into the ocean faster and faster, accelerating the rate of ice loss and pushing sea levels rapidly higher: following the collapse of Larsen B, the rate of flow of some of the glaciers behind it increased sixfold.

The effects of this process are already starkly visible. Since the 1990s the rate of ice loss in Antarctica has more than quadrupled, rising from around 50 billion tonnes a year to more than 200 billion tonnes a year. And this process is accelerating: in the five years between 2012 and 2017 alone the rate of loss tripled.

At Pine Island Bay in West Antarctica, where the Thwaites Glacier flows into the Amundsen Sea, there is growing evidence this process may be the first stage in a much larger transformation. Sometimes dubbed the Doomsday Glacier, Thwaites is a massive river of ice 120 kilometres wide and three times the size of Tasmania. It is so big that if it were to collapse entirely, the ice locked up in it would raise sea levels by 65 centimetres.

Even without the hastening effect of global heating Thwaites is highly dynamic, flowing oceanward at an astonishing 2 kilometres a year. Until very recently its eastern side has been constrained by a massive ice shelf, which obstructs that end of its flow and is in turn restrained by a pinning point where the sea floor rises high enough to hold the shelf in place. Yet as ocean temperatures rise ever higher, warmer water is moving in beneath the shelf and melting the under-side of the glacier, causing Thwaites' grounding line – the point where the ice on the glacier's underside parts company with the sea floor and begins to float – to retreat. This is hollowing the glacier out from below, weakening its structure.

These same processes are at work at many locations around Antarctica. And it only takes a quick glance at a topographical map to understand why they are so concerning. Although we are used to thinking of the continent as roughly circular, or teardrop-shaped, such representations are deceptive, and much of what looks like land is nothing of the sort. It is ice, kilometres thick, the base of which lies hundreds or even thousands of metres below sea level. This is especially true of West Antarctica, where the ice sheet sits in a massive basin bounded on one side by the Transantarctic Mountains and on the other by a scattering of what would otherwise be islands.

Thwaites sits on the seaward edge of this basin. Behind it the ground drops away rapidly, meaning that as the warm water eats away at the underside of the glacier it flows downhill, allowing it to spread further and further beneath the surface of the ice and destabilising the structure above.

Exactly what this will mean for glaciers like Thwaites is not yet clear. Some scientists speculate it might result in a phenomenon called marine ice-cliff instability in which the glacier begins to crumble under its own weight, leading to a runaway collapse of the entire structure. What we do know is that as the ocean around Antarctica warms, the rate at which Thwaites' grounding line is retreating is increasing. On the glacier's eastern side, the rate of retreat has doubled from approximately 600 metres a year between 1992 and 2011 to 1.2 kilometres a year between 2012 and 2017. Even on the western side, where the rate of retreat remains relatively steady at between 600 and 800 metres a year, there are signs of dramatic change. In 2019 NASA researchers discovered a vast cavity, two thirds the size of Manhattan and approximately 300 metres high, below the glacier. Most of the 14 billion tonnes of ice that would once have filled its space is believed to have melted in just three years.

Meanwhile the ice shelf that holds the glacier's eastern section in check is also on the verge of collapse, its structure fatally weakened by a rapidly spreading network of fissures. In 2021 scientists predicted complete collapse was likely to occur within five years. Like the breakup of Larsen B, that process is likely to be rapid. American glaciologist Erin Pettit has compared the shelf to a windshield with a single crack in it. 'You're like, I should get a new windshield. And one day, *bang* – there are a million other cracks there.'

Because the collapse of the West Antarctic Ice Sheet will raise sea levels by over 3 metres, making clearer sense of its dynamics is vital. Yet understanding of the behaviour of Antarctica's ice is constrained by the fact our observations only go back a few decades. In order to fill these holes in our knowledge scientists have turned to the deep past, and in particular the period of warming that took place between 130,000 and 115,000 years ago during the last Interglacial, when global temperatures rose several degrees. In 2019 Chris Turney, professor of Earth Science and pro vice-chancellor (research) at the University of Technology Sydney, led a team that collected blue ice deposits in the Patriot Hills, on the eastern edge of the West Antarctic Ice Sheet. Blue ice occurs in regions where fierce katabatic winds remove upper layers of snow and ice, allowing ancient ice to rise to the surface. The samples Turney's team found revealed a break in the ice-sheet record that coincided with the last interglacial, suggesting an almost complete melting of large parts of West Antarctica had occurred. 'It doesn't look like anything – there's not a hole in it,' Turney says. 'It's just that when you date it you realise there's this massive gap in the record.'

Further evidence this gap in the ice is the result of a near-total collapse of the West Antarctic Ice Sheet has since emerged from an unlikely source. Scientists used DNA from specimens of Turquet's octopus collected from multiple locations around Antarctica to trace the history of the species, using changes in the DNA like a clock to map out movements of different populations of the species over hundreds of thousands of years. This revealed that about 125,000 years ago the populations of Turquet's octopus in the Weddell and Ross seas, which lie on opposite sides of West Antarctica, intermingled – something that would only have been possible if the ice sheet separating them had largely disappeared.

Having established the ice sheet melted, Turney's team then used tiny shards of volcanic glass embedded in the ice immediately before the gap to pinpoint when and how fast the melting occurred. This revealed the ice sheet collapsed very quickly and at under two degrees of warming. Dr Zoë Thomas is a paleoclimatologist at the University of New South Wales and one of the researchers on the Patriot Hills study. She says the fact the melting occurred well before temperatures peaked has frightening implications for today. 'It wasn't a situation where the ice melted gradually as the temperatures rose; it was a very fast response and it happened very early.'

Armed with the knowledge the West Antarctic Ice Sheet collapsed during the last interglacial, Turney, Thomas and some of the others associated with the Patriot Hills study attempted to refine their understanding of these periods of rapid change by looking at the melting that took place after the Last Glacial Maximum 21,000 years ago. To do this they turned to the record left by ancient icebergs. As icebergs break off the edge of Antarctica they are swept away by the currents that encircle the continent. Some find their way into the southern reaches of the Pacific and Indian oceans, but most are propelled up into the South Atlantic near the Antarctic Peninsula. As the bergs pass through this stretch of water often referred to as Iceberg Alley, they begin to melt, releasing dust and other sediment that drifts to the sea floor far beneath.

By using drill cores extracted from the sea floor in Iceberg Alley, Turney, Thomas and their colleagues were able to map the rate of sediment deposit over the past 21,000 years. They found eight distinct periods in which the amount of sediment spiked, suggesting an increase in the number of icebergs, and therefore the rate of ice loss. By cross-referencing these spikes with information about pulses in sea-level rise over the same period, they were able to show that each of these spikes corresponded with a period of rapid sea-level rise lasting several hundred years.

This established that these periods of rapid sea-level rise were caused by ice loss in West Antarctica. But the sediment deposits also made it clear that this ice loss did not begin gradually. Instead, like flipping a switch, the ice sheets shifted abruptly from a state of relative stability to rapid mass loss. This transition from one state to the next took place in less than a decade, and once started, continued for hundreds of years. 'The nature of tipping points is that once the thresholds are passed, you have all these positive feedbacks that accelerate the melt – and that means things happen really, really fast,' says Thomas, who also sees 'concerning parallels' between the past and the present. 'The data suggests we're already past a tipping point. If we're lucky it might be a small one that only results in sustained mass loss for a few decades, but that's kind of irrelevant because any mass loss causes sea-level rise and any sea-level rise, is dangerous.' Dr Michael E. Weber, from the Institute for Geosciences at the University of Bonn and lead author on the paper, is even blunter, saying the findings are 'consistent with a growing body of evidence suggesting the acceleration of Antarctic ice-mass loss in recent decades may mark the beginning of a self-sustaining and irreversible period of ice-sheet retreat and substantial global sea-level rise'.

Trying to get to grips with the idea of this kind of rapid and unstoppable change is not easy, but perhaps one way to grapple with its immensity can be found in the work of the MELT project (Melting at Thwaites grounding zone and its control on sea-level rise), which in the first weeks of 2020 created a 700-metre-deep shaft through the glacier and lowered a submersible into the water beneath.

A section of the resulting footage is on YouTube. In the top half of the image the bottom of the ice shelf moves past overhead, its dirty grey-green flecked with pale specks and fissures oddly reminiscent of videos of the Earth moving beneath an orbiting spacecraft. Between the camera and the ice a glassy layer in the water marks the threshold between the fresh meltwater above and the saltier seawater below; ice crystals and shards float along it, together with the drifting form of an anemone, its delicate tendrils resembling a strange, alien craft escaped from a science-fiction film. As the submersible rises through the shimmer of this halocline, the bottom of the ice shelf comes suddenly, sharply into focus, its surface ridged and scalloped. The tiny forms of fish or other swimming things flick by out of range before some sort of crustacean darts past, its body bleached and overexposed in the light from the submersible. There are several seconds more of the upside-down moonscape of this ice shelf's underside, a final glimpse of another crustacean, and then it is done.

Even without the piano music somebody has laid over it there is a haunting beauty to the video and its brief glimpse of this hidden world. Yet like the 'Earthrise' photo taken by the crew of Apollo 8, it also speaks to something more profound. For these few seconds of video offer a glimpse of planetary change in action. In them we see the inflection point of the climate crisis made suddenly, inescapably tangible – a transition zone not just between ice and water, but between past and future and the deep time of the ancient, slow-moving ice and the hastening fears of our brief lifespans. And in that collapsing of temporalities it is possible to glimpse the rushing collision of the human and the planetary, and also something of the way that inflection reveals the connection between the heedless consumption of the wealthy few and the dislocation and displacement of the rest, and the imbalances of power and corporate greed that allow that to continue.

THE ICE IS THE MEMORY of the world. It contains minute particles of soot and pollen and bubbles of gas that testify to changes in the atmosphere over millions of years. These allow us to trace not just the great cycles and irruptions of the planet's climate, but also the history of human expansion and transformation of the environment.

The ice sheets bear the signature of nuclear tests carried out in the 1950s and 1960s and the smelting of metals and other industrial processes, as well as traces of PCBs. They even offer evidence of early agriculture and the transformation of the landscape by colonisation: ice cores extracted from James Ross Island, just off the northern end of the Antarctic Peninsula, hold soot that corresponds with the burning of forests to open up land for agricultural use by the Māori upon their arrival in New Zealand just over 700 years ago, and may also offer evidence of the destruction wrought by European colonists in Patagonia, several hundred years later. Some scientists even believe that the global pause of the pandemic's first year will be written into ice cores extracted hundreds of thousands of years from now.

But what these cores also record is a period of unusual climactic stability, extending from about 11,700 years ago until the beginning of this century. Known as the Holocene, this window of relative calm made possible the development of agriculture, oceanic and continental trade and expansion, and finally the accumulation and concentration of wealth and power that gave rise to globalised society.

This stability would not have been possible without the Antarctic and its ice. The Antarctic Circumpolar Current and the systems of heat and saline exchange driven by Antarctica's sea ice help drive the conveyor belt of the world's currents, while the continent and its sea ice help reflect the Sun's rays back into space to keep the planet's temperature in equilibrium.

Yet the Antarctic's ice also shows this period of stability is coming to an end. By examining the chemical composition of the bubbles of air trapped inside the ice it is possible to chart the rise and fall of greenhouse gases over hundreds of thousands of years. This shows they move in lockstep, changes in global temperature echoing fluctuations in carbon dioxide.

For most of the Holocene the concentration of carbon dioxide in the atmosphere has remained within a very narrow band. But over recent decades it has risen precipitously. Plotted on a graph the dizzying speed of this process is stark. Just over 11,000 years ago atmospheric carbon dioxide was around 260 parts per million. When the Industrial Revolution began it was 280 parts per million. In 1900 it was under 300 parts per million, by the late 1980s it had reached 350 parts per million, and just thirty years later, in 2017, it passed 400 parts per million. In 2022 it was 417 parts per million, and still rising.

This matters because the last time levels of atmospheric carbon dioxide were this high was four million years ago, during the Pliocene. Global temperatures were 3 degrees warmer than they are today and the collapse of the Greenland, West Antarctic and parts of the East Antarctic ice sheets meant sea levels were more than 20 metres higher. A rise of even a fraction of that amount would devastate contemporary coastlines and displace hundreds of millions of people. And that future is already here. While there may still be time to avert the worst-case scenarios, even the most optimistic predictions suggest the melting underway in Antarctica, Greenland and elsewhere will lead to 10 to 25 centimetres of sea-level rise by 2050, and more than half a metre by 2100 – the collapse of Thwaites alone will add 65 centimetres to sea levels over the next century. Beyond 2100 the uncertainties grow larger: the loss of West Antarctica could raise sea levels by 4 metres, Greenland 7 metres, and East Antarctica many more. In other words the question is no longer whether the ice will melt and sea levels will rise, but how fast and how far. The choices of the past mean there is no way back. Change is coming whatever happens. The challenge that faces us is how to respond to that change.

A planet transformed in this way will be radically less hospitable to human life. But what of the krill? How will they survive in such a radically altered world? If their numbers fall, what will that mean for the Antarctic's ecosystems and the species that depend upon them?

Neither King nor Kawaguchi believe the krill are likely to disappear entirely. But that does not mean they are not deeply concerned. 'Their life cycle is so linked to the dynamics of the sea ice, you have to ask where they're going to hide from predators or find food without it,' says Kawaguchi.

King agrees. 'We know they're robust and durable animals because they've survived previous extinction events when 70 to 90 per cent of marine species disappeared. So it might just be an unprofitable time for them. But I also see potential for massive disruptions in the amount of krill and the predictability of their presence and numbers year to year. And that's a serious problem for the species that rely upon them, because they can't just swim another thousand kilometres if the krill aren't where they're supposed to be – they'll be very unwell or dead before they get home if they try.' He pauses. 'The problem is what we're doing, the change that's taking place, it's just happening so fast. That's a real challenge.'

These questions are also a reminder of the degree to which our own future depends upon the fate of the Antarctic. Geologists call gaps in sedimentary strata unconformities. Their absence is a physical record of discontinuity, of a schism between past and future. The gap in the ancient ice Chris Turney and his team discovered in the Patriot Hills offers an example of the same kind of discontinuity; so too does the moment we are hurtling into right now. What lies on the far side is still unknown. What will the people of the deep future find written in the ice? Will they see a spasm of mass extinction and suffering? Or a story of transformation and survival?

13

HORIZON

'And the state of emergency is also always a state of emergence.'
Homi K. Bhabha

ONE MORNING IN SEPTEMBER 2023 I leave my home and drive to the beach. Although it is early, the day is already unseasonably warm, the sky hazy with smoke from hazard reduction burns to the south and north of the city.

Despite the weather the beach is quiet. Walking to the water's edge I wade out and dive, then stroke outwards until my breath gives out, so I surface with a gasp. Although I have not swum all that far I am already out past the break, so instead of swimming on I turn back towards the beach and tread water slowly. Closer in a few people are waiting for the waves that roll in now and then, behind me three or four swimmers are stroking their way across the bay, but otherwise I am alone.

There is something very particular about looking back towards the shore from deeper water. When I was younger, and my friends and I would slip away from work in the late afternoons to surf, I always loved drifting out beyond the break as evening approached – the way the colour would bleed out of the world, until it was just you and the movement of the swell. Today, though, it seems enough to just float here.

I suppose that in some part of myself I am taking stock. Three and a half years ago, in the first weeks of 2020, my partner, my children and I headed south from Sydney to join my mother and my stepfather and my brother, his partner and their children for a week by the beach. It was vilely hot that week, the air still thick with smoke and the water choked with ash and burned leaves from the bushfires that had devastated the east coast of Australia over previous months, killing thirty-four people and sending the populations of communities further south fleeing onto beaches to await rescue by naval troop carriers.

Against the backdrop of this horrifying conflagration, my family and I were grappling with a much more intimate crisis. After multiple rounds of surgery and years of chemotherapy, my mother's cancer had advanced to a point of no return. Twice in the year prior we had been told the end was imminent and, although she had rallied both times, she was now so weak we all understood the end was near.

On our fourth afternoon, my brother and I noticed a line of pink, soapy scum along the waterline. Realising it was what Australians call 'bio', we decided to return to the beach in the evening to see whether there would be a display of luminescence.

It was almost nine by the time my brother, our partners, and I returned to the beach with the kids. I wasn't sure what to expect, and indeed as we descended onto the beach it seemed I had been right to temper my expectations, for beyond the pale strip of sand, the water was dark. But as my eyes adjusted to the dimness I glimpsed something else, something wondrous. Where the waves broke on the sand, pale blue light filled the water, shifting and flaring as it spilled inwards, coating the beach with sheets of fading radiance. I stopped, amazed, and stared out over the bay. As far as the eye could see, diaphanous skeins of illumination danced in the water, detonating soundlessly in its darkness like lightning moving beneath the surface.

Turning, I saw my children and their cousins were jumping up and down and sprinting here and there, shrieking in delight at the patterns of light that radiated from their feet across the wet sand – glowing traceries that resembled the spreading networks of biological systems, whether veins or the mycorrhizal threads of fungus, lingering for a few seconds before fading. Laughing, my brother and I joined them, stamping our feet and leaping in the air, taking a childlike glee in trying to create the biggest impact.

After a few minutes I noticed figures approaching through the darkness and, stopping for a moment, saw my stepfather helping my mother pick her way across the sand towards us. I went to them and,

taking her arm, led her down to the water's edge. While my stepfather and my brother played with the kids she and I stood, arms linked,
watching the light flicker in the waves and on the sand.

'Have you ever seen it like this before?' she asked me.

I shook my head. 'A few times. But never so much of it at once.'
What I didn't say was that the increasing incidence and growing range
of the organism that causes such displays of luminescence, the single-
celled dinoflagellate *Noctiluca scintillans*, corresponds to rising ocean
temperatures, meaning its uncanny beauty is also a harbinger of accelerating environmental breakdown.

She took my hand and squeezed it, and for a moment I felt not just
how thin she was, but how much she wanted to stay. Three and a half
years is not much against the span of a life, yet the gap between that
night and today is so vast as to seem unbridgeable. My mother is gone,
her body finally failing in late March of 2020, her last weeks played
out against the background of soaring mortality, economic emergency and the panicked lurch towards lockdowns and closed borders
as COVID-19 swept across the planet.

The children who played on that beach lit by the bioluminescent
glow of billions of *Noctiluca* have grown older, and infinitely more
complex, their lives blown off course by the storms of adolescence
and the disruptions of the pandemic, so severely in the case of one of
them that there have been times we did not think we would ever get
her back. I am changed as well: older, less certain, more aware of the
constant proximity of disaster.

The world has also been transformed. A hastening wave of disaster
has made it clear that the first stages of climate breakdown are upon us.
After three years in which floods, heatwaves, fires and storms devastated communities and ecosystems around the world, July and August
2023 were the hottest months ever recorded, exceeding previous
extremes set in 2016 and 2019 by wide margins. In Asia, Europe, North
America and North Africa, records have been shattered over and over

again. Sanbao in China hit 52.2 degrees Celsius; near the Arctic Circle
in Canada temperatures reached almost 38 degrees Celsius; and in
Phoenix in the United States, where temperatures exceeded 43 degrees
Celsius for thirty days straight, hospitals were crowded with people
who had suffered burns from falling onto the pavement. Alongside
another summer of record heat, fires have consumed tens of millions
of hectares of forest in Canada, Russia, Greece, Spain, Algeria and
even Hawaii, while floods and storms have devastated communities
across Asia, Europe and North Africa. Only a few days ago two dams
in Libya burst after a storm dumped nearly half a metre of rain on a
region that normally only sees a millimetre or two a month; the ensuing
flood swept away whole districts and left thousands of people dead.
Meanwhile ocean temperatures have also moved into uncharted territory, rising half a degree above previous records in the North Atlantic
and reaching more than 38 degrees off Florida, with catastrophic
implications for corals and other marine organisms in the region.

Some have taken to calling this wave of rolling disaster the new
normal. But what the world has experienced over the past months and
years is not a new normal – it is just the beginning. Combined with the
legacies of centuries of colonial violence and extractive processes, the
reckless burning of fossil fuels has pushed the planet into a danger-
ously unstable new state. We now live in an Age of Emergency that
will not end in my lifetime.

The brutal reality is that the world has already warmed 1.2
degrees, and 2023's record-breaking temperature anomalies increase
likelihood we will crash through the 1.5 degree guardrail in less than
a decade. Even if implemented in full, the emission reduction targets
announced to date by nations around the world will not prevent this;
instead they place the planet on a path to two degrees of heating. The
level of real-world action falls even further short of what is needed,
committing the world to a temperature increase of 2.5 to three degrees
by the end of the century.

A world that is that much hotter is almost unimaginable. A temperature rise of just 2 degrees will have catastrophic effects on the planet and human life. Heatwaves and extreme weather events will increase significantly. More than half the world's population will be affected by water scarcity, and political instability due to water insecurity will intensify in many regions. Food production will decline markedly, especially in regions such as sub-Saharan Africa, Southeast Asia and Central and South America, contributing to what is described as 'a disproportionately rapid evacuation' of people from tropical countries as extreme heat, water scarcity, decreased food production and social breakdown make them unviable, resulting in population increases of up to 300 per cent in the tropical margins and subtropics. The distribution and incidence of tropical diseases such as malaria and dengue fever will increase significantly as the regions in which the conditions that sustain them expand. The impacts on the non-human world will be even more drastic. Extinction rates, already running at thousands of times the background average, will soar to levels not seen since the last of the mass extinctions that have periodically devastated Earth's ecosystems. Recently-observed collapses in insect populations will accelerate, severely disrupting ecosystems and food production. Coral reefs, already under sustained assault, will all but disappear. Warming and acidifying waters will severely impact the fisheries that provide one-third of the world with their principal source of protein.

Worse yet, with each fraction of a degree of heating the likelihood of sudden and non-linear change increases. In 2008 scientists identified nine global tipping points, boundaries beyond which the process of change becomes self-perpetuating, as a cascade of elements feed on themselves, leading to rapid and irreversible breakdown. In 2022 a new study added seven more regional tipping points, and produced evidence suggesting we may have already pushed the planet past the threshold of five of them. The uncontrolled burning of fossil fuels means we are already at risk of triggering the unstoppable collapse of the Greenland and West Antarctic ice sheets: thawing of the permafrost, the death of reefs in tropical waters, and an abrupt collapse of the North Atlantic subpolar gyre and Labrador Current systems. By 1.5 degrees of warming it becomes increasingly likely these five tipping points will be passed, as will a sixth – the abrupt collapse of the ice in the Barents Sea, causing a rapid inflow of warm water into the Arctic Sea – and four more, including the dieback of the boreal forest and the weakening of the Atlantic Meridonal Overturning Circulation (AMOC), become possible. Beyond 2 degrees the remaining six tipping points move into the realm of possibility.

Exactly when the planet will breach these tipping points is difficult to ascertain. Some, such as the collapse of the West Antarctic Ice Sheet, may well be in motion already, and recent – if controversial – research suggests the shutdown of the AMOC may be much closer than previously thought. But each one that is breached will trigger further changes that will further accelerate global heating. Like dominoes falling this increases the chances of others being exceeded – a cascade of disaster that has the potential to push the planet into a radically less habitable hothouse state.

The effects of this process are already transforming the world. More than half of the 60 million internally displaced people who were forced to flee their homes in 2022 did so as a result of natural disasters such as cyclones, flooding and drought; many more were displaced by conflicts caused or exacerbated by the effects of the climate crisis. And these numbers are only going to continue to rise: the World Bank predicts that by 2050, more than 200 million people will be displaced internally by the effects of the climate crisis, while the Institute for Economics and Peace predicts that without rapid action to address the impacts of climate change, 1.2 billion people will be on the move by 2050. In the words of the United Nations secretary-general, António Guterres, without rapid action to curb emissions and reshape the world economy we face 'a mass exodus of entire populations on a biblical scale', and increased insecurity as a result of displacement and 'ever-fiercer competition for fresh water, land and other resources'.

THIS CRISIS IS SO IMMENSE, so complex and so seemingly intractable, that it can sometimes seem impossible to make sense of. The philosopher Timothy Morton argues climate change is an example of what they dub a hyperobject: a phenomenon so complex and massively distributed through time and space that it exceeds human comprehension. But as the stories in this book suggest, the ocean provides a way of reimagining these questions, of seeing the currents and tidal forces that have borne us here, the way the waves of migration and encounter and exploitation flow across continents and timescales. Indeed in a very real sense the ocean is the original hyperobject: attempting to comprehend its immensity and fluid multiplicity alters us, making it possible to glimpse new continuities and connections.

Simultaneously, though, the ocean reveals that the roots of the crisis we inhabit lie deep in the patterns of violent exploitation and extraction that have shaped the modern world. For those like myself who are the beneficiaries of these historical processes, acknowledging the truth of this violence and its legacies can be confronting, but it is necessary. As the late Sven Lindqvist observes in his interrogation of the racist and genocidal foundations of European imperialism, *Exterminate All the Brutes*, 'It is not knowledge we lack. It is the courage to understand what we know and draw conclusions.'

In other words the path through involves more than just a shift in energy sources. It begins in a reckoning with the past, and demands a far more fundamental reorganisation of the global economy, a shift to a model that operates within planetary boundaries and shares resources for the benefit of all. The technological and economic tools that are necessary to achieve this already exist; what is needed is for those solutions to be put in place.

Such a shift is not impossible. The body of economic and social theory outlining how such a world might operate is extensive. Social experiments exploring sustainable systems are underway in cities and communities around the world. Treaties and agreements to control the spread of plastics, to bring problems such as overfishing and pollution, the burning of fossil fuels and other destructive activities, are gradually being brought into being. There is also increasing recognition of the need for adaptation and support for poorer nations, as the creation of the loss and damage fund at COP27 makes clear. These reforms have not come from nowhere: they are the result of decades of sacrifice by activists, scientists and local and Indigenous communities.

These victories are only a beginning. As recent years have made clear, the forces opposing change are extraordinarily powerful. The influence of fossil fuel companies and other corporations over governments continues, as the increasing use of state power to curtail protest makes clear. In large parts of the world fascism and its bedfellows, racism and white supremacy, are on the march. The wealth of the richest continues to grow, as does the rate at which industrialised society is burning through the planet's reserves. Again, the structures that give rise to these problems did not emerge by chance, they have arisen out of decisions made by governments and others. But while these forces can seem overwhelming, unstoppable, they are not. Only a few years ago, the world was on track for temperature rises of 4 degrees or more. The fact the temperature increases currently predicted are only slightly more than half that is partly the result of a dizzyingly fast uptake of green technologies. But it is also a testament to the environmental movement's tireless efforts to force governments and corporations to alter course, and a reminder that the transformation that is needed will come to pass only through campaigns of mass engagement and civil disobedience.

WHILE WRITING THIS BOOK I have thought a lot about grief. Some of this has been personal – for me as for many others, the past few years of my life have been marked by loss. But I have also contemplated a larger, less personal kind of grief. So much is being lost, and so fast, it is difficult not to feel deranged by it. How do we make sense of the disappearance of coral reefs, of dying kelp and collapsing ecosystems? How do we imagine a world in which the massing life that once inhabited not just the oceans but the earth and the sky is largely gone?

One solution is to simply turn away. The cognitive dissonance of this choice is all around us, as visible in the insistence of politicians that it is possible to keep burning fossil fuels as in the increasingly frantic displays of wealth by the powerful. At a more practical level it is also simply delaying the inevitable: if the past few years have taught us anything, it is that nobody is safe. But it is also to do a kind of violence, for by denying the reality of what is going on we do violence to ourselves, by cauterising our capacity for empathy and grief.

The other alternative, to try to accommodate what is happening, is a far more confronting prospect. The anthropologist and philosopher Deborah Bird Rose, who died in 2018, wrote of the impossibility of bridging the gap between our limited ability to affect what is taking place around ourselves and the cost of facing it. Yet she also recognised that to turn our backs to it was also to turn our backs on ourselves. 'To face others is to become a witness, and to experience our incapacity in this position.' It is also an ethical imperative, a way to 'remain true to the lives within which ours are entangled, whether or not we can effect great change'.

To bear witness in this way is to make ourselves vulnerable, to open ourselves up to loss and sadness. Nonetheless as the philosopher

Thom van Dooren has observed, it is also an act of hope, a refusal to ignore the bonds of care that connect us to the world around us. And, perhaps no less importantly, it embodies a preparedness to absorb the lessons of history and to recognise the reality of the past.

More than that, however, the act of openness creates the possibility of love and joy and – improbably – wonder. And to my surprise that feeling has been one of the central experiences of writing this book. In that process I have realised that however much has been lost, the world still hums with beauty and astonishment. We share the planet with whales that sing across oceans and navigate by watching the stars, with fish that pass ways of knowing across generations, in webs of culture spreading back millions of years, with turtles that follow invisible patterns of magnetism back to the beaches where they were born. To contemplate the strangeness and wonder of these other ways of being is to begin to understand our place in the world very differently, to be reminded that we are not separate, or different, but part of a much larger system of impossible magnificence and complexity.

No less more importantly, it is to recognise that despair is also a form of turning away. A few months ago I spoke to a scientist in Tasmania who is working to regenerate the giant kelp that has been almost wiped out by rising temperatures by selectively breeding specimens that have demonstrated higher thermal tolerance. While we were talking he grew emotional as he conceded it was possible the seemingly unstoppable upward arc of ocean temperatures will soon overwhelm even these more thermally tolerant species. Yet like the scientists working to save coral reefs, he said he did not know what else he could do.

The hope he described is a fragile thing, and difficult to sustain, but it is also an investment in the future, a refusal to give up. No less importantly, it offers a reminder that mourning cannot be an endpoint. Instead grief must be part of a larger recognition that there is no longer any way back, that the only route now is forward. That we must find ways to live in a world on fire. And ways to fight that will ensure the survival of all.

The storm that is upon us will leave nobody untouched. Surviving it demands we build a world that treats everybody – human and non-human – as worthy of life and possibility. That will not happen unless those of us who have benefited from the systems of extraction and subjugation that are destroying the planet learn to see our place in the world differently, to recognise the true cost of the lives we live.

I have times when I think it is possible to see that world taking shape in the distance. Times when it is possible to convince myself we will get there because we have no choice. Because however much is lost, there is still more to save.

I TURN TO LOOK OUT to the horizon, its fading margin between sea and sky a space of grief, but also possibility. Around me the water extends outwards, its embrace holding me, its fluidity connecting me to the planet's systems, to myriad other lives – past, present and future. And putting my face down I start to swim, outwards, towards the unknown.

NOTES & SOURCES

1 Beginnings

The 'inkling of infinity' Capp, Fiona, *That Oceanic Feeling*, Allen & Unwin, Sydney, 2003, p. 11.

'Letting the ocean envelop you' Nestor, James, *Deep: Freediving, renegade science and what the ocean tells us about ourselves*, Profile Books, London, 2014, p. 230.

'Oceanic feeling' Freud, Sigmund, *Civilization and its Discontents*, General Press, New Delhi, 2018, p. 14ff.

The brains of Buddhist monks Danzico, Matt, 'Brains of Buddhist monks scanned in meditation study', *BBC News*, 24 April 2011.

Activities such as surfing and skydiving Nicholls, Wallace J., *Blue Mind: How water makes you happier, more connected and better at what you do*, Little, Brown, London, 2014, pp. 233–5.

When . . . John McPhee sought a metaphor to encompass the vastness of geological time McPhee, John, *Basin and Range*, Farrar, Straus and Giroux, New York, 1982.

Austrian philosopher and writer Ivan Illich Farber, Thomas, *On Water*, Ecco Press, Hopewell, 1994, p. 5.

A galaxy 12.88 billion light years from Earth Jarugula, Sreevani et al., 'Molecular line observations in two dusty star-forming galaxies at Z = 6.9', *Astrophysical Journal*, vol. 921, no. 1, 2021, p. 97.

Enstatite chondrites Piani, Laurette et al., 'Earth's water may have been inherited from material similar to enstatite chondrite meteorites', *Science*, vol. 369, no. 6507, 2020, pp. 1110–13.

Ringwoodite Schmandt, Brandon et al., 'Earth's interior dehydration melting at the top of the lower mantle', *Science*, vol. 344, no. 6189, 2014, pp. 1265–8.

Fossils of microbes dating back 4.3 billion years Dodd, Matthew S. et al., 'Evidence for early life in earth's oldest hydrothermal vent precipitates', *Nature*, vol. 543, no. 7643, 2017, pp. 60–4.

A gentle rain of cometary material Derouin, Sarah, 'Antarctic study shows how much space dust hits Earth every year', *Scientific American*, 29 April 2021.

'Grotesque to Earthly eyes' Kline, Otis Adelbert, *A Vision of Venus*, Project Gutenberg Australia, 2013.

Venus may have continued in a relatively Earthlike state Way, M.J., 'Venusian habitable climate scenarios: Modelling Venus through time and applications to slowly rotating Venus-like exoplanets', *Journal of Geophysical Research: Planets*, 125, 2020, e2019JE006276.

Water is likely to have covered most, if not all, of the planet's surface Dong, J. et al., 'Constraining the volume of Earth's early oceans with a temperature-dependent mantle water storage capacity model', *AGU Advances*, vol. 2, no. 1, 2021, e2020AV000323.

A looming, volcano-scarred ball Maurice, M. et al., 'A long-lived magma ocean on a young Moon', *Science Advances*, vol. 6, no. 28, 10 July 2020.

A layer 50 metres thick Rohling, Eelco J., *The Oceans: A deep history*, Princeton University Press, Princeton, 2017, p. 33.

Sponges and comb jellies Schultz, D.T., 'Chromosomal comparisons reveal comb jellies as the sister group to all other animals', *Nature*, vol. 618, 2023, pp. 110–17.

Skittering around on shore MacNaughton, R.B. et al., 'First steps on land: Arthropod trackways in Cambrian–Ordovician eolian sandstone, southeastern Ontario, Canada', *Geology*, vol. 30, 2002, pp. 391–4.

Java by 1.3 million years ago Matsu'Ura, Shuji et al., 'Age control of the first appearance datum for Javanese *Homo erectus* in the Sangiran area', *Science*, vol. 367, no. 6156, 2020, pp. 210–14.

Neanderthals may have been moving Lawler, Andrew, 'Neanderthals, Stone Age people may have voyaged the Mediterranean', *Science*, 24 April 2018.

The mysterious Denisovans Cooper, A. and Stringer, C.B., 'Did the Denisovans cross Wallace's line?', *Science*, vol. 342, 2013, pp. 321–3.

The Solomon Islands Leavesley, Matthew, 'Late pleistocene complexities in the Bismarck Archipelago', in Lilley, Ian, *Archaeology of Oceania: Australia and the Pacific Isles*, Blackwell Publishing Ltd, Oxford, 2006, pp. 189–204.

The Ryukyu Islands Fujita, Masaki et al., 'Advanced maritime adaptation in the Western Pacific coastal region extends back to 35,000–30,000 years before present', *Proceedings of the National Academy of Sciences*, vol. 113, no. 40, 2016, pp. 11184–9.

The Talaud Islands Bellina, Bérénice et al., *Sea Nomads of Southeast Asia: From the past to the present*, NUS Press, Singapore, 2021, p. 7.

'An animal of the edge' Quoted in Gillis, John R., *The Human Shore: Seacoasts in history*, University of Chicago Press, Chicago, 2012, p. 19.

'The human learning curve accelerated' Gillis, John R., *The Human Shore: Seacoasts in history*, University of Chicago Press, Chicago, 2012, p. 17.

Carved a series of zigzag patterns Thompson, Helen, 'Zigzags on a shell from Java are the oldest human engravings', *Smithsonian Magazine*, 3 December 2014.

Thirty-three beads Kindy, David, 'Are these snail shells the world's oldest known beads?', *Smithsonian Magazine*, 27 September 2021.

Neanderthals were painting shells and making clam necklaces Hoffmann, Dirk L. et al., 'Symbolic use of marine shells and mineral pigments by Iberian Neandertals 115,000 years ago', *Science Advances*, vol. 4, no. 2, 2018, eaar5255.

People in the Bismarck Archipelago Leavesley, Matthew, 'Late pleistocene complexities in the Bismarck Archipelago', in Lilley, Ian, *Archaeology of Oceania: Australia and the Pacific Isles*, Blackwell Publishing Ltd, Oxford, 2006, pp. 189–204.

A web of commerce Paine, Lincoln, *The Sea and Civilization: A maritime history of the world*, Alfred A. Knopf, New York, 2013, pp. 137–8.

Indian sailors were experienced at deep-sea navigation ibid., p. 139.

Before the beginning . . . as far as the Ganges Abulfia, David, *The Boundless Sea: A human history of the oceans*, Allen Lane, London, 2019, pp. 102–3.

Known as the *aster mara* Macfarlane, Robert, *The Old Ways: A journey on foot*, Hamish Hamilton, London, 2012, p. 88.

An ancestral island known as Hawaiki Thompson, Christina, *Sea People: In search of the ancient navigators of the Pacific*, William Collins, London, 2019, p. 166.

A 'constellation linked together by pathways on the ocean' Gladwin, Thomas, *East Is a Big Bird: Navigation and logic on Puluwat Atoll*, Harvard University Press, Cambridge, MA, 2022, p. 34.

'A sea of islands' Epeli, Hau'ofa, *We Are the Ocean: Selected works*, University of Hawaii Press, Honolulu, 2008, p. 33.

Not 'a foot of ground that was not entirely covered with great trees' Crosby, Alfred W., *Ecological Imperialism: The biological expansion of Europe, 900–1900*, Cambridge University Press, Cambridge, 2015, pp. 75–6.

Leaving the settlers in despair ibid., p. 75.

Originally cultivated in New Guinea Horton, M., Langton, P. and Bentley, R.A., 'A history of sugar – the food nobody needs but everybody craves', *The Conversation*, 31 October 2015.

Within five years Crosby, Alfred W., *Ecological Imperialism: The biological expansion of Europe, 900–1900*, Cambridge University Press, Cambridge, 2015, p. 77.

At least 50 kilograms of timber Patel, Raj and Moore, Jason W., *A History of the World in Seven Cheap Things: A guide to capitalism, nature, and the future of the planet*, Verso, London, 2020, p. 17.

'Europe's wealthy ate the sugar' ibid., p. 17.

'Unrestrained mass violence' Adhikari, Mohamed, 'Europe's first settler colonial incursion into Africa: The genocide of Aboriginal Canary Islanders', *African Historical Review*, vol. 49, no. 1, 2017, pp. 1–26.

Replaced with fully racialised slavery of Black Africans French, Howard, *Born in Blackness: Africa, Africans, and the making of the modern world, 1471 to the Second World War*, Liveright, New York, 2021, p. 116.

As low as seven years ibid., p. 202.

Also ravaged African societies ibid., p. 322.

A mere six million Koch, Alexander et al., 'Earth system impacts of the European arrival and great dying in the Americas after 1492', *Quaternary Science Reviews*, vol. 207, 2019, pp. 13–36.

The global dip in temperatures known as the Little Ice Age Lewis, Simon L. and Maslin, Mark A., *The Human Planet: How we created the Anthropocene*, Pelican Books, London, 2018, pp. 179–87.

Even the soil was transformed Mann, Charles C., *1493: How Europe's discovery of the Americas revolutionised trade, ecology and life on Earth*, Granta, London, 2011, pp. 51–3.

Silver extracted from the mines of Potosí Patrick Greenfield, 'How silver turned Potosí into "the first city of capitalism"', *The Guardian*, 21 March 2016.

Almost entirely accounted for by the rise of nations with access to the Atlantic Ocean Acemoglu, Daron et al., 'The rise of Europe: Atlantic trade, institutional change, and economic growth', *American Economic Review*, vol. 95, no. 3, 2005, pp. 546–79.

Close to 70 per cent of export commodities produced in the Americas Inikori, Joseph E., 'Atlantic slavery and the rise of the capitalist global economy', *Current Anthropology*, vol. 61, no. S22, 2020, pp. S159–S171.

The value of enslaved people in the United States Piketty, Thomas, *Capital in the Twenty-First Century*, Harvard University Press, Cambridge, MA, 2017, p. 161.

Helped spur the development of new systems of property rights Acemoglu, Daron et al., 'The rise of Europe: Atlantic trade, institutional change, and economic growth', *American Economic Review*, vol. 95, no. 3, 2005, pp. 546–79.

Sugar . . . allowed workers to be fed cheaply Pomeranz, Kenneth, *The Great Divergence: China, Europe, and the making of the modern world economy*, Princeton University Press, Princeton, 2021.

The coffee houses of the eighteenth century French, Howard, *Born in Blackness: Africa, Africans, and the making of the modern world, 1471 to the Second World War*, Liveright, New York, 2021, pp. 214–16.

This indigenous critique Graeber, David and Wengrow, David, *The Dawn of Everything: A new history of humanity*, Penguin, London, 2022, p. 58.

'As long as the Earth was considered to be alive' Merchant, Carolyn, *The Death of Nature: Women, ecology, and the scientific revolution*, Harper & Row, New York, 1989, p. 3.

'Never take the first plant you find' Kimmerer, Robin Wall, *Braiding Sweetgrass: Indigenous wisdom, scientific knowledge and the teachings of plants*, Penguin, London, 2020, p. 182.

'Viewed from the distance of the Moon' Thomas, Lewis, *The Lives of a Cell: Notes of a biology watcher*, Penguin, New York, 1974, p. 170.

The first-century Roman geographer Pomponius Mela Cartwright, David E., 'On the origins of knowledge of the sea tides from antiquity to the thirteenth century', *Earth Sciences History*, vol. 20, no. 2, 2001, pp. 105–26.

'Some universal truth' Carson, Rachel, *The Edge of the Sea*, Mariner Classics, New York, 1998, p. 250.

2 Bodies

'A feeling for the water' 'A feeling for the water', *ABC News*, 11 November 2013.

The Bajau also possess a series of genetic adaptations Ilardo, Melissa A., 'Physiological and genetic adaptations to diving in sea nomads', *Cell*, vol. 173, no. 3, 2018, pp. 569–80.

This gene appears to have come from the Denisovans Gibbons, Ann, 'Tibetans inherited high-altitude gene from ancient humans', *Science*, 2 July 2014.

The Moken people Thomson, Helen, 'The "sea-nomad" children who see like dolphins', *BBC Future*, 1 March 2016.

'The Tarzanlike figure of the prehominid' Morgan, Elaine, *The Descent of Woman*, Souvenir Press, London, 2011, p. 11.

Our Neanderthal cousins were seafarers Lawler, Andrew, 'Neandertals, Stone Age people may have voyaged the Mediterranean', *Science*, 24 April 2018.

Bony growths in the ear canals of many Neanderthal skulls Trinkaus, Erik et al., 'External auditory exostoses among western Eurasian late Middle and Late Pleistocene humans', *PLOS ONE*, 14 August 2019.

The Sahara dried out rapidly Coffey, Donavyn, 'Could the Sahara ever be green again?', *Live Science*, 27 September 2020.

Hieroglyphs from almost 5000 years ago Carr, Karen Eva, *Shifting Currents: A world history of swimming*, Reaktion Books, London, 2022, pp. 19–21.

Divers have been harvesting pearls ibid., pp. 29–30.

The same sorts of bony growths that are found in the ears of Neanderthals Villotte, Sébastien et al., 'External auditory exostoses and aquatic activities during the Mesolithic and the Neolithic in Europe: Results from a large prehistoric sample', *Anthropologie (Brno)*, 2014, pp. 73–89.

A more than 10,500-year-old willow bark net *Wikipedia*, Wikipedia Foundation, 17 September 2023.

People in northern Eurasia ibid.

From the *Epic of Gilgamesh* to the Torah Carr, Karen Eva, *Shifting Currents: A world history of swimming*, Reaktion Books, London, 2022, p. 51ff.

Plato speaks of those who are 'unlettered and unable to swim' ibid., p. 124, see also p. 102.

A substance 'devilish rather than divine' Sprawson, Charles, *Haunts of the Black Masseur: The swimmer as hero*, Vintage, London, 2013, p. 67.

This aqueousness also took on other, deeper forms Dawson, Kevin, 'Waterscapes and wet bodies: Beach culture in Atlantic Africa and the diaspora, 1444–1888', *Coastal Studies & Society*, vol. 2, no. 1, 2023, pp. 58–81; Sogoba, Mia, 'The cowrie shell: monetary and symbolic value', *Cultures of West Africa*, 4 May 2018.

'Once the children learn to walk by themselves' Dawson, Kevin, 'Africa, 1450 to 1770 CE', in Stein, Stephen K. (ed.), *The Sea in World History: Exploration, travel, and trade, Volume I: Ancient Egypt through the first global age*, ABC-CLIO, Santa Barbara, 2017, pp. 289–95.

The ability to navigate the surf zone Dawson, Kevin, 'A brief history of surfing in Africa and the diaspora', in Malarkey, Sarah (ed.), *Afrosurf,* Mami Wata/Ten Speed Press, Berkeley, 2020, pp. 23–6.

'Dived like cormorants' Dawson, Kevin, *Undercurrents of Power: Aquatic culture in the African diaspora,* University of Pennsylvania Press, Philadelphia, 2018, p. 15.

'The greatest swimmers which there are in the world' Carr, Karen Eva, *Shifting Currents: A world history of swimming,* Reaktion Books, London, 2022, p. 205.

Africans were 'amphibious' Dawson, Kevin, *Undercurrents of Power: Aquatic culture in the African diaspora,* University of Pennsylvania Press, Philadelphia, 2018, p. 15.

European travellers in Africa Dawson, K., 'Enslaved swimmers and divers in the Atlantic world', *Journal of American History,* vol. 92, no. 4, 2006, pp. 1327–55.

The vice-chancellor of the University of Cambridge banned swimming Chaline, Eric, *Strokes of Genius: A history of swimming,* Reaktion Books, London, 2017, p. 103.

Everard Digby's 1587 treatise, *De Arte Natandi* Digby, Everard, *A short introduction for to learne to swimme. Gathered out of Master Digbies Booke of the Art of Swimming. And translated into English for the better instruction of those who vnderstand not the Latine tongue. By Christofer Middleton,* Early English Books Online Text Creation Partnership, Ann Arbor, 2011, quod.lib.umich.edu/e/eebo/A20436 .0001.001.

Digby included a series of woodcuts 'The Art of Swimming (1587)', *Public Domain Review,* 29 August 2014.

'Performing on the Way many Feats and Activity' Means, Howard, *Splash: 10,000 years of swimming,* Allen & Unwin, Sydney, 2020, p. 75.

Teaching naval and marine cadets to swim Chaline, Eric, *Strokes of Genius: A history of swimming,* Reaktion Books, London, 2017, pp. 111–12.

Byron . . . was notorious for pretending to be vague about exactly how far he had swum Sprawson, Charles, *Haunts of the Black Masseur: The swimmer as hero,* Vintage, London, 2013, p. 122.

Magellan observed Micronesians swimming overarm Means, Howard, *Splash: 10,000 years of swimming,* Allen & Unwin, Sydney, 2020, p. 97.

An account of Easter Islanders doing the same ibid.

It had begun to be adopted by at least some American colonists Carr, Karen Eva, *Shifting Currents: A world history of swimming,* Reaktion Books, London, 2022, pp. 305–7.

'Smooth and graceful' Dawson, K., 'Enslaved swimmers and divers in the Atlantic world', *Journal of American History,* vol. 92, no. 4, 2006, pp. 1327–55.

'White' swimming Dawson, Kevin, *Undercurrents of Power: Aquatic culture in the African diaspora,* University of Pennsylvania Press, Philadelphia, 2018.

'Totally un-European' Means, Howard, *Splash: 10,000 years of swimming,* Allen & Unwin, Sydney, 2020, p. 99.

Celebrated the reassertion of English superiority ibid., p. 100.

Cavill may have picked up the technique earlier Osmond, Gary and Phillips, Murray G., 'The bloke with a stroke: Alick Wickham, the "crawl" and social memory', *Journal of Pacific History*, vol. 39, no. 3, 2004, pp. 309–24.

'None of the black people' Webb, Matthew, *The Art of Swimming*, Ward, Lock and Tyler, London, 1875, p. 64.

In Japan Means, Howard, *Splash: 10,000 years of swimming*, Allen & Unwin, Sydney, 2020, pp. 236–7, and Sprawson, Charles, *Haunts of the Black Masseur: The swimmer as hero*, Vintage, London, 2013, p. 287.

In Burroughs' original novel Burroughs, Edgar Rice, *Tarzan of the Apes*, Project Gutenberg Australia, 2008.

'Cultural resilience served as a mechanism of resistance' Dawson, Kevin, *Undercurrents of Power: Aquatic culture in the African diaspora*, University of Pennsylvania Press, Philadelphia, 2018, p. 6.

'They transgressed *haole* expectations and categories in the waves' Walker, Isaiah Helekunihi, 'Hui Nalu, beachboys, and the surfing boarder-lands of Hawai'i', *Contemporary Pacific*, vol. 20, no. 1, 2008, pp. 89–113.

The fate of the women was even more horrific Taylor, Rebe, *Unearthed: The Aboriginal Tasmanians of Kangaroo Island*, Wakefield Press, Adelaide, 2008, p. 39.

'Baby died on way. Child on her back' ibid., p. 41.

'Down below there are no colours' Fagotto, Matteo, 'Swimming under the ice: "There's nothing. You're completely alone."', *The Observer*, 15 March 2020.

Ocean swimming brings us into contact with the sublime Young, Damon, 'Why swimming is sublime', *The Guardian*, 7 February 2014.

Thomas De Quincey intuits a connection between swimming and taking drugs Sprawson, Charles, *Haunts of the Black Masseur: The swimmer as hero*, Vintage, London, 2013, p. 135.

'Addiction is liminal too' Ziolkowski, Thad, *The Drop: How the most addictive sport can help us understand addiction and recovery*, Harper Wave, New York, 2021.

'Swimming can become an ecological meditation for the Anthropocene' Mentz, Steve, *Ocean*, Object Lessons, Bloomsbury, London, 2020, p. 131.

'Floating brings the Atlantic to light' Tolbert, Makshya, 'Becoming Water: Black memory in slavery's afterlives', *Emergence*, 17 February 2022.

3   Migrations

May total 10 billion tonnes or more Courage, Katherine Harmon, 'Greatest migration on Earth happens every day', *Scientific American*, 1 August 2022.

May play a bigger part in ocean circulation than the wind Pennisi, Elizabeth, 'Tiny shrimp may be mixing ocean water as much as the wind and waves', *Science*, 18 April 2018.

It even helps regulate the climate Kelly, Thomas B. et al., 'The importance of mesozooplankton diel vertical migration for sustaining a mesopelagic food web', *Frontiers in Marine Science*, vol. 6, 2019.

A loggerhead turtle released from an aquarium in Cape Town Michelmore,

Karen, 'Loggerhead turtle's journey tracked 37,000 km from Cape Town in South Africa to Australia', *ABC News*, 7 March 2020.

The distance to the Moon and back seven times over Safina, Carl, *Eye of the Albatross: Visions of hope and survival*, Holt Paperbacks, New York, 2003.

'A great migration' deep beneath the South Atlantic Milligan, Rosanna J. et al., 'Evidence for seasonal cycles in deep-sea fish abundances: A great migration in the deep SE Atlantic?', *Journal of Animal Ecology*, vol. 89, no. 7, 2020, pp. 1593–603.

When the coast myall, or kai'arrewan, flowered Clark, Anna, *The Catch: The story of fishing in Australia*, National Library of Australia, Canberra, 2017, p. 26.

Regard the salmon that course through their coastal waters and rivers as gift-bearing relatives Mueller, Martin Lee, *Being Salmon, Being Human: Encountering the wild in us and us in the wild*, Chelsea Green Publishing, White River Junction, 2017.

The storks that nest in Europe in the summer spend their winters on the Moon Gorman, Alice, 'When birds migrated to the moon', *The MIT Press Reader*, 1 December 2020.

British zoologist and geographer Philip Sclater Greer, Kirsten, 'Geopolitics and the avian imperial archive: The zoogeography of region-making in the nineteenth-century British Mediterranean', *Annals of the Association of American Geographers*, vol. 103, no. 6, 2013, pp. 1317–31.

As the European powers continued to expand Jacobs, Nancy J., 'Africa, Europe and the birds between them', *Eco-Cultural Networks and the British Empire: New views on environmental history*, edited by James Beattie et al., Bloomsbury, London, 2015, pp. 92–120.

Sailors occasionally reported great flocks of swallows Beilby, Ralph and Bewick, Thomas, *History of British Birds, Vol. I: Containing the history and description of land birds.* Printed by Sol. Hodgson, for Beilby & Bewick: sold by them, and G.G. and J. Robinson, London, 1798. *Eighteenth Century Collections Online*, p. xviii.

'More . . . than can be bred in any one district' White, Gilbert, *The Natural History of Selborne*, Penguin Classics, London, 1977.

Chain Home commander Sir Edward Fennessy An engineer and a key figure in the development of radar technology, Fennessy was also responsible for recruiting the young Arthur C. Clarke into the Royal Air Force. Clarke would later write his seminal 1945 paper about communication satellites while working on ground-approach radar.

The . . . 'angels' . . . were the ghosts of fallen soldiers Clarke, David R., 'Radar Angels', drdavidclarke.co.uk, 2011.

The biologist David Lack Anderson, Ted R., *The Life of David Lack: Father of evolutionary ecology*, Oxford University Press, Oxford, 2013.

Huge flocks of starlings Fox, Anthony D. and Beasley, Patrick D.L., 'David Lack and the birth of radar ornithology', *Archives of Natural History*, vol. 37, no. 2, 2010, pp. 325–32.

The wandering glider, or globe skimmer dragonfly Hobson, Keith A., 'Isotopic

evidence that dragonflies (*Pantala flavescens*) migrating through the Maldives come from the northern Indian Subcontinent', *PLOS ONE*, vol. 7, no. 12, 2012, e52594.
'All streams and waters are completely filled with them' Roberts, Callum, *The Unnatural History of the Sea: The past and future of humanity and fishing*, Gaia Thinking, London, 2007, p. 50.
'So full of salmon' ibid., p. 52.
Columns of fish 8 to 10 kilometres long and 5 to 7 kilometres wide ibid., p. 124.
Huge schools of massive bluefin tuna Monbiot, George, *Feral: Rewilding the land, sea and human life*, Penguin, London, 2014, pp. 231–2.
'Hardly has [the herring] issued from the egg' Hérubel, Marcel A., *Sea Fisheries: Their treasures and toilers*, T. Fisher Unwin, London, 1912, p. 25.
Along the west coast of Africa Ragnhild, Overå, 'Institutions, mobility and resilience in the Fante migratory fisheries in West Africa', *Transactions of the Historical Society of Ghana*, no. 9, 2005, pp. 103–23; Abobi, Seth Mensah and Alhassan, Elliot Haruna, 'A review of fisheries-related human migration in the Gulf of Guinea', *Coastal Management*, vol. 18, no. 1, 2015.
Each year in India Mondal, Madhuri, 'On the Road: Tracing the current patterns of fishworkers' migration in India', *The Bastion*, 4 September 2020.
The annual dance of the fifteen or so species of eels 'Evolution of the eel a matter of continental drift', in 'On the origin and demise of coasts', *World Ocean Review*, 2017.
Special proteins known as cryptochromes Hore, Peter J. and Mouritsen, Henrik, 'The quantum nature of bird migration', *Scientific American*, vol. 326, no. 4, 2022.
Some birds . . . possess an inbuilt magnetic map of the entire planet Kishkinev, Dmitry et al., 'Navigation by extrapolation of geomagnetic cues in a migratory songbird', *Current Biology*, vol. 31, no. 7, 2021, pp. 1563–9; Holland, Richard and Kishkinev, Dmitry, 'Birds use massive magnetic maps to migrate – and some could cover the whole world,' *The Conversation*, 14 February 2021.
The journey might be a relic of an earlier age Carr, Archie, 'Transoceanic migrations of the green turtle', *Bioscience*, vol. 14, no. 8, 1964, pp. 49–52.
Turtles appear to develop magnetic 'maps' Lohmann, Kenneth J., 'Sea turtles: Navigating with magnetism', *Current Biology*, vol. 17, no. 3, 2007, R102–104.
The hawksbill turtles that breed in the Chagos Archipelago Hays, Graeme C. et al., 'Travel routes to remote ocean targets reveal the map sense resolution for a marine migrant', *Journal of the Royal Society Interface*, vol. 19, no. 190, 2022.
Leatherback and loggerhead turtles Avens, Larisa et al., 'Use of multiple orientation cues by juvenile loggerhead sea turtles *Caretta caretta*', *Journal of Experimental Biology*, vol. 206, no. 23, 2003; Dodge, Kara, 'Leatherback sea turtles use mysterious "compass sense" to migrate hundreds of miles', *The Conversation*, 11 March 2015.
Forty-five sperm whales stranded themselves on beaches in the North Sea Vanselow, Klaus Heinrich et al., 'Solar storms may trigger sperm whale strandings: Explanation approaches for multiple strandings in the North Sea in 2016', *International Journal of Astrobiology*, vol. 17, no. 4, 2018, pp. 336–44.

Strandings of gray whales seem to correspond with the incidence of sunspots Granger, Jesse et al., 'Gray whales strand more often on days with increased levels of atmospheric radio-frequency noise', *Current Biology*, vol. 30, no. 4, 2020, pp. 155–6.

Humpbacks . . . seem to orient themselves Horton, Travis W. et al., 'Straight as an arrow: Humpback whales swim constant course tracks during long-distance migration', *Biology Letters*, vol. 7, no. 5, 2011, pp. 674–9.

Some whales even seem to be able to perceive subtle permutations in the Earth's gravity Horton, Travis W. et al., 'Route fidelity during marine megafauna migration', *Frontiers in Marine Science*, vol. 4, article 422, 2017.

Salmon imprint on the odour of the stream or lake in which they hatch Dingle, Hugh, *Migration: The biology of life on the move*, Oxford University Press, Oxford, 2014, p. 138.

Sea lampreys are so disturbed by the scent of dead lampreys Nuwer, Rachel, 'I smell sex and death: Manipulating invasive sea lampreys with odor', *Audubon Magazine*, 8 August 2011; Kelly, Erin, 'Researchers use lamprey's sense of smell to aid control effort', *Virginia Institute of Marine Science News*, 17 July 2013.

Shearwaters . . . will fly straight back Abolaffio, Milo et al., 'Olfactory-cued navigation in shearwaters: Linking movement patterns to mechanisms', *Scientific Reports*, vol. 8, no. 1, 2018, pp. 11590–9; Gagliardo, Anna et al., 'Oceanic navigation in Cory's shearwaters: Evidence for a crucial role of olfactory cues for homing after displacement', *Journal of Experimental Biology*, vol. 216, 2013, pp. 2798–805.

Petrels and albatross . . . rely on scent to find their way to and from feeding and nesting sites Nevitt, Gabrielle A. et al., 'Evidence for olfactory search in wandering albatross, *Diomedea exulans*', *Proceedings of the National Academy of Sciences*, vol. 105, no. 12, 2008, pp. 4576–81.

Albatross and related species . . . learn to recognise seasonal changes in the concentration of these chemical cues Nevitt, Gabrielle A., 'Sensory ecology on the high seas: The odor world of the procellariiform seabirds', *Journal of Experimental Biology*, vol. 211, 2008, pp. 1706–13.

Their movements allow researchers to infer changes in the distribution of life at those depths McMahon, Clive R. et al., 'Finding mesopelagic prey in a changing Southern Ocean', *Scientific Reports*, vol. 9, no. 1, 2019, pp. 19011–13.

'It's almost as if we had 287 mini-submarines' 'Unique new insights into the life of southern elephant seals', *IMOS News*, 16 June 2016.

The small size and weight of the transmitters 'Animals on the air', *ICARUS: Global Monitoring with Animals*, 2023, icarus.mpg.de/28874/sensor-animals -tracking.

ICARUS may even be able to provide an early warning system for earthquakes and tsunamis 'Animals' early warning system', *ICARUS: Global Monitoring with Animals*, 2023, icarus.mpg.de/28810/animals-warning-sensors.

Sonar, radar, GPS, even early satellite tracking harnesses Macdonald, Helen, *Falcon*, Reaktion Books, London, 2006, pp. 155–6.

'Military superiority is built on knowing where everything is' ibid., p. 153.

Free-swimming species at the equator are declining in number Chaudhary, Chhaya et al., 'Global warming is causing a more pronounced dip in marine species richness around the equator', *Proceedings of the National Academy of Sciences*, vol. 118, no. 15, 2021, e2015094118.

Marine animals are shifting polewards at an average of 6 kilometres a year Lenoir, Jonathan et al., 'Species better track warming in the oceans than on land', *Nature Ecology & Evolution*, vol. 4, no. 8, 2020, pp. 1044–59.

Reducing diversity and simplifying foodwebs Chown, Steven L., 'Marine food webs destabilised', *Science*, vol. 369, no. 6505, 2020, pp. 770–1.

Off Sydney . . . coral is beginning to appear Tatham, Harriet, 'Sydney growing its own coral reef with help from tropical fish finding warmer waters', ABC Radio, Sydney, 13 September 2019.

Warming waters have allowed king crabs . . . to ascend the continental slope Smith, Kathryn, 'The march of the king crabs: A warning from Antarctica', *The Conversation*, 21 July 2015.

On the Great Barrier Reef's Raine Island . . . hatching rates of eggs Pike, David A. et al., 'Nest inundation from sea-level rise threatens sea turtle population viability', *Royal Society Open Science*, vol. 2, no. 7, 2015.

The arrival of migrating shearwaters in Tasmania and elsewhere Readfearn, Graham, 'From Alaska to Australia, anxious observers fear mass shearwater deaths', *The Guardian*, 24 November 2019.

Floating pumice Kilvert, Nick, 'Seabirds were so famished they ate pumice stones before mass "wreck"', *ABC Science*, 25 March 2021.

Even blue whales . . . are being affected Yong, Ed, 'Is the world's largest animal too reliant on the past?', *The Atlantic*, 25 February 2019.

The number of people displaced globally *Global Trends: Forced Displacement in 2022*, Statistics and Demographics Section, UNHCR Global Data Service, Copenhagen, 2023; '2022 Year in Review: 100 million displaced, "a record that should never have been set"', *UN News*, 26 December 2022.

Nearly 61 million people were displaced internally Internal Displacement Monitoring Centre, *GRID 2023*, IDMC, Geneva, 2022.

Subsidence and hastening erosion caused . . . are forcing Indigenous communities from their homes Ilhardt, Julia, '"It was sad having to leave": Climate crisis splits Alaskan town in half', *The Guardian*, 9 June 2021.

In the Sahel region Bough, Moussa, 'Sahel internal displacement tops 2 million as violence surges', *UNHCR News*, 22 January 2021.

'Just like a slavery boat' Gladstone, Rick, 'Stepping over the dead on a migrant boat', *New York Times*, 5 October 2016.

'I can't sleep. I cry every time I see a picture of him' Blackall, Molly, '"Where is my child? Only God knows": Migrant parents losing babies during perilous crossings to Europe', *iNews*, 31 July 2021.

'Just as bombs follow oil' Klein, Naomi, *On Fire: The burning case for a green new deal*, Allen Lane, London, 2019, pp. 160–3.

The detention camps operated by the Australian Government in Nauru and elsewhere Farrell, Paul, Evershed, Nick and Davidson, Helen, 'The Nauru files: Cache of 2000 leaked reports reveal scale of abuse of children in Australian offshore detention', *The Guardian*, 10 August 2016.

4   Echo

A specially designed hydrophone placed at the deepest point of the Mariana Trench Dziak, Robert et al., 'Ambient sound at Challenger Deep, Mariana Trench', *Oceanography*, vol. 30, no. 2, 2017, pp. 186–97.

Sound can even travel between oceans Fraser, I. and Morash, P., 'It's a small world: Underwater sound transmission from the southern Indian Ocean to the western North Atlantic', *Canadian Acoustics*, vol. 20, 1992, pp. 67–8.

The tides and currents produce turbulence that results in extremely low frequency infrasound Sand, O. and Karlsen, H.E., 'Detection of infrasound and linear acceleration in fishes', *Philosophical Transactions of the Royal Society B: Biological Sciences*, vol. 355, no. 1401, 2000, pp. 1295–8.

Even the Earth hums Webb, Spahr C., 'The Earth's "hum" is driven by ocean waves over the continental shelves', *Nature*, vol. 445, no. 7129, 2007, pp. 754–6.

'I heard the high whistle of one type, *moi*' Konesni, Bennett, 'Songs of the Lalaworlor: Musical labor on Ghana's fishing canoes', Worksongs.org, 2008.

Fishers working the waters west of the Japanese island of Kyushu Takemura, Akira et al., 'Underwater calls of the Japanese marine drum fishes (*Sciaenideae*)', *Bulletin of the Japanese Society of Scientific Fisheries*, vol. 44, no. 2, 1978, pp. 121–5.

The waters of the Batticaloa Lagoon 'The singing fish of Batticaloa', *BBC Sounds*, 28 July 2014.

'Like a man idly playing on the keys of a piano' 'In clear print it is stated, land of the singing fish', iam.lk, Vimeo, 7 September 2012.

American submariners quickly learned to use the racket of the shrimp Ball, Desmond and Tanner, Richard, *The Tools of Owatatsumi: Japan's ocean surveillance and coastal defence capabilities*, ANU Press, Canberra, 2015, p. 38.

A history of interfering with military sonar W.K., 'Sea creatures trouble sonar operators: New enzyme', *New York Times*, 2 February 1947.

Even humble plankton make noise Selivanovsky, D.A. et al., 'Acoustical investigation of phytoplankton', *ICES Journal of Marine Science*, vol. 53, no. 2, 1996, pp. 313–6; Freeman, Simon E. et al., 'Photosynthesis by marine algae produces sound, contributing to the daytime soundscape on coral reefs', *PLOS ONE*, vol. 13, no. 10, 2018, e0201766.

The discovery in corals of genes associated with the reception and production of sound 'Could corals use sound to communicate?', *EurekAlert!*, 27 April 2021, and Vermeij, M.J.A. et al., 'Coral larvae move toward reef sounds', *PLOS ONE*, vol. 5, no. 5, 2010, e10660.

Ga and Ewe fishers sing as they work Konesni, Bennett, 'Songs of the Lalaworlor: Musical labor on Ghana's fishing canoes', Worksongs.org, 2008; Otchere, Eric Debrah, 'In a world of their own': Memory and identity in the fishing songs of a migrant Ewe community in Ghana', *African Music*, vol. 10, no. 3, 2017, pp. 7–22; and Lalouschek, Adu and Wondergem, Alex, *Meet Ghana's Singing Fishermen*, Inthelife.tv, 26 November 2017.

The Mukkuvar fisherpeople who live along the shores of Kerala and Tamil Nadu Pampackal, Thomas Mathew, 'Indigenous knowledge and ecological concerns: A case study from India', *Concilium*, 15 January 2021.

The Narungga people sang to the dolphins and sharks Clark, Anna, 'Friday essay: Traps, rites and kurrajong twine – the incredible ingenuity of Indigenous fishing knowledge', *The Conversation*, 1 September 2023; Roberts, A. et al., 'They planned their calendar . . . they set up ready for what they wanted to feed the tribe: A first-stage analysis of Narungga fish traps on Yorke Peninsula, South Australia', *Journal of Island and Coastal Archaeology*, vol. 11, no. 1, 2016, pp. 1–25.

The Yuin cooperated with killer whales Clode, Danielle, *Killers in Eden: The story of a rare partnership between men and whales*, Museum Victoria, Melbourne, 2011.

The people of the Marshall Islands and other Pacific cultures use chants Schwartz, Jessica A., 'How the sea is sounded: Remapping Indigenous sound-ings in the Marshallese diaspora'; Steingo, Gavin and Sykes, Jim (eds), *Remapping Sound Studies*, Duke University Press, Durham, 2020, pp. 77–106.

'Merging the body with this rhythmic sea' Ingersoll, Karin Amimoto, *Waves of Knowing: A seascape epistemology*, Duke University Press, Durham, 2016, p. 129.

Sounds of unfathomable sadness and pain Rediker, Marcus, *The Slave Ship: A human history*, John Murray, London, 2008, p. 150.

'The rattling of chains, smacking of whips' Cugoano, Ottobah, 'Narrative of the enslavement of Ottobah Cugoano, a native of Africa; published by himself, in the Year 1787', electronic edition, *Documenting the American South*, 1999.

'They frequently sing' Rediker, Marcus, *The Slave Ship: A human history*, John Murray, London, 2008, p. 282.

He 'had never found it anything joyous' Epstein, Dena J., *Sinful Tunes and Spirituals: Black folk music to the Civil War*, University of Illinois Press, Champaign, 1977, p. 8.

'Slow airs, of a pathetic nature' Butterworth, William, *Three Years Adventures, of a Minor, in England, Africa, the West Indies, South Carolina and Georgia*, Thomas Inchbold, Leeds, 1831, pp. 93–4.

Conditions on the ships were so cruel 'The Middle Passage', History of Slavery, *National Museums Liverpool*, online resource.

'Pregnant with as many dead as living under the sentence of death' Glissant, Édouard and Wing, Betsy (trans.), *Poetics of Relation*, University of Michigan Press, Ann Arbor, 1997, p. 6.

'The sea is history' Walcott, Derek, *Collected Poems, 1948–1984*, Farrar, Straus & Giroux, New York, 1986.

'Like a train of gunpowder, ignited at one end' Rediker, Marcus, *The Slave Ship: A human history*, John Murray, London, 2008, p. 281.

'Songs of joy superseded the vociferations of discontent' Butterworth, William,

Three Years Adventures of a Minor, in England, Africa, the West Indies, South Carolina and Georgia, Thomas Inchbold, Leeds, 1831, p. 131.

Boys carried messages back and forth ibid., p. 120.

The women on the ship sang songs that told 'the History of their Lives' Rediker, Marcus, *The Slave Ship: A human history*, John Murray, London, 2008, p. 284; Romilly, Samuel, *The Life of Sir Samuel Romilly* (3rd edn), John Murray, London, 1842, p. 269.

A remarkable resemblance to musical structures employed in Senegambian music Green, Toby, *A Fistful of Shells: West Africa from the rise of the slave trade to the Age of Revolution*, Penguin, London, 2019, p. 398.

'An intellectual production that embodied many of the pathways . . . criss-crossing the ocean' ibid., p. 396.

An 'alchemy of pain' Smith, Zadie, *Changing My Mind: Occasional Essays*, Penguin, London, 2011.

'I had never heard anything like it' McQuay, Bill and Joyce, Christopher, 'It took a musician's ear to decode the complex song in whale calls', *NPR*, 6 August 2015.

'Few experiences have had a deeper effect on me' Ocean Alliance, 'Songs of the humpback whale', Whale.org, 2021.

Controversial dolphin researcher John Lilly Riley, Christopher, 'The dolphin who loved me: The NASA-funded project that went wrong', *The Guardian*, 8 June 2014.

'Humpback whales . . . produce a series of beautiful and varied sounds' Payne, R.S. and McVay, S., 'Songs of humpback whales', *Science*, vol. 173, no. 3997, 1971, pp. 585–97.

The record stretched the mind 'to encompass alien artforms' Rothenberg, David, *Thousand Mile Song: Whale music in a sea of sound*, Basic Books, New York, 2008, p. 18.

'There was a burst of realisation that the world could change its relation to wildlife . . .' Rothenberg, David, *Thousand Mile Song: Whale music in a sea of sound*, Basic Books, New York, 2008, p. 45.

The whales employ a variety of sounds Whitehead, Hal and Rendell, Luke, *The Cultural Lives of Whales and Dolphins*, University of Chicago Press, Chicago, 2015, pp. 77–9; Allen, Jenny A. et al., 'Network analysis reveals underlying syntactic features in a vocally learnt mammalian display, humpback whale song', *Proceedings of the Royal Society B*, vol. 286, no. 1917, 2019, 20192014; and Whale Trust, 'Song structure and composition', WhaleTrust.org, n.d.

Nor is the structure of the songs random Allen et al., ibid.

The whales seem to rely upon rhyme Guinee, Linda N. and Payne, Katharine B., 'Rhyme-like repetitions in songs of humpback whales', *Ethology*, vol. 79, no. 4, 1988, pp. 295–306.

A system of cultural transmission that spans the entire Pacific Schulze, Josephine N. et al., 'Humpback whale song revolutions continue to spread from the central into the eastern South Pacific', *Royal Society Open Science*, vol. 9, no. 8, 2022, 220158.

'When you swim up next to a singing whale' Payne, Roger, *Among Whales*, Scribner, New York, 1995, p. 145.

Bowhead whales . . . sing almost constantly across the winter months Hickey, Hannah, 'Bowhead whales, the "jazz musicians" of the Arctic, sing many different songs', *University of Washington News*, 3 April 2018; Bright, Richard, 'The changing acoustic environment of the Arctic', *Interalia Magazine*, January 2021.

'Humpbacks are like jazz singers' Land, Sherry, 'New population of blue whales discovered with help of bomb detectors', *UNSW News*, 8 June 2021.

French bioacoustician Dr Emmanuelle Leroy Leroy, Emmanuelle C. et al., 'Multiple pygmy blue whale acoustic populations in the Indian Ocean: Whale song identifies a possible new population', *Scientific Reports*, vol. 11, 2021, 8762.

Seasonal changes in the pitch of the whales' songs Leroy, Emmanuelle C. et al., 'Long-term and seasonal changes of large whale call frequency in the Southern Indian Ocean', *Journal of Geophysical Research: Oceans*, vol. 123, no. 11, 2018, pp. 8568–80.

At least some right whale populations sing Crance, Jessica L. et al., 'Song production by the North Pacific right whale, *Eubalaena japonica*', *Journal of the Acoustical Society of America*, vol. 145, no. 6, pp. 3467–79; Daley, Jason, 'Listen to the first known song of the North Pacific right whale', *Smithsonian Magazine*, 24 June 2019.

Some populations even enjoy getting high Nuwer, Rachel, 'Dolphins seem to use toxic pufferfish to get high', *Smithsonian Magazine*, 30 December 2013.

Signature whistles Morell, Virginia, 'Dolphins can call each other, not by name, but by whistle', *Science*, 20 February 2013.

Dolphins . . . are able to associate particular sounds with particular objects Mustill, Tom, *How to Speak Whale: A voyage into the future of animal communication*, William Collins, London, 2022, pp. 117–18.

Wild dolphins taught to associate a particular whistle with a specific species of seaweed Hodson, Hal, 'Dolphin whistle instantly translated by computer', *New Scientist*, 26 March 2014.

They are even capable of learning to communicate with other species Cosentino, Mel et al., 'I beg your pardon? Acoustic behaviour of a wild solitary common dolphin who interacts with harbour porpoises', *Bioacoustics*, vol. 31, no. 5, 2022, pp. 517–34.

African elephants . . . use a special alarm call Ngama, Steeve et al., 'How bees deter elephants: Beehive trials with forest elephants (*Loxodonta africana cyclotis*) in Gabon', *PLOS ONE*, vol. 11, no. 5, 2016, e0155690.

Codas seem to help reinforce social bonds Oliveira, Cláudia et al., 'Sperm whale codas may encode individuality as well as clan identity', *Journal of the Acoustical Society of America*, vol. 139, no. 5, 2016, pp. 2860–9.

An ongoing process of cultural evolution Whitehead, Hal and Rendell, Luke, *The Cultural Lives of Whales and Dolphins*, University of Chicago Press, Chicago, 2015, pp. 153–4.

The opening of the Drake Passage Steeman, Mette E. et al., 'Radiation of extant cetaceans driven by restructuring of the oceans', *Systematic Biology*, vol. 58, no. 6, 2009, pp. 573–85.

'Highly developed cross-modal sensory processing' Marino, Lori, 'The brain:

Evolution, structure and function', in Herzing, Denise L. and Johnson, Christine
M., *Dolphin Communication and Cognition: Past, present, and future*, MIT Press,
Cambridge, MA, 2015, pp. 12–13.

May be the origin of the legend of Davy Jones Whitehead, Hal and Rendell,
Luke, *The Cultural Lives of Whales and Dolphins*, University of Chicago Press,
Chicago, 2015, p. 150.

Dolphins … can 'see' objects the size of golf balls from 100 metres away
Kloepper, Laura N. et al., 'A view askew: Bottlenose dolphins improve echolocation precision by aiming their sonar beam to graze the target', Acoustical Society
of America, 170th Meeting, 20 October 2015; White, Thomas I., *In Defense of
Dolphins: The new moral frontier*, Blackwell Publishing, Malden, 2007, p. 21.

Sonar also renders the world around dolphins transparent White, ibid.

They may well be able to tell when other individuals are ill or injured or pregnant
ibid., pp. 22–3.

'I have never again observed the dolphins escort the boat in the same manner'
Herzing, Denise L. and Johnson, Christine M., *Dolphin Communication and Cognition:
Past, present, and future*, MIT Press, Cambridge, MA, 2015, pp. 31–2.

'Individuality is reduced close to zero' Norris, Ken, *Spinner Days: The life and times
of the spinner dolphin*, Avon Books, New York, 1993, p. 186.

Mass strandings also mostly involve matrilineal toothed whales Whitehead,
Hal and Rendell, Luke, *The Cultural Lives of Whales and Dolphins*, University of
Chicago Press, Chicago, 2015, p. 259.

Simultaneously themselves and part of a larger, social or collective self White,
Thomas I., *In Defense of Dolphins: The new moral frontier*, Blackwell Publishing,
Malden, 2007, p. 181.

Since the introduction of powerful low- and mid-frequency naval sonar
Alvarado-Rybak, Mario et al., '50 years of cetacean strandings reveal a concerning
rise in Chilean Patagonia', *Scientific Reports*, vol. 10, no. 1, 2020, p. 9511.

Of 136 mass strandings of beaked whales reported between 1876 and 2004
D'Amico, Angela et al., 'Beaked whale strandings and naval exercises', *Aquatic
Mammals*, vol. 35, no. 4, 2009, pp. 452–72.

Strandings in Britain, Greece, the Canary Islands and the Pacific Jepson,
Paul D. et al., 'What caused the UK's largest common dolphin (*Delphinus delphis*)
mass stranding event?', *PLOS ONE*, vol. 8, no. 4, 2013, e60953; Frantzis, A.,
'Does acoustic testing strand whales?', *Nature*, vol. 392, no. 6671, 1998, p. 29; and
Simonis, Anne E. et al., 'Co-occurrence of beaked whale strandings and naval
sonar in the Mariana Islands, Western Pacific', *Proceedings of the Royal Society B*,
vol. 287, no. 1921, 2020, 20200070.

Haemorrhages in their ears and brains Nevala, Amy E., 'The sound of sonar and
the fury about whale strandings', *Oceanus*, 15 February 2008.

'Envision a football squeezed to the size of a ping-pong ball' Balcomb, Ken,
'Letter to US Navy concerning high power military sonars', Ocean Mammal
Institute, 23 February 2001.

Shipping noise has increased dramatically Jones, N., 'Ocean uproar: Saving marine life from a barrage of noise', *Nature*, vol. 568, no. 7751, pp. 158–61; Erbe, Christine et al., 'The marine soundscape of the Perth Canyon', *Progress in Oceanography*, vol. 137, 2015, pp. 38–51.

A constant barrage of noise so loud it can be heard up to 4000 kilometres away Robbins, Jim, 'Oceans are getting louder, posing potential threats to marine life', *New York Times*, 22 January 2019.

Whales . . . suffer damage to their ears and hearing loss Jones, N, 'Ocean uproar: Saving marine life from a barrage of noise', *Nature*, vol. 568, no. 7751, pp. 158–61.

Humpback and bowhead whales exposed to the roar of ships Lemos, Leila S. et al., 'Effects of vessel traffic and ocean noise on gray whale stress hormones', *Scientific Reports*, vol. 12, no. 1, 2022, 18580.

Fish . . . raise their voices like humans Anthes, Emily, 'When fish shout', *New Yorker*, 10 November 2014.

The problems don't stop there de Soto, Natacha Aguilar et al., 'Anthropogenic noise causes body malformations and delays development in marine larvae', *Scientific Reports*, vol. 3, no. 1, 2013, 2831; Engås, Arill and Løkkeborg, Svein, 'Effects of seismic shooting and vessel-generated noise on fish behaviour and catch rates', *Bioacoustics*, vol. 12, no. 2–3, 2002, pp. 313–16.

The spiny chromis Nedelec Sophie, L. et al., 'Limiting motorboat noise on coral reefs boosts fish reproductive success', *Nature Communications*, vol. 13, no. 1, 2022, 2822.

Seismic surveys kill lobsters and crabs Clark, Chris, 'A two-year FRDC-funded study is investigating different seismic survey methods that could protect fisheries while giving oil and gas explorers the data they need', *Fish*, vol. 30, no. 2, 2022.

Cephalopods André, Michel et al., 'Low-frequency sounds induce acoustic trauma in cephalopods', *Frontiers in Ecology and the Environment*, vol. 9, no. 9, 2011, pp. 489–93.

Scallop larvae exposed to sonic pulses develop abnormally de Soto, Natacha Aguilar et al., 'Anthropogenic noise causes body malformations and delays development in marine larvae', *Scientific Reports*, vol. 3, no. 1, 2013, p. 2831.

Blue mussels exposed to ship noise Kershaw, Francine, 'New science: noise seriously impacts marine invertebrates', *NRDC News*, 29 July 2016.

The noise from sonic blasts also kills zooplankton up to 1.2 kilometres away McCauley, Robert D. et al., 'Widely used marine seismic survey air gun operations negatively impact zooplankton', *Nature Ecology & Evolution*, vol. 1, no. 7, 2017, p. 195; Tollefson, Jeff, 'Air guns used in offshore oil exploration can kill tiny marine life', *Nature*, vol. 546, no. 7660, 2017, pp. 586–7.

In the Arctic Bright, Richard, 'The changing acoustic environment of the Arctic', *Interalia Magazine*, January 2021.

Climate change is 'altering the acoustic fabric of the planet' Sueur, Jérôme et al., 'Climate change is breaking earth's beat', *Trends in Ecology and Evolution*, vol. 34, no. 11, 2019, pp. 971–3.

A superpod of more than 300 sperm whales Rodrigo, Malaka, 'Researchers miss out on sperm whale superpod in Sri Lanka amid pandemic', *Mongabay*, 6 May 2020.
The 'war of the soul' Mustakeem, Sowande' M., *Slavery at Sea: Terror, sex, and sickness in the Middle Passage*, University of Illinois Press, Champaign, 2016, p. 187ff.
A blue whale's heart beats just two times a minute Liverpool, Layal, 'A blue whale's heart beats just twice a minute when it dives for food', *New Scientist*, 25 November 2019; Kendall-Bar, Jessica, 'Listen to the heart rate of a blue whale', YouTube, 5 July 2022.
'The first time I ever recorded the songs of humpback whales at night' Payne, Roger, *Among Whales*, Scribner, New York, 1995, p. 145.

5 Beings

A mere eight weeks McGrouther, Mark, 'Which species of fish has the shortest lifespan?', Australian Museum, 30 March 2020.
The mysterious Greenland shark Miquel Sureda, Anfres, 'Near-blind shark is world's longest-lived vertebrate', *Nature*, 11 August 2016.
Rainbowfish learn . . . three times as fast as rats and twice as fast as dogs Brown, Culum, 'Fish intelligence, sentience and ethics', *Animal Cognition*, vol. 18, no. 1, 2015, pp. 1–17.
When tested again almost a year later Balcombe, Jonathan, *What a Fish Knows: The inner lives of our underwater cousins*, Farrar, Straus and Giroux, New York, 2016, p. 110.
Tilapia ibid., pp. 112–13.
Gobies Brown, Culum, 'Fish intelligence, sentience and ethics', *Animal Cognition*, vol. 18, no. 1, 2015, pp. 1–17.
The foundation of culture ibid.
Archerfish Brown, Culum and Laland, Kevin, 'Social learning in fishes', in Brown, Culum et al., *Fish Cognition and Behavior*, Wiley Blackwell, Oxford, 2011, p. 269.
Guppies Brown, Culum, 'Fish intelligence, sentience and ethics', *Animal Cognition*, vol. 18, no. 1, 2015, pp. 1–17.
French grunts ibid., p. 9.
Young female bluehead wrasse Brown, Culum and Laland, Kevin, 'Social learning in fishes', in Brown, Culum et al., *Fish Cognition and Behavior*, Wiley-Blackwell, Oxford, 2011, p. 244.
Cultural traditions are quickly lost Brown, Culum, 'Fish intelligence, sentience and ethics', *Animal Cognition*, vol. 18, no. 1, 2015, p. 9.
Port Jackson sharks Brown, Culum, 'The social lives of sharks', *Australasian Science*, vol. 39, no. 2, 2018, pp. 28–9.
Yellowtail amberjacks Schmitt, R. and Strand, S., 'Cooperative foraging by yellowtail, *Seriola lalandei (Carangidae)*, on two species of fish prey', *Copeia*, no. 3, 1982, pp. 714–17.
Some species of cichlids Dey, Cody J. et al., 'Direct benefits and evolutionary transitions to complex societies', *Nature Ecology & Evolution*, vol. 1, no. 5, 2017, p. 137.

Sticklebacks Brown, Culum, 'Fish intelligence, sentience and ethics', *Animal Cognition*, vol. 18, no. 1, 2015, p. 11.

Atlantic cod . . . took to stealing food from a feeding device Millot, Sandie et al., 'Innovative behaviour in fish: Atlantic cod can learn to use an external tag to manipulate a self-feeder', *Animal Cognition*, vol. 17, no. 3, 2014, pp. 779–85.

As Helen Macdonald has observed Macdonald, Helen, *Vesper Flights: New and collected essays*, Jonathan Cape, London, 2020, p. 5.

Groupers and morays in combination Bshary, R. et al., 'Interspecific communicative and coordinated hunting between groupers and giant moray eels in the Red Sea', *PLOS Biology*, vol. 4, no. 12, 2006, e431.

Pointing in magpies Kaplan, Gisela, 'Pointing gesture in a bird: Merely instrumental or a cognitively complex behavior?', *Current Zoology*, vol. 57, no. 4, 2011, pp. 453–67.

Some dogs seem to be able to learn to understand Worsley, Hannah K. et al., 'Cross-species referential signalling events in domestic dogs (*Canis familiaris*)', *Animal Cognition*, vol. 24, no. 4, 2018, pp. 457–65.

The ability of the trout . . . is almost identical to that of chimpanzees Vail, Alexander L. et al., 'Fish choose appropriately when and with whom to collaborate', *Current Biology*, vol. 24, no. 17, 2014, pp. R791–R793.

Groupers have learned to team up with octopuses Bayley, D.T.I. and Rose, A., 'Multi-species co-operative hunting behaviour in a remote Indian Ocean reef system', *Marine and Freshwater Behaviour and Physiology*, vol. 53, no. 1, 2020, pp. 35–42.

The wrasse are capable . . . of recognising and remembering more than 100 clients Bshary, Redouan, 'Machiavellian intelligence in fishes', in Brown, Culum et al. (eds), *Fish Cognition and Behaviour*, Wiley-Blackwell, Oxford, 2011, p. 284; Chojnacka, Dominika et al., 'Relative brain and brain part sizes provide only limited evidence that Machiavellian behaviour in cleaner wrasse is cognitively demanding', *PLOS ONE*, vol. 10, no. 8, 2015, e0135373.

They only ever bite non-predatory species Bshary, Redouan, 'Machiavellian intelligence in fishes', in Brown, Culum et al. (eds), *Fish Cognition and Behaviour*, Wiley-Blackwell, Oxford, 2011, p. 283ff.

Cleaner wrasse have been observed visiting the cleaning stations of other wrasse ibid., pp. 286–7.

The wrasse are capable of recognising themselves in a mirror Masanori, Kohda et al., 'If a fish can pass the mark test, what are the implications for consciousness and self-awareness testing in animals?', *PLOS Biology*, vol. 17, no. 2, 2019, p. e3000021.

Manta rays placed in a tank with a mirror Blaszczak-Boxe, Agata, 'Manta rays are first fish to recognise themselves in a mirror', *New Scientist*, 21 March 2016.

Diana Reiss . . . is similarly sceptical Preston, Elizabeth, 'A "self-aware" fish raises doubts about a cognitive test', *Quanta Magazine*, 12 December 2018.

A continuum of self-cognisance, 'ranging from self-referencing to self-awareness to self-consciousness' Bekoff, Marc and Sherman, Paul W., 'Reflections on animal selves', *Trends in Ecology & Evolution*, vol. 19, no. 4, 2004, pp. 176–80.

Damselfish . . . have intricate patterns on their faces Losey, George S., 'Crypsis and communication functions of UV-visible coloration in two coral reef damselfish, *Dascyllus aruanus* and *D. reticulatus*', *Animal Behaviour*, vol. 66, no. 2, 2003, pp. 299–307.

Longer-lived species tend to have ultraviolet filters in their eyes Scales, Helen, *Eye of the Shoal: A fishwatcher's guide to life, the ocean and everything*, Bloomsbury Sigma, London, 2018, p. 101.

The eyes of salmon ibid., p. 98.

Gobies and flounders Balcombe, Jonathan, *What a Fish Knows: The inner lives of our underwater cousins*, Farrer, Straus and Giroux, New York, 2016, p. 27.

Four-eyed fish ibid., p. 28.

Sharks . . . are believed to be able to hear prey from up to 250 metres away Hackner, Stacy, 'Question of the week: How do sharks hear?', Researchers in Museums, University College London, 23 April 2014.

Some species of fish . . . rely upon infrasound Sand, O. and Karlsen, H.E., 'Detection of infrasound and linear acceleration in fishes', *Philosophical Transactions of the Royal Society B: Biological Sciences*, vol. 355, no. 1401, 2000, pp. 1295–8.

In great whites . . . 14 per cent of their total brain mass is devoted to olfaction 'Smell and taste', *Biology of Sharks and Rays*, elasmo-research.org.

Known as *schreckstoff* Jabr, Ferris, 'Scary stuff: Fright chemicals identified in injured fish', *Scientific American*, 23 February 2012.

The distinctive shape of the hammerhead shark's head and the . . . toothed snouts of sawfish Whitehead, Darryl and Collin, Shaun, 'The functional roles of passive electroreception in non-electric fishes', *Animal Biology*, vol. 54, no. 1, 2004, pp. 1–25.

Knifefish . . . use electrical signals to organise complex hierarchies Buehler, Jake, 'Knifefish use electric fields to develop complex social hierarchy', *New Scientist*, 17 February 2021.

Juvenile Doederlein's cardinalfish Bottesch, Michael et al., 'A magnetic compass that might help coral reef fish larvae return to their natal reef', *Current Biology*, vol. 26, no. 24, 2016, R1266–R1267.

American eels Dodson, Julian and Castonguay, Martin, 'Tracking American eels on the open sea to crack the mystery of their migration', *The Conversation*, 28 October 2015.

'If a lion could talk' Wittgenstein, Ludwig, *Philosophical Investigations: The collected works of Ludwig Wittgenstein*, electronic edition, InteLex Corp, 1998, p. 223.

Thomas Nagel . . . argued consciousness is essentially subjective Nagel, Thomas, 'What is it like to be a bat?', *Philosophical Review*, vol. 83, no. 4, 1974, pp. 435–50.

Open-source software Van Den Berg, Cedric P. et al., 'Quantitative colour pattern analysis (QCPA): A comprehensive framework for the analysis of colour patterns in nature', *Methods in Ecology and Evolution*, vol. 11, no. 2, 2020, pp. 316–32; for more see Ferreira, Becky, 'Scientists created open source tools to see in animal vision', *Motherboard*, 4 December 2019.

'There exist, alongside ours, endless other worlds' Peña-Guzmán, David M., *When Animals Dream: The hidden world of animal consciousness*, Princeton University Press, Princeton, 2022, p. 9.

'unfamiliar . . . uncanny, haunting' Haraway, Donna, *Staying with the Trouble: Making kin in the Chthulucene*, Duke University Press, Durham, 2016, p. 103; see also Paulson, Steven, 'Making kin: An interview with Donna Haraway', *Los Angeles Review of Books*, 6 December 2019.

'Science can be a way of forming intimacy and respect' Kimmerer, Robin Wall, *Braiding Sweetgrass: Indigenous wisdom, scientific knowledge and the teachings of plants*, Penguin, London, 2020, p. 253.

Trout given injections of acetic acid Braithwaite, Victoria, *Do Fish Feel Pain?*, Oxford University Press, Oxford, 2010, pp. 67–9.

Somewhere between 790 billion and 2.3 trillion wild fish were caught and killed annually 'Estimated numbers of individuals in annual global capture tonnage (FAO) of fish species (2007–2016)', Fishcount.org.uk, 2016.

As many as 167 billion fish were killed each year in aquaculture 'Estimated numbers of individuals in global aquaculture production (FAO) of fish species (2017)', Fishcount.org.uk, 2017.

Almost all of these fish die slowly and painfully 'Humane slaughter', Fishcount.org.uk, 2019.

Fish caught in nets suffer horribly as well 'Estimating the number of fish caught in global fishing each year', ibid.

The agonising effects of the bends Salleh, Anna, 'Even fish get the bends, and it's "worse than in humans"', *ABC Science*, 14 December 2018.

Fish are specifically excluded from animal welfare legislation in many countries Mood, A. and Brooke, P. 'Towards a strategy for humane fishing in the UK', Fishcount.org.uk, November 2019.

We routinely downplay the mental abilities and capacity for suffering of the land-based animals we eat Bastian, Brock et al., 'Don't mind meat? The denial of mind to animals used for human consumption', *Personality and Social Psychology Bulletin*, vol. 38, no. 2, 2012, pp. 247–56.

'It is a pain to be in pain' Gruen, Lori, 'The moral status of animals', in Zalta, Edward N. (ed.), *The Stanford Encyclopaedia of Philosophy*, Summer 2021.

Over the past fifty years humans have wiped out at least half the fish on the planet *Living Blue Planet Report: Species, habitats and human well-being*, WWF, Gland, 2015.

The loss of migratory behaviours that have existed for thousands of generations De Luca, Giancarlo et al., 'Fishing out collective memory of migratory schools', *Journal of the Royal Society Interface*, vol. 11, no. 95, 2014, 20140043.

6   Beaches

'Vast heaps of the largest Oyster Shells I ever saw' Cook, James, *The Journals of Captain Cook*, Penguin Classics, London, 2003, p. 130.

'Great quantity' of oysters of 'amazing size' Attenbrow, Val, *Sydney's Aboriginal Past: Investigating the archaeological and historical records*, UNSW Press, Sydney, 2002, p. 67.

A man might '[tie] his boat to a stake, then [commence] to dredge for six months' McAfee, Dominic et al., 'The value and opportunity of restoring Australia's lost rock oyster reefs', *Restoration Ecology*, vol. 28, no. 2, 2020, pp. 304–14; see also Lergessner, J. G., *Oysterers of Moreton Bay*, Schuurs Publications, Bribie Island, 2008.

In some places the crust rises and falls by more than 50 centimetres Lau, Harriet C.P. and Schindelegger, Michael, 'Solid earth tides', in Green, Mattias and Duarte, João C. (eds), *A Journey Through Tides*, Elsevier, Amsterdam, 2022, pp. 365–87.

400 million years ago a day lasted almost three hours less than it does today Pannella, Giorgio, 'Paleontological evidence on the Earth's rotational history since early Precambrian', *Astrophysics and Space Science*, vol. 16, no. 2, 1972, pp. 212–37.

This gradual slowing may even have paved the way for complex life Klatt, J.M. et al., 'Possible link between Earth's rotation rate and oxygenation', *Nature Geoscience*, vol. 14, no. 8, 2021, pp. 564–70; Klatt, Judith, 'A long day for microbes', *Max Planck News*, 2 August 2021.

The Bay of Fundy National Oceanic and Atmospheric Administration, 'Where is the highest tide?', *Tides & Currents*, 27 September 2023.

The Muslim astrologer Abu Ma'shar al Balkhi Cartwright, David E., 'On the origins of knowledge of the sea tides from antiquity to the thirteenth century', *Earth Sciences History*, vol. 20, no. 2, 2001, pp. 105–26.

The best time for the Allied landings at Normandy on D-Day Aldersey-Williams, Hugh, *Tide: The science and lore of the greatest force on Earth*, Penguin, London, 2017, pp. 261–2.

People in Southern Europe etched detailed lunar calendars into mammoth tusks and antlers 'The oldest lunar calendars', *Solar System Exploration Research Virtual Institute News*, 4 October 2011.

While stones at Wurdi Youang Davey, Melissa, 'Australian rock formation could be older than Stonehenge, researchers say', *The Guardian*, 13 October 2016.

'Abstract' time Postone, Moishe, *Time, Labor, and Social Domination: A reinterpretation of Marx's Critical Theory*, Cambridge University Press, Cambridge, 1993, p. 203.

Abstract time is 'a central part of the deep structure of environmental damage' Adam, Barbara, *Timescapes of Modernity: The environment and invisible hazards*, Routledge, London, 2005, p. 9.

'The concept of everywhen unsettles the way historians and archaeologists have conventionally treated time' McGrath, Ann, Rademaker, Laura and Troy, Jakelin, *Everywhen: Australia and the language of deep history*, NewSouth Publishing, Sydney, 2023, p. 15.

Mirning woman Iris Burgoyne Burgoyne, Iris, *The Mirning: We are the whales*, Magabala Books, Broome, 2000.

Palawa stories transcribed in the 1830s Hamacher, Duane et al., 'The Archaeology of Orality: Dating Aboriginal Tasmanian oral traditions to the Late Pleistocene', *Journal of Archaelogial Sciences*, vol. 159, 2023, 105819.

'**Immediately . . . the earth crumbled away**' Nunn, Patrick D., *The Edge of Memory: Ancient stories, oral tradition and the post-glacial world*, Bloomsbury Sigma, London, 2018, p. 79.

Kulin and Bunurong traditions ibid., pp. 76–9.

Shorelines may have retreated as much as 500 metres a decade ibid., p. 104.

The Narungga people of the Yorke Peninsula ibid., pp. 69–71.

A king tide on Muralag Island Dinnen, Richard, 'Rising sea level unearths Torres Strait burial site', *TropicNow*, 20 September 2022.

On the islands of Saibai and Boigu high tides are already spilling over seawalls Bunch, Aaron, 'Torres Strait flooding fears as sea-level peak imminent', *South Coast Register*, 19 February 2023.

A virtual copy of their nation Shepherd, Tory, 'Could a digital twin of Tuvalu preserve the island nation before it's lost to the collapsing climate?', *The Guardian*, 30 September 2022.

Pacific Island nations 'refuse to be the proverbial canaries in the world's coal mine' 'PM Bainimarama remarks at the Climate Reality Blue Pacific event', Ministry of Foreign Affairs, Fiji, 7 October 2021.

Representatives of fossil fuel companies at the summit outnumbered Pacific Island negotiators twelve to one Bainimarama, Frank, 'Pacific Island negotiators are outnumbered by fossil fuel reps at COP26 by more than 12 to 1', Twitter, 11 November 2021.

'**Moving outside of Tuvalu will not solve any climate change issues**' Roy, Eleanor Ainge, 'One day we'll disappear: Tuvalu's sinking islands', *The Guardian*, 16 May 2019.

'**You tell them / we don't want to leave**' Jetñil-Kijiner, Kathy, *Iep Jaltok: Poems from a Marshallese daughter*, University of Arizona Press, Tucson, 2017, pp. 66–7.

'**Slow violence**' Nixon, Rob, *Slow Violence and the Environmentalism of the Poor*, Harvard University Press, Harvard, 2011, p. 2.

A place 'where we see ourselves reflected in somebody else's otherness' Dening, Greg, *Readings/Writings*, Melbourne University Press, Melbourne, 1998, p. 87.

'**The islanded must understand that to live long and well, they need to take care**' Teaiwa, T., 'To island', in Jetñil-Kijiner, K., Santos Perez, C. and Kava, L. (eds), *Indigenous Pacific Islander Eco-Literatures*, University of Hawaii Press, Honolulu, 2022, p. 127.

7   Deep

Our eyes compress gradations of light Widder, Edith, *Below the Edge of Darkness: A memoir of exploring light and life in the deep sea*, Text Publishing, Melbourne, 2021, p. 74.

Humboldt squid . . . produce complex displays of red and white light Kubota, Taylor, 'Stanford researcher investigates how squid communicate in the dark', *Stanford News Service*, 23 March 2020.

Known as *te lapa* Widder, Edith, *Below the Edge of Darkness: A memoir of exploring light and life in the deep sea*, Text Publishing, Melbourne, 2021, pp. 119–22; see also George, Marianne, 'Polynesian navigation and *te lapa* – "The flashing"', *Time and Mind*, vol. 5, no. 2, 2012, pp. 135–73.

The name refers to Hades, the Greek god of the underworld Jamieson, Alan, *The Hadal Zone: Life in the deepest oceans*, Cambridge University Press, Cambridge, 2015, p. 18.

Below about 8500 metres . . . their cells begin to lose integrity Gerringer, Mackenzie, 'The deepest-dwelling fish in the sea is small, pink and delicate', *The Conversation*, 1 February 2018.

What Macfarlane calls the underland Macfarlane, Robert, *Underland: A deep time journey*, Hamish Hamilton, London, 2019.

Some West African cultures believed the resisting tension of the ocean's surface Mack, John, *The Sea: A cultural history*, Reaktion Books, London, 2011, p. 92.

Challenger's **discovery of tiny shells . . . more than 7 kilometres down** Jamieson, Alan, *The Hadal Zone: Life in the deepest oceans*, Cambridge University Press, Cambridge, 2015, p. 11.

'An endless variety of sculptured pillow forms' Ballard, Robert D., 'Notes on a major oceanographic find', *Oceanus*, vol. 20, no. 3, 1977.

Over cycles of approximately 10 to 20 million years the entire volume of the ocean passes through them 'Importance: Seawater chemistry', Discovering Hydrothermal Vents section, Woods Hole Oceanographic Institution website.

Governments consigned . . . obsolete chemical weapons to the depths Curry, Andrew, 'Weapons of war litter the ocean floor', *Hakai Magazine*, 10 November 2016; Wilkinson, Ian, 'Chemical weapon munitions dumped at sea: An interactive map', James Martin Center for Nonproliferation Studies, 1 August 2017.

A 'Chernobyl in slow motion on the sea floor' Luhn, Alec, 'Russia's "slow-motion Chernobyl" at sea', *BBC Future Planet*, 2 September 2020.

Almost 70,000 tonnes of nuclear waste Sjöblom, Kirsti-Liisa and Linsley, Gordon, 'Sea disposal of radioactive wastes: The London Convention 1972', *International Atomic Energy Agency Bulletin*, vol. 36, 1994, pp. 12–16, and *Inventory of radioactive waste disposals at sea*, International Atomic Energy Agency, IAEA, Vienna, August 1999, IAEA-TECDOC-1105.

Plans to dispose of up to 750,000 cubic metres of nuclear waste Busby, Martha, 'Undersea nuclear waste dump off Cumbria would imperil marine life, experts warn', *The Guardian*, 29 July 2022.

The 2014 leak of radioactive material at a waste disposal facility . . . in New Mexico Tollefson, Jeff, 'US seeks waste-research revival', *Nature*, vol. 507, no. 7490, 2014, pp. 15–16.

The Japan Agency for Marine-Earth Science and Technology (JAMSTEC) Deep-sea Debris Database Deep-sea Debris Database, jamstec.go.jp; Sanae, Chiba, 'Shopping bags in the abyss: Addressing the deep-sea plastic crisis', Nippon .com, 3 October 2018.

The number of such objects exceeds 300 per square kilometre Sanae, Chiba et al., 'Human footprint in the abyss: 30-year records of deep-sea plastic debris', *Marine Policy*, vol. 96, 2018, pp. 204–12.

Whales and birds are consuming microplastics in large quantities Rivers-Auty,

Jack et al., 'The one-two punch of plastic exposure: Macro- and micro-plastics induce multi-organ damage in seabirds', *Journal of Hazardous Materials*, vol. 442, 2023, e130117.

As much as 99.8 per cent of the more than 11 million tonnes of plastic that enters the ocean This figure is probably a significant underestimate. Studies suggest that in 2010 between 4.8 and 12.7 million tonnes of plastic entered the ocean, but between 2010 and 2022 global plastic production increased by 50 per cent. (Jambeck, Jenna R. et al., 'Plastic waste inputs from land into the ocean', *Science*, vol. 347, no. 6223, 2015, pp. 768–71; Koelmans, Albert A. et al., 'All is not lost: Deriving a top-down mass budget of plastic at sea', *Environmental Research Letters*, vol. 12, no. 11, 2017, 114028.)

This process may transport more than 400,000 tonnes of plastic into the deep ocean every year Imbler, Sabrina, 'In the ocean, it's snowing microplastics', *New York Times*, 3 April 2022.

Trillions of tiny plastic fragments . . . accumulate in the canyons and trenches Kane, Ian and Clare, Michael, 'Seafloor currents sweep microplastics into deep-sea hotspots of ocean life', *The Conversation*, 1 May 2020; Kane, Ian A. et al., 'Seafloor microplastic hotspots controlled by deep-sea circulation', *Science*, vol. 368, no. 6495, 2020, pp. 1140–5.

Microplastics are now ubiquitous in ocean sediments Woodall, Lucy C. et al., 'The deep sea is a major sink for microplastic debris', *Royal Society Open Science*, vol. 1, no. 4, 2014, 140317.

Plastic microfilaments coating the surface of deep-sea corals ibid.

Radioactive carbon-14 from the detonation of nuclear bombs in the 1940s and 1950s Levy, Adam, '"Bomb carbon" has been found in deep-ocean creatures', *Scientific American*, 15 May 2019.

Radioactive caesium from the Fukushima nuclear disaster Jamieson, Alan, *The Hadal Zone: Life in the deepest oceans*, Cambridge University Press, Cambridge, 2015, p. 269.

PCBs . . . have simply migrated into deeper waters Jamieson, Alan J. et al., 'Bioaccumulation of persistent organic pollutants in the deepest ocean fauna', *Nature Ecology & Evolution*, vol. 1, no. 3, 2017, p. 51.

The different isotopes contained in their shells 'Seafloor fossils provide clues to climate change', *ScienceDaily*, 8 November 2009; Keating-Bitonti, Caitlin and Chang, Lucy, 'Here's how scientists reconstruct Earth's past climates', *Smithsonian Magazine*, 23 March 2014.

Microbes that had become trapped as it was deposited Pennisi, Elizabeth, 'Scientists pull living microbes, possibly 100 million years old, from beneath the sea', *Science*, 28 July 2020.

Oil and gas companies are continuing to sink hundreds of millions of dollars a day into new and expanded production Carrington, Damian and Taylor, Matthew, 'Revealed: The "carbon bombs" set to trigger catastrophic climate breakdown', *The Guardian*, 11 May 2022.

The United Nations Environment Programme (UNEP) found that the fossil fuel projects then planned SEI et al., *The Production Gap Report 2021*, Stockholm Environment Institute, International Institute for Sustainable Development, ODI, E3G and United Nations Environment Programme, 2021.

This projected new production continues to expand Carrington, Damian and Taylor, Matthew, 'Revealed: The "carbon bombs" set to trigger catastrophic climate breakdown', *The Guardian*, 11 May 2022.

Not only does the project pose risks to ancient rock art on the Peninsula Ritchie, Nick and Edney-Browne, Alex, *Deep-Sea Disaster: Why Woodside's Burrup Hub project is too risky to proceed*, Greenpeace Australia, Sydney, June 2022.

Trillions of such nodules lie scattered across the . . . abyssal plain Hein, James R. et al., 'Deep-ocean polymetallic nodules as a resource for critical materials', *Nature Reviews Earth & Environment*, vol. 1, 2020, pp. 158–69.

Twice as much nickel and more than three times as much cobalt ibid.

The fragile détente between the two superpowers Bennett, M. Todd, 'Détente in Deep Water: The CIA mission to salvage a sunken Soviet submarine and US–USSR relations, 1968–1975', *Intelligence and National Security*, vol. 33, no. 2, 2018, pp. 196–210.

It also helped stimulate real research Thulin, Lila, 'During the Cold War, the CIA secretly plucked a Soviet submarine from the ocean floor using a giant claw', *Smithsonian Magazine*, 10 May 2019.

Production of nickel will need to double and production of cobalt will need to increase fivefold Hund, Kirsten et al., *Minerals for Climate Action: The mineral intensity of the clean energy transition*, World Bank, Washington, D.C., 2020.

Mineral production will need to rise sixfold *The Role of Critical Minerals in Clean Energy Transitions*, International Energy Agency, Paris, May 2021.

Huge Chinese-owned mines that supply nickel to Tesla Timmerman, Antonia, 'The dirty road to clean energy: How China's electric vehicle boom is ravaging the environment', *Rest of World*, 28 November 2022.

More than three quarters of the world's terrestrial reserves of cobalt are held by just four countries Garside, M., 'Reserves of cobalt worldwide in 2022, by country', *Statista*, 18 February 2023.

Run-off from the massive Moa mine Watts, Jonathan, 'How the race for cobalt risks turning it from miracle metal to deadly chemical', *The Guardian*, 18 December 2019; Diaz Rizo, O. et al., 'Chromium, cobalt and nickel contents in urban soils of Moa, northeastern Cuba', *Bulletin of Environmental Contamination and Toxicology*, vol. 86, no. 2, 2011, pp. 189–93.

An ongoing human and ecological tragedy Searcey, Dionne, Forsythe, Michael and Lipton, Eric, 'A power struggle over cobalt rattles the clean energy revolution', *New York Times*, 20 November 2021; see also Kara, Siddharth, *Cobalt Red: How the blood of the Congo powers our lives*, St Martin's Press, New York, 2023.

'Across twenty-one years of research into slavery and child labour' Kara, ibid., p. 5.

'A battery in a rock' The Metals Company, 'A battery in a rock', Metals.co/nodules/, 2021.

The din of the mining machines could travel up to 500 kilometres McVeigh, Karen, 'One deep sea mine could send noise 500 km across the ocean: Report', *The Guardian*, 8 July 2022.

Between 70 and 90 per cent of the species recovered were new to science Amon, Diva J. et al., 'Assessment of scientific gaps related to the effective environmental management of deep-seabed mining', *Marine Policy*, vol. 138, 2022, 105006.

It is impossible to make a coherent assessment of the potential costs and benefits Shackleford, Scott et al., 'Deep seabed mining plans pit renewable energy demand against ocean life in a largely unexplored frontier', *The Conversation*, 18 January 2023.

Developing the knowledge base necessary will take decades Amon, Diva J. et al., 'Assessment of scientific gaps related to the effective environmental management of deep-seabed mining', *Marine Policy*, vol. 138, 2022, 105006.

'Everybody in Brooklyn can . . . say, "I don't want to harm the ocean"' Lipton, Eric, 'Secret data, tiny islands and a quest for treasure on the sea floor', *New York Times*, 29 August 2022.

The places of Belgium and Nauru . . . were filled by executives from seabed-mining corporations Casson, Louisa et al., *Deep Trouble: The murky world of the deep sea mining industry*, Greenpeace International, Amsterdam, 2020.

A project to mine black smokers Groch, Sherryn, '"It's bloody El Dorado": Undersea riches to be reaped by the few', *Sydney Morning Herald*, 9 September 2023.

Google, Samsung, BMW and Volvo 'Major companies call for moratorium on deep-sea mining', Radio New Zealand, 1 April 2021.

Twenty-one nations . . . called for a precautionary pause on mining Mehta, Angeli, 'Policy watch: After fraught global meeting, future of deep-sea mining still hangs in balance', *Reuters*, 4 August 2023.

Apple has already announced plans to employ only recycled cobalt 'Apple will use 100 per cent recycled cobalt by 2025', *Apple Newsroom*, 14 April 2023.

An accelerated rollout of public transport, bikes and e-bikes IPCC, *Climate Change 2022: Impacts, adaptation, and vulnerability – Summary for policymakers*, 2022, p. 38.

Nor did he see anything that remotely resembled 'corporate commitments to human rights standards' Kara, Siddharth, *Cobalt Red: How the blood of the Congo powers our lives*, St Martin's Press, New York, 2023, p. 5.

The significance of the discovery of deep time Griffiths, Tom, 'Travelling in deep time: La longue durée in Australian History', *Australian Humanities Review*, no. 18, June 2000.

95 per cent of the ocean biosphere Cario, Anäis et al., 'Exploring the deep marine biosphere: Challenges, innovations and opportunities', *Frontiers in Earth Science*, vol. 7, 2019, p. 225.

8 Cargo

Almost 30,000 ocean-going ships 'Facts & Figures', portofrotterdam.com, n.d.

Most synthetic fibres begin their life as oil *Preferred Fiber & Materials: Market report 2021*, Textile Exchange, 2021.

Your clothes Sabanoglu, Tuba, 'Share in world exports of the leading clothing exporters in 2021, by country', *Statitsta*, 5 December 2022.

Your shoes Smith, P., 'Leading 10 footwear producers worldwide from 2016 to 2021, by country', *Statista*, 10 October 2022.

Over the past fifty years international maritime trade has more than quadrupled United Nations Conference on Trade and Development, *Review of Maritime Transport 2022: Navigating stormy waters*, UNCTAD, New York, 2022, p. 3.

The first began in the years immediately after World War II Kaukiainen, Yrjö, 'The role of shipping in the "second stage of globalisation"', *International Journal of Maritime History*, vol. 26, no. 1, 2014, pp. 64–81; see also Campling, Liam and Colás, Alejandro, *Capitalism and the Sea: The maritime factor in the making of the modern world*, Verso, London, 2021, pp. 240–50.

International trade grew more than twice as fast as manufacturing production Levinson, Marc, *The Box: How the shipping container made the world smaller and the world economy bigger*, 2nd edn, Princeton University Press, Princeton, 2016, p. 15.

Biodiversity loss from deforestation has also accelerated Feng, Yu et al., 'Doubling of annual forest carbon loss over the tropics during the early twenty-first century', *Nature Sustainability*, vol. 5, no. 5, 2022, pp. 444–51; Pacheco, P. et al., *Deforestation Fronts: Drivers and responses in a changing world*, WWF, Gland, 2021.

The insatiable demand for cardboard Shaer, Matthew, 'Where does all the cardboard come from? I had to know', *New York Times*, 28 November 2022.

'The twin signatures of this era have been the mass export of products' Klein, Naomi, *This Changes Everything: Capitalism vs the climate*, Allen Lane, London, 2014, p. 20.

'Coffins of remote labour-power' Sekula, Allan and Burch, Noel, 'The forgotten space: Notes for a film', *New Left Review*, no. 69, 2011, pp. 78–9.

Enough to end extreme poverty seventy times over Hickel, Jason et al., 'Imperialist appropriation in the world economy: Drain from the Global South through unequal exchange, 1990–2015', *Global Environmental Change*, vol. 73, 2022.

81 per cent of the growth in material use . . . results from increased consumption in developed countries Hickel, Jason, *Less Is More: How degrowth will save the world*, William Heinemann, London, 2020.

'An extraordinary apparatus of management' Cowen, Deborah, *The Deadly Life of Logistics: Mapping violence in global trade*, University of Minnesota Press, Minneapolis, 2014, p. 20.

The 'almost inconceivable global logistics' Logistics Art Project, *Logistics*, logisticsartproject.com, 2021.

Their dream is to follow the parts of the pedometer 'Logistics', an 857-hour movie, tracks a pedometer from shop back to factory', Hacker News, news .ycombinator.com, 13 August 2022.

Capitalism aspires to annihilate time and space Campling, Liam and Colás, Alejandro, *Capitalism and the Sea: The maritime factor in the making of the modern world*, Verso, London, 2021, pp. 213–4.

Trapped on their vessels Pauksztat, B. et al., *Seafarers' Experiences During the Covid-19 Pandemic*, World Maritime University, Malmö, 2020.

The profits of shipping companies soared 'Threat of hard landing hangs over soaring shipping industry', *Financial Times*, 8 September 2022; 'Operating profit of container shipping companies between 2010 and 2022', Statista, 13 July 2023.

Waves of migration Khalili, Laleh, *Sinews of War and Trade: Shipping and capitalism in the Arabian Peninsula*, Verso Books, London, 2020, p. 135.

Climate protesters disrupted trains bound for Sydney's Port Botany Chung, Laura, 'Activists slam new anti-protest laws as "knee-jerk" after three days of protests', *Sydney Morning Herald*, 24 March 2022.

In 2022 alone there were seven major oil spills 'Seven large oil spills reported in 2022', *Safety4Sea*, 11 January 2023.

A tanker that was unloading at a refinery in Peru Collins, Dan, 'Oil spill at sea: Who will pay for Peru's worst environmental disaster?', *The Guardian*, 7 March 2022.

The explosion of the refinery ship FPSO *Trinity Spirit* Obukata, Joseph, 'Vessel tragedy: Untold story of how 2 crew members were arrested, arraigned', *Sun Nigeria*, 30 March 2022.

More than 3000 containers were reported lost at sea *Safety and Shipping Review 2021*, Allianz Global Corporate & Specialty SE, Munich, 2021.

The number of containers being lost each year is increasing ibid.

The loss of almost 29,000 rubber ducks in the northern Pacific Hohn, Donovan, *Moby-Duck: The true story of 28,800 bath toys lost at sea and of the beachcombers, oceanographers, environmentalists, and fools, including the author, who went in search of them*, Scribe, Melbourne, 2011.

Around 3500 seafarers were taken hostage by Somali pirates 'Deaths of seafarers in Somali pirate attacks soar', *Reuters*, 21 June 2011.

In the Gulf of Guinea Igwe, Uchenna, 'Murder, kidnapping and arson: Nigerian pirates switch targets from ships to shore', *The Guardian*, 17 January 2023.

Piracy is also on the rise in the Straits of Singapore, and continues to be an issue off Peru ibid.

Cyberattacks on shipping companies *Safety and Shipping Review 2021*, Allianz Global Corporate & Specialty SE, Munich, 2021.

The majority come from Asia United Nations Conference on Trade and Development, *Review of Maritime Transport 2022*, New York, 2022; 'Seafarer supply, quinquennial, 2015 and 2021', unctadstat.unctad.org, 2023.

Several hundred seafarers are killed or go missing every year Placek, Martin, 'Number of seafarers reported killed and missing globally from 2015 to 2019', *Statista*, 30 September 2022.

At least 60,000 premature deaths a year Rutherford, Dan and Miller, Josh, 'Silent but deadly: The case of shipping emissions' [blog entry], International Council on Clean Transportation, 22 March 2019.

This poisonous fug has helped shield the planet from the heating effect of fossil fuels Hausfather, Zeke and Forster, Piers, 'Analysis: How low-sulphur shipping rules are affecting global warming', *Carbon Brief*, 3 July 2023.

Shipping exhaust is speeding up the melting of the ice in the Arctic Comer, Bryan et al., *The International Maritime Organization's Proposed Arctic Heavy Fuel Oil Ban: Likely impacts and opportunities for improvement*, International Council on Clean Transportation, Washington, D.C., 2020.

Yet these impacts pale in comparison to shipping's contribution to global heating United Nations Conference on Trade and Development, *Review of Maritime Transport 2022: Navigating stormy waters*, UNCTAD, New York, 2022.

Representatives of shipbuilders, mining and fossil fuel companies, shipping companies and lobbyists Collett-White, Rich, 'Cook Islands UN negotiator paid $700K by shipping industry lobby group', *DeSmog*, 17 September 2021.

'It's mind-boggling how corrupt and unaccountable the IMO is' Darby, Megan, 'It's mind-boggling how corrupt and unaccountable the IMO is, hidden under a veneer of sheer tedium', Twitter, 15 May 2019.

'I was fishing with my grandfather' Harvey, Fiona, 'Tony de Brum obituary', *The Guardian*, 11 October 2017.

'Time is short, and it is not our friend' ibid.

De Brum and his delegation Apuzzo, Matt and Hurts, Sarah, 'Tasked to fight climate change, a secretive UN agency does the opposite', *New York Times*, 3 June 2021.

These targets were revised upwards again in 2023 'Revised GHG reduction strategy for global shipping adopted', *IMO Newsroom*, 7 July 2023.

They do demand significant reductions in the emissions produced by individual ships de Beukelaer, Christiaan and Smith, Tristan, 'Why the shipping industry's increased climate ambition spells the end for its fossil fuel use', *The Conversation*, 14 July 2023; Smith, Tristan and Shaw, Alison, 'An overview of the discussions from IMO MEPC 80 and frequently asked questions', *UMAS*, London, 7 July 2023.

Only a third of the world's major shipping companies *Ready, Set, Decarbonize! Are shipowners committed to a net zero future?*, Mærsk Mc-Kinney Møller Center for Zero Carbon Shipping, Copenhagen, 31 May 2022.

Maersk committed to reaching net zero by 2050 'A.P. Moller-Maersk aims at having carbon neutral vessels commercially viable by 2030 and calls for strong industry involvement', *Maersk News*, 4 December 2018.

In 2022 the company brought that target forward to 2040 'A.P. Moller-Maersk accelerates net zero emission targets to 2040 and sets milestone 2030 targets', *Maersk News*, 12 January 2022.

Maersk has also ordered nineteen 'zero-emission' ships 'A.P. Moller-Maersk continues green transformation with six additional large container vessels', *Maersk News*, 5 October 2022.

CMA CGM has . . . placed orders for eighteen container ships Li, Martina, 'CMA CGM books another dozen 13,000 TEU methanol-powered ships', *Load Star*, 2 February 2023.

COSCO has a dozen ultra-large methanol-ready container ships on order 'COSCO orders 12 ultra-large, green methanol containerships for $2.9B', *Maritime Executive*, 28 October 2022.

Pacific Basin Shipping . . . has announced it intends to transition its fleet 'Pacific Basin selects methanol for next gen dry bulkers', *Maritime Executive*, 27 October 2022.

In 2023 less than a million tonnes was produced worldwide 'Renewable Methanol', The Methanol Institute, methanol.org/renewable/.

Methanol has several attractions as an alternative to HFO Placek, Martin, 'Energy density of maritime fuels in 2020, by fuel type', *Statista*, 3 February 2023; Delbert, Caroline, 'Green ammonia could power the future – if we can get over the risks', *Popular Mechanics*, 25 February 2020; and Butterworth, Paul, 'Green ammonia fuel faces three big challenges', *CRU*, 17 February 2022.

Sourcing green methanol is likely to be a significant challenge IRENA and Methanol Institute, *Innovation Outlook: Renewable Methanol*, International Renewable Energy Agency, Abu Dhabi, 2021.

Just powering the ships Maersk already has on order 'Cargo owners for zero emission vessels unites 19 global brands for 2040 zero-carbon ocean shipping ambition', *Shipping Australia Limited*, 23 September 2022; 'Maersk signs first green methanol deal in step towards dropping fossil fuels', *Reuters*, 19 August 2021.

Potentially serious implications for food and water security and biodiversity Monbiot, George, 'Why are we feeding crops to our cars when people are starving?', *The Guardian*, 30 June 2022.

By 2050 biofuels will require three times the land they require today 'Biofuels can provide up to 27% of world transportation fuel by 2050, IEA report says: IEA "roadmap" shows how biofuel production can be expanded in a sustainable way, and identifies needed technologies and policy actions', *International Energy Agency News*, 20 April 2011.

Around seventy per cent of the ammonia currently produced is used as fertiliser *Ammonia Technology Roadmap: Towards more sustainable nitrogen fertiliser production*, International Energy Agency, 2021.

Ammonia production is responsible for around 1.8 per cent of the world's emissions *Ammonia: Zero-carbon fertiliser, fuel and energy store*, The Royal Society, London, February 2020.

Although it does not produce carbon dioxide when combusted, burning ammonia *Managing Emissions from Ammonia-Fuelled Vessels: An overview of regulatory drivers, emission types, sources, scenarios, reduction technologies, and solutions*, Mærsk Mc-Kinney Møller Center for Zero Carbon Shipping, Copenhagen, March 2023; *Ammonia: Zero-carbon fertiliser, fuel and energy store*, ibid.

Its energy density is also only around a third that of HFO Placek, Martin, 'Energy density of maritime fuels in 2020, by fuel type', *Statista*, 3 February 2023.

Hydrogen . . . is significantly less energetically dense than either HFO or ammonia Placek, Martin, 'Energy density of maritime fuels in 2020, by fuel type', *Statista*, 3 February 2023.

Almost half the voyages along the crucial route between China and the United States Mao, Xiaoli et al., *Refueling assessment of a zero-emission container corridor between China and the United States: Could hydrogen replace fossil fuels?*, Working Paper 2020-05, International Council on Clean Transportation, 2020.

Installation of rotor sails on ships operated by Berge Bulk 'Anemoi Marine to provide rotor sails for Berge Bulk ships', *Ship Technology*, 14 July 2022.

Windcoop . . . has plans for a sail-powered container ship capable of carrying 60 to 68 containers Windcoop, 'The activist shipping company', wind.coop, 2023; 'Sustainable transportation: Windcoop is raising funds to build a sailing container ship', *Business Solutions Atlantic France News*, 12 October 2022.

Will require the investment of hundreds of billions of dollars Berti, Adele, 'The financial cost of low-carbon transformation in shipping', *Ship Technology*, 23 April 2020.

9   Traces

At least 414 million pieces – or 238 tonnes – of plastic waste on Cocos's beaches Lavers, J.L. et al., 'Significant plastic accumulation on the Cocos (Keeling) Islands, Australia', *Scientific Reports*, vol. 9, no. 1, 2019, 7102.

170 trillion particles of microplastic Eriksen, Marcus et al., 'A growing plastic smog, now estimated to be over 170 trillion plastic particles afloat in the world's oceans: Urgent solutions required', *PLOS ONE*, vol. 18, no. 3, 2023, e0281596.

Where he died three years later Smith, F., 'Borneo's first "White Rajah": New light on Alexander Hare, his family and associates', *Borneo Research Bulletin*, vol. 44, 2013, pp. 93–131.

'Could not help but liken the situation to what I imagine life would be like for slaves' Kerr, Alan, *Report on Visit to Cocos*, 14–19 November, Department of Territories, 1971.

'Forced, underpaid labour' *Report of the Special Committee on the Situation with Regard to the Implementation of the Declaration on the Granting of Independence to Colonial Countries and People, Volume IV*, Official Records of the General Assembly of the United Nations, Twenty-Ninth Session, Supplement No. 23 (a/9623/Rev.1), pp. 112–16.

He 'did their thinking for them' Mowbray, Martin, 'Decolonisation and community development on the Cocos (Keeling) Islands', *Community Development Journal*, vol. 32, no. 4, 1997, p. 321.

The 'oppressive' glare . . . of the islands' 'calcareous beaches' Darwin, Charles, *Voyage of the Beagle*, Penguin Classics, London, 1989, p. 333.

'Exceedingly limited' ibid., p. 334ff.

'There is to my mind a considerable degree of grandeur' ibid., p. 338.

Capt. FitzRoy at the distance of but little more than a mile from the shore Darwin, Charles, *Charles Darwin's Beagle Diary*, 12 April 1836, darwin-online .org.uk.

Cautioned the younger man not to 'flatter yourself that you will be believed' Jones, Steve, *Coral: A pessimist in paradise*, Abacus, London, 2008, p. 32.

In the aftermath of the American nuclear tests in the Marshall Islands ibid., pp. 37–8.

The blobs some believe are the remnants of Theia Voosen, Paul, 'Remains of impact that created the Moon may lie deep within Earth', *Science*, 23 March 2021.

This convergence of military interests and scientific research Jones, Steve, *Coral: A pessimist in paradise*, Abacus, London, 2008, pp. 38–40.

The United States detonated sixty-seven nuclear devices Kimball, Daryl G. and Rostampour, Chris, 'US, Marshall Islands grapple with nuclear legacy', *Arms Control Today*, November 2022.

'While it is true that these people do not live the way that Westerners do' Rust, Susanne, 'How the US betrayed the Marshall Islands, kindling the next nuclear disaster', *Los Angeles Times*, 10 November 2019.

The Marshallese are still grappling with the toxic legacy of the nuclear violence Abella, Maveric K.I.L. et al., 'Background gamma radiation and soil activity measurements in the northern Marshall Islands', *Proceedings of the National Academy of Sciences*, vol. 116, no. 31, 2019, pp. 15425–34.

'The Tomb' . . . is at risk of cracking Rust, Susanne, 'How the US betrayed the Marshall Islands, kindling the next nuclear disaster', *Los Angeles Times*, 10 November 2019.

Hitherto unknown information about the speed and direction of the winds and currents Bruno, Laura A., 'The bequest of the nuclear battlefield: Science, nature, and the atom during the first decade of the Cold War', *Historical Studies in the Natural Sciences*, vol. 33, no. 2, 2003, pp. 237–60.

Bears this radioactivity west . . . to deposit it on the reefs of Cocos Andrews, Allen H. et al., 'Bomb-produced radiocarbon in the western tropical Pacific Ocean: Guam coral reveals operation-specific signals from the Pacific proving grounds', *Journal of Geophysical Research: Oceans*, vol. 121, no. 8, 2016, pp. 6351–66.

Plastic is now believed to be present in the bodies of more than 90 per cent of all pelagic birds Wilcox, Chris et al., 'Threat of plastic pollution to seabirds is global, pervasive, and increasing', *Proceedings of the National Academy of Sciences*, vol. 112, no. 38, 2015, pp. 11899–904.

More than half a million crabs were dying Lavers, Jennifer L. et al., 'Entrapment in plastic debris endangers hermit crabs', *Journal of Hazardous Materials*, vol. 387, 2020, 121703.

Sex ratios have already been so drastically affected Welch, Craig, '99% of these sea turtles are turning female: Here's why', *National Geographic*, 8 January 2018.

A base for US drones Whitlock, Craig, 'US, Australia to broaden military ties amid Pentagon pivot to SE Asia', *Washington Post*, 27 March 2012.

Upgrades to the runway on West Island Kuper, Stephen, 'Government awards contract for upgrade to Cocos (Keeling) Island infrastructure', *Defence Connect*, 3 February 2020.

'Labor rejects the notion' 'Sweeping changes to mandatory detention announced', *ABC News*, 29 July 2008.

In 2012 . . . seventy-three boats were intercepted on or near Cocos Project SafeCom, 'Keeping perspective: Australia's boat arrivals', safecom.org.au, 2009–12.

The mass of the objects made . . . outweighs the biomass of every living thing Elhacham, Emily et al., 'Global human-made mass exceeds all living biomass', *Nature*, vol. 588, no. 7838, 2020, pp. 442–4.

Humans are . . . the most powerful geological force on Earth Cooper, Anthony H. et al., 'Humans are the most significant global geomorphological driving force of the 21st century', *Anthropocene Review*, vol. 5, no. 3, 2018, pp. 222–9.

The sudden proliferation of plastic, or phosphates, or . . . spheroidal carbonaceous particles Subramanian, Meera, 'Humans versus earth: The quest to define the Anthropocene', *Nature*, vol. 572, no. 7768, 2019, pp. 168–70.

The bones of the tens of billions of chickens that humans consume every year Bennett, Carys E. et al., 'The broiler chicken as a signal of a human reconfigured biosphere', *Royal Society Open Science*, vol. 5, no. 12, 2018, 180325.

In mid-2023 the Anthropocene Working Group selected a sediment core from Crawford Lake Witze, Alexandra, 'This quiet lake could mark the start of a new Anthropocene epoch', *Nature*, 11 July 2023.

'Placing the Anthropocene at this time' Biello, David, 'Mass deaths in Americas spark new CO2 epoch', *Scientific American*, 11 March 2015.

'When I think about the question, *what is a plantation*' Haraway, Donna et al., 'Anthropologists are talking – about the Anthropocene', *Ethnos*, vol. 81, no. 3, pp. 535–64.

We inhale and ingest tens of thousands of these particles Prata, Joana Correia et al., 'Environmental exposure to microplastics: An overview on possible human health effects', *Science of the Total Environment*, vol. 702, 2020, 134455.

10 Reef

More than 90 per cent of reefs . . . bleached Hughes, Terry, Schaffelke, Britta and Kerry, James, 'How much coral has died in the Great Barrier Reef's worst bleaching event?', *The Conversation*, 29 November 2016; Hughes, Terry and Kerry, James, 'Back-to-back bleaching has now hit two-thirds of the Great Barrier Reef', *The Conversation*, 12 April 2017.

'We've never seen anything like this scale of bleaching before' Slezak, Michael, 'Great Barrier Reef: 93% of reefs hit by coral bleaching', *The Guardian*, 20 April 2016.

Healthy corals produce energy at many times the rate of most terrestrial plants Adams, Melanie S. et al., 'Coral reef productivity and diversity: Contributions from enhanced photosynthesis via demand for carbohydrate from the host', *Marine Ecology*, vol. 41, no. 6, 2020, e12618.

More than 60 per cent of corals were affected Berkelmans, R. and Oliver, J.K., 'Large-scale bleaching of corals on the Great Barrier Reef', *Coral Reefs*, vol. 18, no. 1, 1999, pp. 55–60, cited in Veron, J.E.N., *A Reef in Time: The Great Barrier Reef from beginning to end*, Belknap Press, Harvard, 2008.

Elsewhere the effects were even worse Wilkinson, C.P., *The 1997–1998 Mass Bleaching Event Around the World*, Australian Institute of Marine Science Research, 1998.

'Terminal' Knaus, Christopher and Evershed, Nick, 'Great Barrier Reef at "terminal stage": Scientists despair at latest coral bleaching data', *The Guardian*, 10 April 2017.

25 per cent of reefs up and down the entire 2200-kilometre expanse Readfearn, Graham, 'Great Barrier Reef's third mass bleaching in five years the most widespread yet', *The Guardian*, 7 April 2020.

The average distance between severely bleached reefs and only mildly affected reefs Dietzel, Andreas et al., 'The spatial footprint and patchiness of large-scale disturbances on coral reefs', *Global Change Biology*, vol. 27, no. 19, 2021, pp. 4825–38.

The rate of coral recruitment . . . fell by almost 90 per cent Hughes, Terry P. et al., 'Global warming impairs stock-recruitment dynamics of corals', *Nature*, vol. 568, no. 7752, 2019, pp. 387–90.

Only 2 per cent of the Reef had escaped being bleached Hughes, Terry et al., 'Emergent properties in the responses of tropical corals to recurrent climate extremes', *Current Biology*, vol. 31, no. 23, 2021, 5393–5399.

Coral recruitment had declined by more than 70 per cent Cheung, Mandy W.M. et al., 'Cumulative bleaching undermines systemic resilience of the Great Barrier Reef', *Current Biology*, vol. 31, no. 23, 2021, pp. 5835–5392e4.

99 per cent of reefs worldwide will be exposed to frequent thermal stress Dixon, Adele M. et al., 'Future loss of local-scale thermal refugia in coral reef ecosystems', *PLOS Climate*, vol. 1, no. 2, 2022, p. e0000004.

Thirteen to twenty-six times more likely to survive Quigley, K.M. et al., 'Assessing the role of historical temperature regime and algal symbionts on the heat tolerance of coral juveniles', *Biology Open*, vol. 9, no. 1, 2020, bio047316.

The team later placed hundreds of these selectively bred juveniles on the Reef 'Next generation corals undergo first field tests on the Great Barrier Reef', *Australian Institute of Marine Science News*, 2 July 2019.

Experimental crossbreeding of different species of *Acropora* Chan, Wing Yan et al., 'Interspecific hybridization may provide novel opportunities for coral reef restoration', *Frontiers in Marine Science*, vol. 5, 2018, p. 1.

Inoculated corals . . . with three different species of microalgae Quigley, K.M.

et al., 'Assessing the role of historical temperature regime and algal symbionts on the heat tolerance of coral juveniles', *Biology Open*, vol. 9, no. 1, 2020, bio047316.

Symbionts are replaced with strains of microalgae that have been 'hardened' Buerger, P. et al., 'Heat-evolved microalgal symbionts increase coral bleaching tolerance', *Science Advances*, vol. 6, no. 20, 2020, eaba2498.

A 2019 report commissioned by the Australian Government Hardisty, P. et al., *Reef Restoration and Adaptation Program: Investment case*, Reef Restoration and Adaptation Program, Australian Institute of Marine Science, 2019.

Fleets of ships equipped with mist machines Tollefson, Jeff, 'Can artificially altered clouds save the Great Barrier Reef?', *Nature*, vol. 596, no. 7873, 2021, pp. 476–8.

'We feel surprise' Darwin, Charles, *Voyage of the Beagle*, Penguin Classics, London, 1989, p. 342.

Information about weather patterns and water temperatures stretching back to the eighteenth century Zinke, J. et al., 'Corals record long-term Leeuwin current variability including Ningaloo Niño/Niña since 1795', *Nature Communications*, vol. 5, no. 1, 2014, p. 3607.

Evidence about river flows and rainfall up to 6000 years ago Lough, J.M. et al., 'Evidence for suppressed mid-Holocene northeastern Australian monsoon variability from coral luminescence', *Paleoceanography*, vol. 29, no. 6, 2014, pp. 581–94.

The ways in which the Reef has responded to changing conditions over the past 30,000 years Webster, Jody M. et al., 'Response of the Great Barrier Reef to sea-level and environmental changes over the past 30,000 years', *Nature Geoscience*, vol. 11, no. 6, 2018, pp. 426–32.

11 Net

'Let me not return with an empty boat' Amadou Souaré . . . says that a better spell than the Prayer of Seven *Waws* is this one: You draw in the sand a pentagonal star. Look, I'm showing you. You say a prayer for each of the triangles here, and here, and here, and here, and here and then you say a prayer for the center – here – and then you take a little bit of sand from each part, and then you blow on it, and then you take it aboard the boat with you. It's the same prayer that the Prophet Muhammad said, and that the Wolof, and Serer, and Bambara, and white people say. It goes: "There is no God but God and I pray to God and I pray to the Prophet. If there are a lot of fish, we will share it. If there are no fish, I will still get my part. Let me not return with an empty boat." This way you get God's blessing for your trip.' Badkhen, Anna, *Fisherman's Blues: A West African community at sea*, Riverhead Books, New York, 2018, p. 31.

Fishing employs around half a million of the country's 8.5 million people Food and Agriculture Organization of the United Nations, *Aquaculture for Food Security, Livelihood and Nutrition in Sierra Leone*, FAO, Rome, 2019.

In Ghana . . . overfishing has pushed the country's once-bountiful fisheries to the point of collapse Environmental Justice Foundation, *A Human Rights Lens on* *the Impacts of Industrial Illegal Fishing and Overfishing on the Socio-Economic Rights of Small-Scale Fishing Communities in Ghana*, EJF, London, 2021.

Almost half the fish caught off West Africa are caught illegally Belhabib, Dyhia et al., 'The fisheries of Africa: Exploitation, policy, and maritime security trends', *Marine Policy*, vol. 101, 2019, pp. 80–92.

Illegal fishing siphons an estimated US$9.4 billion out of West Africa every year Daniels, Alfonso et al., *Fishy Networks: Uncovering the companies and individuals behind illegal fishing globally*, Financial Transparency Coalition, Boston, 2022.

The passage out across the Atlantic to the Canaries Jones, Sam, '"I'd never seen a boat come in with so many bodies": The mortal cost of the Atlantic migrant route', *The Guardian*, 8 August 2021.

As much as 40 per cent of all IUU fishing worldwide Daniels, Alfonso et al., *Fishing Networks: Uncovering the companies and individuals behind illegal fishing globally*, Financial Transparency Coalition, Boston, 2022.

Populations of large fish . . . have been reduced by about 90 per cent Myers, Ransom A. and Worm, Boris, 'Rapid worldwide depletion of predatory fish communities', *Nature*, vol. 423, no. 6937, 2003, pp. 280–3.

Close to 95 per cent of the world's fish stocks are being fished to their limits or beyond Food and Agriculture Organization of the United Nations, *The State of World Fisheries and Aquaculture 2022: Towards blue transformation*, FAO, Rome, 2022, p. 46.

Almost half of the 1439 stocks it assessed were overfished Minderoo Foundation, *Global Fishing Index: Assessing the sustainability of the world's fisheries*, Perth, 2023.

Sharks and rays have been particularly hard-hit Camhi, Merry D. et al., *The Conservation Status of Pelagic Sharks and Rays: Report of the IUCN shark specialist group pelagic shark red list workshop*, International Union for the Conservation of Nature and Natural Resources, Gland, 2007; Yan, Helen F. et al., 'Overfishing and habitat loss drive range contraction of iconic marine fishes to near extinction', *Science Advances*, vol. 7, no. 7, 2021, eabb6026; and Sherman, C. Samantha et al., 'Half a century of rising extinction risk of coral reef sharks and rays', *Nature Communications*, vol. 14, no. 1, 2023, 15.

'These sharks and rays have evolved over 450 million years' Readfearn, Graham, '"Extinction crisis" of sharks and rays to have devastating effect on other species, study finds', *The Guardian*, 18 January 2023.

In addition to the 40 million people employed as fishers Food and Agriculture Organisation of the United Nations, *The State of World Fisheries and Aquaculture 2022: Towards blue transformation*, Rome, FAO, 2022.

The impact of the meteorite that wiped out the dinosaurs on terrestrial life 'Lifelines: Daniel Pauly', *Nature*, vol. 421, no. 6918, 2003, p. 23.

In the North Sea . . . catches dropped by two-thirds or more Roberts, Callum, *The Unnatural History of the Sea: The past and future of humanity and fishing*, Gaia Thinking, London, 2007, p. 164.

'The greatest human transformation of marine habitats ever seen' ibid., p. 162.
As developed nations expanded and modernised their fleets Campling, Liam and Colás, Alejandro, *Capitalism and the Sea: The maritime factor in the making of the modern world*, Verso, London, 2021, pp. 196–7.
From 1950 to 1980 the reach of industrial fisheries Pauly, D. and Jacquet, Jennifer, *Vanishing Fish: Shifting baselines and the future of global fisheries*, David Suzuki Institute/Greystone Books, Vancouver/Berkeley, 2019, p. 5; Food and Agriculture Organisation of the United Nations, *The State of World Fisheries and Aquaculture 2022: Towards blue transformation*, FAO, Rome, 2022.
'A poem, a stink, a grating noise' Steinbeck, John, *Cannery Row*, Penguin, Camberwell, 2009, p. 5.
Happened with the anchoveta Pauly, D. and Jacquet, Jennifer, *Vanishing Fish: Shifting baselines and the future of global fisheries*, David Suzuki Institute/Greystone Books, Vancouver/Berkeley, 2019, pp. 2–3.
Haddock and rosefish ibid., p. 207ff.
So thick in the water it was difficult to row a boat through them E Magazine Editors, 'A run on the banks', *E–The Environmental Magazine*, 20 July 2004.
Historical reconstructions Roberts, Callum, *The Unnatural History of the Sea: The past and future of humanity and fishing*, Gaia Thinking, London, 2007, p. 219.
Chinese authorities provided false figures Watson, Reg and Pauly, Daniel, 'Systematic distortions in world fisheries catch trends', *Nature*, vol. 414, no. 6863, 2001, pp. 534–6.
The total distance travelled by industrial fishers Tickler, David et al., 'Far from home: Distance patterns of global fishing fleets', *Science Advances*, vol. 4, no. 8, 2018; Palomares, Maria L.D. and Pauly, Daniel, 'On the creeping increase of vessels' fishing power', *Ecology and Society*, vol. 24, no. 3, 2019, p. 31.
In the Indian Ocean . . . the catch per unit effort fell 78 per cent Zeller, D. et al., 'Trends in Indian Ocean marine fisheries since 1950: Synthesis of reconstructed catch and effort data', *Marine and Freshwater Research*, vol. 74, no. 4, 2023, pp. 301–19.
US$35.4 billion Sumaila, R. et al., 'Updated estimates and analysis of global fisheries subsidies', *Marine Policy*, vol. 109, 2019, 103695.
High-seas fishing fleets globally received subsidies amounting to almost US$4.2 billion Sala, Enric et al., 'The economics of fishing the high seas', *Science Advances*, vol. 4, no. 6, 2018, eaat2504.
The expansion of fishing activity . . . in the northwestern Indian Ocean World Wide Fund for Nature, WWF, *Unregulated Fishing on the High Seas of the Indian Ocean: The impacts on, risks to, and challenges for sustainable fishing and ocean health*, Gland, 2020.
Today China fields the world's largest distant-water fishing fleet Gutiérrez, Miren et al., *China's Distant-Water Fishing Fleet: Scale, impact and governance*, Overseas Development Institute, London, 2020.
In 2020 the Ecuadorian Government called in the United States Coast Guard

Oceana, *Oceana Finds 300 Chinese Vessels Pillaging the Galapagos for Squid*, September 2020; Myers, Steven Lee et al., 'How China Targets the Global Fish Supply', *New York Times*, 26 September 2022.

The Ecuadorian Navy captured a Chinese trawler Sanchez, Wilder Alejandro, 'A growing concern: Chinese illegal fishing in Latin America', *CIMSEC*, 19 September 2017.

'Do double-duty as maritime militias' Sinclair, Michael, 'The national security imperative to tackle illegal, unreported, and unregulated fishing', Brookings.edu, 25 January 2021.

Helping push their populations closer and closer to complete collapse Environmental Justice Foundation, *The 'People's' Fishery on the Brink of Collapse: Small pelagics in landings of Ghana's industrial trawl fleet*, EJF, London, 2021.

US$50 billion a year Sumaila, U.R. et al., 'Illicit trade in marine fish catch and its effects on ecosystems and people worldwide', *Science Advances*, vol. 6, no. 9, 2020, eaaz3801.

The benefit . . . flows overwhelmingly to wealthier countries McCauley, Douglas J. et al., 'Wealthy countries dominate industrial fishing', *Science Advances*, vol. 4, no. 8, 2018, eaau2161.

Video of Taiwanese fishers executing at least four men Urbina, Ian, *The Outlaw Ocean: Crime and survival in the last untamed frontier*, Bodley Head, London, 2019, p. 320ff; see also Morris, James X., 'The dirty secret of Taiwan's fishing industry', *The Diplomat*, 18 May 2018.

'I thought I was going to die' Environmental Justice Foundation, *Pirates and Slaves: How overfishing in Thailand fuels human trafficking and the plundering of our oceans*, EJF, London, 2015.

'The worst was when I saw one of my co-workers fall into the sea' ibid.

'We were beat frequently by the Thai crew' United Nations Inter-Agency Project on Human Trafficking, *Exploitation of Cambodian Men at Sea*, UNIAP, Bangkok, 22 April 2009.

Close to 20 per cent have experienced conditions of slavery Tickler, David et al., 'Modern slavery and the race to fish', *Nature Communications*, vol. 9, no. 1, 2018, pp. 4643–9.

Almost two fifths have been the victims of human trafficking Environmental Justice Foundation, *Slavery at Sea: The continued plight of trafficked migrants in Thailand's fishing industry*, EJF, London, 2014.

The catch per unit effort has declined by 97 per cent ibid.

'Ecosystem decline and slavery exist in a vicious cycle' 'Overfishing and pirate fishing perpetuate environmental degradation and modern-day slavery in Thailand', *Environmental Justice Foundation News*, 25 February 2015.

Illegal tuna fishing costs island nations upwards of US$200 million a year MRAG Asia Pacific, *The Quantification of Illegal, Unreported and Unregulated (IUU) Fishing in the Pacific Islands Region – a 2020 Update: A report prepared for the Pacific Islands Forum Fisheries Agency*, Toowong, 2021.

Multiple instances of fraud and illegal fishing Stop Illegal Fishing, *Illegal Fishing? Evidence and Analysis*, Gaborone, 2017.

The Indo-Pacific Partnership for Maritime Domain Awareness White, Edward, 'Quad and the WTO focus on fishing', *Financial Times*, 8 August 2022.

Farmed fish and pork fed on West African fishmeal is being sold by . . . Europe and the UK Cashion, Tim et al., 'Most fish destined for fishmeal production are food-grade fish', *Fish and Fisheries*, vol. 18, no. 5, 2017, pp. 837–44; Greenpeace Africa, *Feeding a Monster: How European aquaculture and animalfeed industries are stealing food from West African communities*, Greenpeace Africa and Changing Markets Foundation, 2021.

12 South

A total biomass of somewhere between 300 and 500 million tonnes Meyer, Bettina et al., 'Successful ecosystem-based management of Antarctic krill should address uncertainties in krill recruitment, behaviour and ecological adaptation', *Communications Earth & Environment*, vol. 1, no. 1, 2020, pp. 1–12; Bar-On, Yinon M. et al., 'The biomass distribution on Earth', *Proceedings of the National Academy of Sciences*, vol. 115, no. 25, 2018, pp. 6506–11.

Close to five times the biomass of every living wild mammal, bird and reptile combined Bar-On et al., ibid.; Greenspoon, Lior et al., 'The global biomass of wild mammals', *Proceedings of the National Academy of Sciences*, vol. 120, no. 10, 2023, e2204892120.

Video taken within swarms OceanX, 'Trying to pilot a sub through a "krill swarm"', *OceanX*, YouTube, 4 January 2019.

70 per cent of the world's krill population is concentrated in the region below the southwest Atlantic Meyer, Bettina et al., 'Successful ecosystem-based management of Antarctic krill should address uncertainties in krill recruitment, behaviour and ecological adaptation', *Communications Earth & Environment*, vol. 1, no. 1, 2020, pp. 1–12.

30 per cent of the body mass of some penguins Bruno, David and Saucède, Thomas, *Biodiversity of the Southern Ocean*, Elsevier, London, 2015.

Blue whales Connor, Steve, 'The anatomy of a whale', BBCEarth.com, n.d.

The milk of humpbacks is even richer Kingdon, Amorina, 'Humpback moms need a quiet place to nurse', *Hakai Magazine*, 6 March 2019.

'So fearless we killed as ma[n]y as we chose' Cook, James, *The Journals of Captain Cook*, Penguin Classics, London, 2003, p. 442.

The Nazis sent a secret mission to Antarctica Fater, Luke, 'Hitler's secret Antarctic expedition for whales', *Atlas Obscura*, 6 November 2019; Epstein, Charlotte, *The Power of Words in International Relations: Birth of an anti-whaling discourse*, MIT Press, Cambridge, MA, 2008.

The collapse in whale numbers did not lead to an explosion in krill Yong, Ed, 'The enormous hole that whaling left behind', *The Atlantic*, 3 November 2021.

Krill faeces and shell casings sequester some 23 million tonnes of carbon dioxide Cavan, E.L. et al., 'The importance of Antarctic krill in biogeochemical cycles', *Nature Communications*, vol. 10, no. 1, 2019, p. 4742; Cavan, Emma et al., *Antarctic Krill: Powerhouse of the Southern Ocean*, WWF Australia, Sydney, 2022.

Their hearts beat faster Ward, Ashley, 'When krill host social gatherings other ocean animals thrive', *Popular Science*, 8 March 2022.

In 2022 the reported catch was just over 415,000 tonnes CCAMLR Secretariat, *Fishery Report 2022:* Euphausia superba *in Area 48*, 17 March 2023.

This upward trend looks likely to continue Dickie, Gloria, 'In Antarctica, does a burgeoning krill fishery threaten wildlife?', *Reuters*, 25 February 2022.

The value of the carbon sequestered by the krill Cavan, Emma et al., *Antarctic Krill: Powerhouse of the Southern Ocean*, WWF Australia, Sydney, 2022.

Krill populations in the southwest Atlantic Sector have fallen by 70 to 80 per cent Atkinson, Angus et al., 'Long-term decline in krill stock and increase in salps within the Southern Ocean', *Nature*, vol. 432, no. 7013, 2004, pp. 100–3.

A 2022 report by . . . Dr Emma Cavan Cavan et al., *Antarctic Krill*, WWF Australia, Sydney, 2022.

In 2017 and 2018 both the winter maximum and summer minimum Scott, Michon, 'Understanding climate: Antarctic sea ice extent', *NOAA News*, 14 March 2023.

By July, when the ice would usually have expanded to around 16.4 million square kilometres Readfearn, Graham, '"Something weird is going on": Search for answers as Antarctic sea ice stays at historic lows', *The Guardian*, 29 July 2023.

A five-sigma event Alvaro, Alexandra, 'Antarctic sea-ice levels dive in "five-sigma event", as experts flag worsening consequences for planet', *ABC News*, 24 July 2023.

The record-breaking sea ice anomalies in 2022 and 2023 herald a more profound transformation of the Antarctic environment Purich, Ariaan and Doddridge, Edward W., 'Record low Antarctic sea-ice coverage indicates a new sea-ice state', *Communications Earth & Environment*, vol. 4, 2023, 314.

Thousands of penguin chicks drowned Fretwell, Peter T., Boutet, Aude and Ratcliffe, Norman, 'Record low 2022 Antarctic sea ice led to catastrophic breeding failure of emperor penguins', *Communications Earth & Environment*, vol. 4, 2023, 273.

The combined effects of warming water and shrinking sea ice Piñones, Andrea and Fedorov, Alexey V., 'Projected changes of Antarctic krill habitat by the end of the 21st century: Changes in Antarctic krill habitat', *Geophysical Research Letters*, vol. 43, no. 16, 2016, pp. 8580–9.

The krill's habitat will contract southwards and deteriorate significantly Devi, Veytia et al., 'Circumpolar projections of Antarctic krill growth potential', *Nature Climate Change*, vol. 10, no. 6, 2020, pp. 568–75; Veytia, Devi and Cornery, Stuart, 'Climate change threatens Antarctic krill and the sea life that depends on it', *The Conversation*, 19 May 2020.

Populations of salps . . . are increasing Atkinson, Angus et al., 'Long-term decline in krill stock and increase in salps within the Southern Ocean', *Nature*, vol. 432, no. 7013, 2004, pp. 100–3.

Adult krill are mostly unaffected by . . . increases in ocean pH Ericson, Jessica A. et al., 'Near-future ocean acidification does not alter the lipid content and fatty acid composition of adult Antarctic krill', *Scientific Reports*, vol. 9, no. 1, 2019, 12375.

Fears of a tipping point 'Krill face deadly cost of ocean acidification', *Australian Antarctic Division News*, 13 October 2012.

In mid-March of 2022 Antarctica experienced an unprecedented heatwave Bergstrom, Dana, Robinson, Sharon and Alexander, Simon, 'Record-smashing heatwaves are hitting Antarctica and the Arctic simultaneously. Here's what's driving them, and how they'll impact wildlife', *The Conversation*, 22 March 2022.

Temperatures on the Antarctic Peninsula have risen by more than 3 degrees Turner, John and Bracegirdle, Thomas, 'Antarctica and climate change', *British Antarctic Survey*, 30 June 2022.

'Antarctic climatology has been rewritten' Di Battista, Stefano, 'It is impossible, we would have said until two days ago', Twitter, 18 March 2022.

'Upended our expectations about the Antarctic climate system' Samenow, Jason and Patel, Kasha, 'It's 70 degrees warmer than normal in eastern Antarctica. Scientists are flabbergasted', *Washington Post*, 18 March 2022.

East Antarctica's Conger Ice Shelf abruptly collapsed Lu, Donna, 'Satellite data shows entire Conger ice shelf has collapsed in Antarctica', *The Guardian*, 25 March 2022.

The rate of flow of some of the glaciers behind it increased sixfold Scambos, T.A., 'Glacier acceleration and thinning after ice shelf collapse in the Larsen B embayment, Antarctica', *Geophysical Research Letters*, vol. 31, no. 18, 2004, L18402.

The rate of ice loss in Antarctica has more than quadrupled Shepherd, Andrew et al., 'Mass balance of the Antarctic ice sheet from 1992 to 2017', *Nature*, vol. 558, no. 7709, 2018, pp. 219–22; 'Ramp-up in Antarctic ice loss speeds sea-level rise', *NASA News*, 13 June 2018.

Thwaites is a massive river of ice International Thwaites Glacier Collaboration, 'Thwaites Glacier facts', thwaitesglacier.org, 2023.

In 2019 NASA researchers discovered a vast cavity Rasmussen, Carol, 'Huge cavity in Antarctic glacier signals rapid decay', *JPL News*, 30 January 2019.

'You're like, I should get a new windshield' Voosen, Paul, 'Key Antarctic ice shelf is within years of failure', *Science*, 13 December 2021.

An almost complete melting of large parts of West Antarctica Turney, Chris S.M. et al., 'Early last interglacial ocean warming drove substantial ice mass loss from Antarctica', *Proceedings of the National Academy of Sciences*, vol. 117, no. 8, 2020, pp. 3996–4006.

DNA from specimens of Turquet's octopus Lau, Sally C.Y. et al., 'Genomic evidence for West Antarctic ice sheet collapse during the last interglacial period', *bioRxiv*, 31 January 2023; Readfearn, Graham, 'Clue to rising sea levels lies in DNA of 4-year-old octopus, scientists say', *The Guardian*, 5 February 2023.

By using drill cores extracted from the sea floor in Iceberg Alley Weber, Michael E. et al., 'Decadal-scale onset and termination of Antarctic ice-mass loss during the last deglaciation', *Nature Communications*, vol. 12, no. 1, 2021, 6683.

The signature of nuclear tests carried out in the 1950s and 1960s Eisenstadt, Abigail, 'Radioactive chlorine from nuclear bomb tests still present in Antarctica', Phys.org, 16 October 2019.

The burning of forests . . . by the Māori McConnell, Joseph R. et al., 'Hemispheric black carbon increase after the 13th-century Māori arrival in New Zealand', *Nature*, vol. 598, no. 7879, 2021, pp. 82–85; see also Newnham, Rewi M., 'Black carbon attribution', *Nature*, vol. 612, no. 7941, 2022, E18–19; Newnham, Rewi, 'Research linking soot in Antarctic ice exclusively with early Māori fires was flawed: There were other sources elsewhere', *The Conversation*, 22 December 2022; McConnell, Joseph R. et al., 'Reply to: Black carbon attribution', *Nature*, vol. 612, no. 7941, 2022, E20–E21.

The global pause of the pandemic's first year 'Glacial ice will likely hold records of the COVID-19 pandemic, researchers say', *Ohio State News*, 7 May 2020.

The last time levels of atmospheric carbon dioxide were this high was four million years ago Shulmeister, James, 'Climate explained: What the world was like the last time carbon dioxide levels were at 400ppm', *The Conversation*, 8 July 2020.

Global temperatures were 3 degrees warmer Rovere, Alessio et al., 'Higher than present global mean sea level recorded by an early Pliocene intertidal unit in Patagonia (Argentina)', *Communications Earth & Environment*, vol. 1, no. 1, 2020, 68.

The melting underway in Antarctica, Greenland and elsewhere IPCC, *Climate Change 2021: The physical science basis, 2021* (Chapter 9: Ocean, Cryosphere and Sea Level Change), IPCC, 2021.

13 Horizon

July and August 2023 were the hottest months ever recorded World Meteorological Organization, 'Earth had hottest three-month period on record, with unprecedented sea surface temperatures and much extreme weather', *World Meteorological Organization*, 6 September 2023.

Records have been shattered over and over again Copernicus Climate Change Service, 'Surface air temperature for July 2023', *Copernicus Climate Change Service News*, 2023.

Ocean temperatures have also moved into uncharted territory Copernicus Climate Change Service, 'Record-breaking North Atlantic Ocean temperatures contribute to extreme marine heatwaves', *Copernicus Climate Change Service News*, 6 July 2023.

Even if implemented in full Rojelj, Jorei et al., 'Credibility gap in net-zero climate targets leaves world at high risk', *Science*, vol. 3380, no. 6649, 2023, pp. 1014–16.

A world that is that much hotter is almost unimaginable IPCC, *Climate Change 2022: Impacts, Adaptation and Vulnerability*, IPCC, 2022.

'**A disproportionately rapid evacuation**' Hoegh-Guldberg, O. et al., 'Impacts of 1.5°C global warming on natural and human systems', in IPCC, *Global Warming of 1.5°C: An IPCC Special Report on the impacts of global warming of 1.5°C above pre-industrial levels and related global greenhouse gas emission pathways, in the context of strengthening the global response to the threat of climate change, sustainable development, and efforts to eradicate poverty*, IPCC, Geneva, 2018.

Scientists identified nine global tipping points Lenton, T.M. et. al., 'Tipping points in the Earth's climate system', *PNAS*, vol. 105, no. 6, 2008, pp. 1786–93.

Seven more regional tipping points McKay, David I. Armstrong et al., 'Exceeding 1.5°C global warming could trigger multiple climate tipping points', *Science*, vol. 377, no. 6611, 2022, eabn7950.

The shutdown of the AMOC may be much closer than previously thought Ditlevsen, Peter and Ditlevsen, Susanne, 'Warning of a forthcoming collapse of the Atlantic meridional overturning circulation', *Nature Communications*, vol. 14, 2023, 4252.

More than half of the 60 million internally displaced people Kristy Siegfried, 'Climate change and displacement: the myths and the facts', *UNHCR Australia*, 15 November 2023.

And these numbers are only going to continue to rise Clement, Viviane et al., *Groundswell Part 2: Acting on internal climate migration*, the World Bank, Washington, 2021; Institute for Economics and Peace, *Ecological Threat Register 2020: Understanding ecological threats, resilience and peace*, IEP, Sydney, 2020.

'**A mass exodus of entire populations on a biblical scale**' Secretary-General, United Nations, 'Secretary-General's remarks to the Security Council debate on "Sea-level rise: Implications for international peace and security"', 14 February 2023.

'**It is not knowledge we lack**' Linqvist, Sven, *'Exterminate All the Brutes': One man's odyssey into the heart of darkness and the origins of European genocide*, Granta Books, 2018, p. 172.

'**To face others is to become a witness**' Rose, Deborah Bird, 'Slowly – writing in the Anthropocene', *TEXT*, no. 20, October 2013, p. 8.

As the philosopher Thom van Dooren has observed Van Dooren, Thom, *A World in a Shell: Snail stories for a time of extinction*, MIT Press, Cambridge, MA, 2022, pp. 193–5.

TEXT & IMAGE CREDITS

Text

1   Alice Oswald, *Nobody*, Jonathan Cape, London, 2019. Republished by permission of the author.

39 Natalie Diaz, *Postcolonial Love Song*, Faber & Faber, London, 2020. Republished by permission of the author and Faber & Faber.

65 Christina Sharpe, *In the Wake: On blackness and being*, Duke University Press, Durham, 2016. Republished by permission of the author and Duke University Press.

91 Ronald Johnson, *Ark*, Flood Editions, Chicago, 2013. Republished by permission of Flood Editions.

125 Peter Godfrey-Smith, *Other Minds: The octopus and the evolution of intelligent life*, William Collins, London, 2017. Republished by permission of the author.

153 Evelyn Araluen, 'This all come back now', from Mykaela Saunders, *This All Come Back Now*, University of Queensland Press, St Lucia, 2022. Republished by permission of the author.

185 William Beebe, *Half Mile Down*, Harcourt, Brace and Company, New York, 1934, p. 225.

223 Hesiod, *Works and Days*, line 641.

253 Oodgeroo Noonuccal, *My People*, John Wiley & Sons, Brisbane, 2020. Republished by permission of John Wiley & Sons.

285 Jorie Graham, *Sea Change*, Ecco Press, New York, 2009. Republished by permission of the author.

315 Anna Badkhen, *Fisherman's Blues: A West African community at sea*, Riverhead, New York, 2018. Republished by permission of Penguin Random House LLC.

341 Jean Sprackland, *Tilt*, Jonathan Cape, London, 2007. Republished by permission of the author.

371 Homi K. Bhabha, 'Remembering Fanon: Self, psyche and the colonial condition', foreword to Frantz Fanon, *Black Skin, White Masks*, Pluto Press, London, 1986. Republished by permission of Pluto Press.

Images

ACKNOWLEDGEMENTS

I would first like to thank the people whose efforts made this book a reality, beginning with everybody at Penguin Random House Australia, and particularly my publisher, Meredith Curnow, and editor, Rachel Scully. Meredith's friendship and belief in this project over many years has been a gift, and I am grateful to her for her input and engagement across multiple drafts. Similarly, I am deeply indebted to Rachel, whose rigour and attention to detail helped make this a far better book than it would otherwise have been. I would also like to thank the team at Scribe Publications UK, and especially my publisher, Molly Slight, whose suggestions played an important role in helping the book find its final shape. My thanks also to my agent, Matthew Turner, at Rogers, Coleridge and White, whose clarity and commitment to this project played a big part in bringing it into being, and to my former publisher, Ben Ball, who believed in it when it was still just an idea.

I am in the debt of the many, many people who gave up their time to speak with me or share their thoughts and expertise. My particular thanks to Heidi Alleway, Zohra Aly, Jilda Andrews, Julia Baird, Kate Barclay, Chris Bellman, Calisha Bennett, Scott Bennett, Patrick Bradley, Culum Brown, Neal Cantin, Emma Cavan, Anna

Clark, Danielle Clode, John Cooper, Megan Cope, Jem Cresswell, Alexandra Crosby, James Q. Davies, Ceridwen Dovey, David Dyer, Graham Edgar, Tess Egan, Mike Emslie, David Farrier, Hunter Forbes, Rebecca Giggs, Jenna Guffogg, Duane Hamacher, Siobhan Hannan, Robert Harcourt, Lesley Head, Caspar Henderson, Kathryn Heyman, Ove Hoegh-Guldberg, Peter Horn, Mark Horstman, Terry Hughes, Craig Humphrey, Rebecca Huntley, Lisa Viliamu Jameson, Alan Jamieson, Sam Jones, Simon Jones, Alex Jordan, Amara Kalone, So Kawaguchi, Rob King, Skye Krichauff, Matthew Lamb, Jorne Langelaan, Jennifer Lavers, Uncle Bunna Lawrie and the Yerkala Mirning Elders, Cayne Layton, Emmanuelle Leroy, Lucy Manne, Justin Marshall, Joy McCann, John McCarthy, Jane McCredie, Robert Macfarlane, Jessica Meeuwig, Dhana Merritt, Yessie Mosby, Anita Nedosyko, Amanda Nettelbeck, Patrick Nunn, Tim O'Hara, Julie Oswald, Bruce Pascoe, Daniel Pauly, Graham Readfearn, Tracey Rogers, Gene Raymond Ross, Zoe Sadokierski, Richard Short, Mariela Soto, Danielle Southcott, Jared Thomas, Zoë Thomas, Kirsten Tranter, Chris Turney, Alex Vail, Thom van Dooren, Madeleine van Oppen, Charlie Veron, Jody Webster and Christopher Wright. I am also grateful to Margaret Morgan for welcoming me to Rotterdam, and to Bill Hancock for his incredibly generous hospitality during my time in London.

Several sections of this book began life as articles and essays. My thanks to the editors who commissioned and edited those pieces, and in particular Nick Feik, Michael Williams and Patrick Witton at *The Monthly*, Gail MacCallum at *Cosmos*, Catriona Menzies-Pike at *Sydney Review of Books* and Cameron Muir, Kirsten Wehner and Jenny Newell, the editors of *Living with the Anthropocene: Love, loss and hope in the face of environmental crisis*. Their trust in my work played a vital role in helping me believe it might be possible for me to write this book. Although it was later abandoned, a much earlier version of this volume was also part of a doctoral project at the University of Technology Sydney. I owe a great deal to my former supervisor, the late Ross Gibson, for his input and inspiring suggestions during that process. His death in early 2023 was a huge loss, and he is greatly missed. I am also grateful to Libby Robin for her input to the same project.

Many people were generous enough to read sections of the manuscript and offer comments and suggestions or support of other kinds, including Jenny Allen, Line Bay, Duncan Campbell, Neal Cantin, Danielle Celermajer, Sophie Cunningham, Christiaan De Beukelaer, Joëlle Gergis, Shane Gero, Jane Rawson, Eelco Rohling, David Schlosberg, Jock Serong, Helen Sullivan, Rashid Sumaila, Geordie Williamson and Tim Winton. My thanks to all of them. I am also hugely grateful to Delia Falconer, Billy Griffiths and David Ritter, all of whom took the time to read the entire manuscript and offer thoughtful and sometimes stringent critiques of parts of it, a process that improved the finished book enormously. But my deepest thanks go to my friend Ashley Hay, who read multiple versions of what would become this book over many months, and helped sustain my faith in the project when it wavered.

Making a living as a writer is always challenging, and becomes more so with each passing year. I am incredibly grateful to the Copyright Agency for awarding me their Non-Fiction Fellowship for 2020. That decision quite literally changed my life, and it would not have been possible to write this book without that support. I am also extremely grateful to Penny Clive and Detached Cultural Organisation for their financial support of my trip to the Cocos Islands in 2020.

Since 2020 I have been fortunate enough to have been an Honorary Associate at the Sydney Environment Institute at the University of Sydney. Being a part of that community has been incredibly rewarding, and access to the university's resources, and in particular the library, has been invaluable.

And finally, my thanks to my family, and in particular my partner, Mardi McConnochie, and my children, Annabelle and Theo. This book has taken me away from them on many occasions; I'm forever grateful to Mardi for making that possible, and for everything else, and to Annabelle and Theo for making both our lives richer in so many ways.

ABOUT THE AUTHOR

James Bradley is a writer and critic. His books include the novels *Wrack*, *The Deep Field*, *The Resurrectionist*, *Clade* and *Ghost Species*, a book of poetry, *Paper Nautilus*, and *The Penguin Book of the Ocean*. His essays and articles have appeared in *The Monthly*, *The Guardian*, *Sydney Review of Books*, *Griffith Review*, *Meanjin*, the *Weekend Australian* and the *Sydney Morning Herald*. In 2012 he won the Pascall Prize for Australia's Critic of the Year, and he has been shortlisted twice for the Bragg Prize for Science Writing and nominated for a Walkley Award. He lives in Sydney.